Process Systems Analysis and Control

The Series

Process Systems Analysis and Control

Donald R. Coughanowr

Associate Professor of Chemical Engineering
Purdue University

Lowell B. Koppel

Associate Professor of Chemical Engineering
Purdue University

McGraw-Hill Book Company

New York
St. Louis
San Francisco
Toronto
London
Sydney

Process Systems Analysis and Control

Library of Congress Catalog Card Number
64-23273
5 6 7 8 9 – M P – 9 8
13210

To our wives,
Effie and Barbara

Preface

During the last decade, an increasing number of chemical engineering undergraduates, in an increasing number of schools, has been receiving some instruction in process dynamics or control or both. A required course covering both topics for chemical engineering seniors has been taught at Purdue since 1960. It has become evident to us from this teaching experience, and from the growth of the science of process control, that a fundamentally based textbook which takes the student from the basic mathematics to the design application is needed. This book is written to meet that need.

In its present form, the book represents the third revision of a set of class notes which has evolved for six semesters in the Purdue University School of Chemical Engineering. These notes were used (1) as a text for a required senior course, (2) for part of the text material and as reference material in an elective graduate course entitled "Advanced Process Control," and (3) as the text for a graduate-level extension course, given to practicing chemical engineers with little or no previous experience in control, at the Purdue Calumet campus. Even though we have found the material to be satisfactory in each of these applications, the book is written primarily as an undergraduate textbook.

A knowledge of calculus, unit operations, and, to a lesser extent, complex numbers and thermodynamics is presumed on the part of the student. No previous knowledge of Laplace transforms or differential equations is required. In certain later chapters, which would not normally be used for undergraduates, more advanced mathematical preparation is desirable. Some examples are complex variables in Chap. 21, partial differential equations in Chap. 25, and Fourier series in Chap. 29.

Chapter 1 is intended to meet one of the problems we have consistently faced in presenting this material to chemical engineering students, that is, one of perspective. The methods of analysis used in the control area are so different from the previous experience of the students that the material comes to be regarded as a sequence of special mathematical techniques, rather than as an integrated design approach to a class of real and practically significant industrial problems. Therefore, this chapter presents an overall, albeit superficial, look at a simple control system design problem. The body of the text covers the following topics:

1. Laplace transforms, Chaps. 2 to 4.
2. Transfer functions and responses of open-loop systems, Chaps. 5 to 8
3. Basic techniques of closed-loop control and systems engineering, Chaps. 9 to 13
4. Stability, Chap. 14
5. Root-locus methods and design techniques, Chaps. 15 to 17
6. Frequency-response methods and design, Chaps. 18 to 21
7. Industrial process control, Chaps. 22 to 24
8. Advanced process dynamics, Chap. 25
9. Phase-plane analysis of nonlinear systems, Chaps. 26 to 28
10. Describing-function analysis of nonlinear systems, Chap. 29
11. The analog computer in systems engineering and process control, Chaps. 30 to 32

The root-locus material is essentially independent of the remainder of the text and may be omitted without loss in continuity.

It has been our experience that the book contains sufficient material for a one-semester undergraduate course and for the first half of a subsequent one-semester graduate course. In a typical 3-semester-hour senior course we have covered Chaps. 1 to 20, Chaps. 23 and 24, and Chaps. 30–32.

The authors wish to acknowledge the support and encouragement of the Purdue University School of Chemical Engineering for fostering the evolution of this text in its curriculum and for providing the clerical staff and supplies for the several editions of the class notes. For their cooperative assistance in typing, duplicating, assembling, and the myriad other tasks associated with the preparation of this work, we thank Kay Bossung, Lois Christopher, Carol Rawles, and Charlotte Whiteman. Numerous aids and suggestions were provided by our graduate students, particularly P. R. Latour, D. A. Mellichamp, N. E. Moore, and H. H. Orent.

Donald R. Coughanowr

Lowell B. Koppel

Contents

1 *An Introductory Example*

In this chapter we consider an illustrative example of a control system. Our goal is to introduce some of the basic principles and problems involved in process control and to give the reader an early look at an overall problem typical of those we shall face in later chapters.

The System   A liquid stream at temperature T_i is available at a constant flow rate of w in units of mass per time.   It is desired to heat this stream to a higher temperature T_R.   The proposed heating system is shown in Fig. 1.1.   The fluid flows into a well-agitated tank which is equipped with a heating device.   It is assumed that the agitation is sufficient to ensure that all fluid in the tank will be at the same temperature T. Heated fluid is removed from the bottom of the tank at the flow rate w as the product of this heating process.   Under these conditions, the mass of fluid retained in the tank remains constant in time, and the temperature of the effluent fluid is the same as that of the fluid in the tank.   For a satisfactory design this temperature must be T_R.   The specific heat of the fluid C is assumed to be constant, independent of temperature.

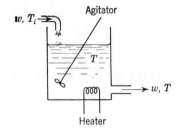

Fig. 1.1   Agitated heating tank.

Steady-state Design   A process is said to be at steady state when none of the variables are changing with time.   At the desired steady state, an energy balance around the heating process may be written as follows:

$$q_s = wC(T_s - T_{i_s}) \tag{1.1}$$

where q_s is the heat input to the tank and the subscript s is added to indicate a steady-state design value.   Thus, for example, T_{i_s} is the normally anticipated inlet temperature to the tank.   For a satisfactory design, the steady-state temperature of the effluent stream T_s must equal T_R.   Hence

$$q_s = wC(T_R - T_{i_s}) \tag{1.2}$$

However, it is clear from the physical situation that, if the heater is set to deliver only the constant input q_s, then if process conditions change, the tank temperature will also change from T_R.   A typical process condition which may change is the inlet temperature T_i.

An obvious solution to the problem is to design the heater so that its energy input may be varied as required to maintain T at or near T_R.

Process Control   It is necessary to decide how much the heat input q is to be changed from q_s to correct any deviations of T from T_R.   One solution would be to hire a process operator, who would be responsible for controlling the heating process.   The operator would observe the temperature in the tank, presumably with a measuring instrument such as a thermocouple or thermometer, and compare this temperature with T_R.   If T is less than T_R, he would increase the heat input and vice versa. As he became experienced at this task, he would learn just how much to change q for each situation.   However, this relatively simple task can be easily and less expensively performed by a machine.   The use of machines for this and similar purposes is known as *automatic process control.*

The Unsteady State   If a machine is to be used to control the process, it is necessary to decide in advance precisely what changes are to be made in the heat input q for every possible situation which might occur.   We cannot rely upon the judgment of the machine as we could upon that of the operator.   Machines do not think; they simply perform a predetermined task in a predetermined manner.

To be able to make these control decisions in advance, it is necessary to know how the tank temperature T changes in response to changes in T_i and q. This necessitates writing the *unsteady-state*, or *transient*, energy balance for the process. The input and output terms in this balance are the same as those used in the steady-state balance, Eq. (1.1). In addition, there is a transient accumulation of energy in the tank, which may be written

$$\text{Accumulation} = \rho V C \frac{dT}{dt} \qquad \text{energy units/time}^1$$

where ρ = fluid density
$\quad V$ = volume of fluid in the tank
$\quad t$ = independent variable, time
By the assumption of constant and equal inlet and outlet flow rates, the term ρV, which is the mass of fluid in the tank, is constant. Since

$$\text{Accumulation} = \text{input} - \text{output}$$

we have

$$\rho V C \frac{dT}{dt} = wC(T_i - T) + q \tag{1.3}$$

Equation (1.1) is the steady-state solution of Eq. (1.3), obtained by setting the time derivative to zero. We shall make use of Eq. (1.3) presently.

Feedback Control As discussed above, the controller is to do the same job that the human operator was to do, except that the controller is told in advance *exactly* how to do it. This means that the controller will use the existing values of T and T_R to adjust the heat input according to a predetermined formula. Let the difference between these temperatures, $T_R - T$, be called *error*. Clearly, the larger this error, the less we are satisfied with the present state of affairs and vice versa. In fact we are completely satisfied only when the error is exactly zero.

Based on these considerations, it is natural to suggest that the controller should change the heat input by an amount *proportional* to the error. Thus, a plausible formula for the controller to follow is

$$q(t) = wC(T_R - T_{i_s}) + K_c(T_R - T) \tag{1.4}$$

where K_c is a (positive) constant of proportionality. This is called *proportional control*. In effect, the controller is instructed to maintain the heat input at the steady-state design value q_s as long as T is equal to T_R [cf. Eq. (1.2)], i.e., as long as the error is zero. If T deviates from T_R, causing an error, the controller is to use the magnitude of the error to change the heat input proportionally. The reader should satisfy himself that this

---

[1] A rigorous application of the first law of thermodynamics would yield a term representing the transient change of internal energy with temperature at constant pressure. Use of the specific heat, at either constant pressure or constant volume, is an adequate engineering approximation for most liquids and will be applied extensively in this text.

change is in the right direction. We shall reserve the right to vary the parameter K_c to suit our needs. This degree of freedom forms a part of our instructions to the controller.

The concept of using information about the deviation of the system from its desired state to control the system is called *feedback* control. Information about the state of the system is "fed back" to a controller which utilizes this information to change the system in some way. In the present case, the information is the temperature T and the change is made in q. When the term $wC(T_R - T_{i_s})$ is abbreviated to q_s, Eq. (1.4) becomes

$$q = q_s + K_c(T_R - T) \tag{1.4a}$$

Transient Responses   Substituting Eq. (1.4a) into Eq. (1.3) and rearranging yield

$$\tau_1 \frac{dT}{dt} + \left(\frac{K_c}{wC} + 1\right) T = T_i + \frac{K_c}{wC} T_R + \frac{q_s}{wC} \tag{1.5}$$

where

$$\tau_1 = \frac{\rho V}{w}$$

The term τ_1 has the dimensions of time and is known as the *time constant* of the tank. We shall study the significance of the time constant in more detail in Chap. 5. At present, it suffices to note that it is the time required to fill the tank at the flow rate w. T_i is the inlet temperature, which we have assumed is a function of time. Its normal value is T_{i_s}, and q_s is based upon this value. Equation (1.5) describes the way in which the tank temperature changes in response to changes in T_i and q.

Suppose that the process is proceeding smoothly at steady-state design conditions. At a time arbitrarily called zero, the inlet temperature which was at T_{i_s} suddenly undergoes a permanent rise of a few degrees to a new value $T_{i_s} + \Delta T_i$, as shown in Fig. 1.2. For mathematical convenience, this disturbance is idealized to the form shown in Fig. 1.3. The equation for the function $T_i(t)$ of Fig. 1.3 is

$$T_i(t) = \begin{cases} T_{i_s} & t < 0 \\ T_{i_s} + \Delta T_i & t > 0 \end{cases} \tag{1.6}$$

This type of function is known as a step function and is used extensively in the study of transient response because of the simplicity of Eq. (1.6). The justification for use of the step change is that the response of T to this function will not differ significantly from the response to the more realistic disturbance depicted in Fig. 1.2.

To determine the response of T to a step change in T_i, it is necessary to substitute Eq. (1.6) into (1.5) and solve the resulting differential equation

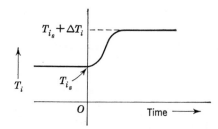

Fig. 1.2 Inlet temperature versus time.

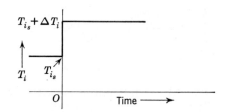

Fig. 1.3 Idealized inlet temperature versus time.

for $T(t)$. Since the process is at steady state at (and before) time zero, the initial condition is

$$T(0) = T_R \tag{1.7}$$

The reader can easily verify (and should do so) that the solution to Eqs. (1.5), (1.6), and (1.7) is

$$T = T_R + \frac{\Delta T_i}{K_c/wC + 1} \left(1 - e^{-(K_c/wC+1)t/\tau_1}\right) \tag{1.8}$$

This system *response*, or tank temperature versus time, to a step change in T_i is shown in Fig. 1.4 for various values of the adjustable control parameter K_c. The reader should compare these curves with Eq. (1.8) to satisfy himself that they are correct, particularly in respect to the relative positions of the curves at the new steady states.

It may be seen that the higher K_c is made, the "better" will be the control, in the sense that the new steady-state value of T will be closer to T_R. At first glance, it would appear desirable to make K_c as large as possible. A little reflection will show that large values of K_c are likely to cause other problems. For example, note that we have considered only one type of disturbance in T_i. Another possible behavior of T_i with time is shown in Fig. 1.5. Here, T_i is fluctuating about its steady-state value. A typical response of T to this type of disturbance in T_i, *without control action*, is shown in Fig. 1.6. The fluctuations in T_i are delayed and "smoothed" by the large volume of liquid in the tank, so that T does not fluctuate as much as T_i. Nevertheless, it should be clear from Eq. (1.4a) and Fig. 1.6 that a

Fig. 1.4 Tank temperature versus time for various values of K_c.

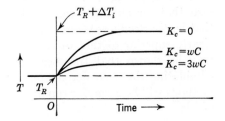

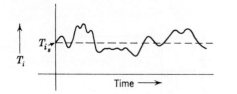

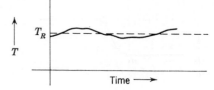

Fig. 1.5 A fluctuating behavior of T_i.

Fig. 1.6 The response to a fluctuating T_i.

control system with a high value of K_c will have a tendency to "overadjust." In other words, it will be too *sensitive* to disturbances which would tend to disappear in time *even without control action*. This will have the undesirable effect of *amplifying* the effects of these disturbances and causing excessive wear on the control system.

The dilemma may be summarized as follows: In order to obtain accurate control of T, despite "permanent" changes in T_i, we must make K_c larger (see Fig. 1.4). However, as K_c is increased, the system becomes oversensitive to spurious fluctuations in T_i. (These fluctuations as depicted in Fig. 1.5 are called *noise*.) The reader is cautioned that there are additional effects, produced by changing K_c, which have not been discussed here for the sake of brevity but which may be even more important. This will be one of the major subjects of interest in later chapters. The two effects mentioned are sufficient to illustrate the problem.

Integral Control   A considerable improvement may be obtained over the proportional control system by adding integral control. The controller is now instructed to change the heat input by an additional amount proportional to the time integral of the error. Quantitatively, the heat input function is to follow the relation

$$q(t) = q_s + K_c(T_R - T) + K_R \int_0^t (T_R - T)\, dt \tag{1.9}$$

This control system is to have two adjustable parameters, K_c and K_R.

The response of the tank temperature T to a step change in T_i, using a control function described by (1.9), may be derived by solution of Eqs. (1.3), (1.6), (1.7), and (1.9). Curves representing this response, which the reader is asked to accept, are given for various values of K_R at a fixed value of K_c in Fig. 1.7. The value of K_c is a moderate one, and it may be seen that for all three values of K_R the steady-state temperature is T_R; that is, *the steady-state error is zero*. From this standpoint, the response is clearly superior to that of the system with proportional control only. It may be shown that the steady-state error is zero for *all* $K_R > 0$, thus eliminating the necessity for high values of K_c. (In subsequent chapters, methods will be given for rapidly constructing response curves such as those of Fig. 1.7.)

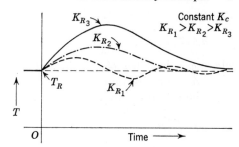

Fig. 1.7 **Tank temperature versus time—step input for proportional and integral control.**

It is clear from Fig. 1.7 that the responses for $K_R = K_{R_2}$ and $K_R = K_{R_1}$ are better than that for $K_R = K_{R_3}$ because T returns to T_R faster, but it may be difficult to choose between K_{R_2} and K_{R_1}. Thus, while the response for K_{R_2} "settles down" sooner, it also has a higher maximum error. The choice might depend upon the particular use for the heated stream. This and related questions form the study of *optimal* control systems. We shall touch only briefly on this important subject in this introductory text and do so more to point out the existence of the problem than to solve it.

To recapitulate, the curves of Fig. 1.7 give the transient behavior of the tank temperature, in response to a step change in T_i, when the tank temperature is controlled according to Eq. (1.9). They show that the addition of integral control in this case eliminates steady-state error and allows use of moderate values of K_c.

More Complications   At this point, it would appear that the problem has been solved in some sense. A little further probing will shatter this illusion.

It has been assumed in writing Eqs. (1.4a) and (1.9) that the controller receives instantaneous information about the tank temperature T. From a physical standpoint, some measuring device such as a thermocouple will be required to measure this temperature. Furthermore, the tempera-ture of a thermocouple inserted in the tank may or may not be the same as the temperature of the fluid in the tank. This can be demonstrated by writing the energy balance for a typical thermocouple installation, such as the one depicted in Fig. 1.8. Assuming that the junction is at a uniform temperature T_m and neglecting any conduction of heat along the thermo-couple lead wires, the net rate of input of energy to the thermocouple junc-

Fig. 1.8 **Thermocouple installation for
heated-tank system.**

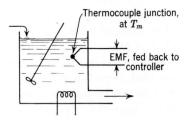

tion is

$$hA(T - T_m)$$

where h = heat-transfer coefficient between fluid and junction
   A = area of junction
The rate of accumulation of energy in the junction is

$$mC_m \frac{dT_m}{dt}$$

where C_m = specific heat of junction
   m = mass of junction
Combining these in an energy balance,

$$\tau_2 \frac{dT_m}{dt} + T_m = T \tag{1.10}$$

where $\tau_2 = mC_m/hA$ is the time constant of the thermocouple. Thus, changes in T are not instantaneously reproduced in T_m. A step change in T causes a response in T_m similar to the curve of Fig. 1.4 for $K_c = 0$ [see Eq. (1.5)]. This is analogous to the case of placing a mercury thermometer in a beaker of hot water. The thermometer does not instantaneously rise to the water temperature. Rather, it rises in the manner described.

Since the controller will receive values of T_m (possibly in the form of a thermoelectric voltage) and *not* values of T, Eq. (1.9) must be rewritten as

$$q = q_s + K_c(T_R - T_m) + K_R \int_0^t (T_R - T_m)\, dt \tag{1.9a}$$

The *apparent error* is given by $(T_R - T_m)$, and it is this quantity upon which the controller acts, rather than the true error $(T_R - T)$. The response of T to a step change in T_i is now derived by simultaneous solution of (1.3), (1.6), (1.9a), and (1.10), with initial conditions

$$T(0) = T_m(0) = T_R \tag{1.11}$$

Equation (1.11) implies that, at time zero, the system has been at rest at T_R for some time, so that the thermocouple junction is at the same temperature as the tank.

The solution to this system of equations is represented in Fig. 1.9 for a particular set of values of K_c and K_R. For this set of values, the effect of the thermocouple delay in transmission of the temperature to the controller is primarily to make the response somewhat more oscillatory than that shown in Fig. 1.7 for the same value of K_R. However, if K_R is increased somewhat over the value used in Fig. 1.9, the response is that shown in Fig. 1.10. The tank temperature oscillates with *increasing* amplitude and will continue to do so until the physical limitations of the heating system are

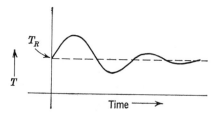

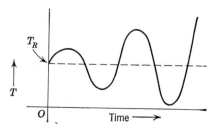

Fig. 1.9 **Tank temperature versus time with measuring lag.**

Fig. 1.10 **Tank temperature versus time for increased K_R.**

reached. The control system has actually caused a deterioration in performance. Surely, the uncontrolled response for $K_c = 0$ in Fig. 1.4 is to be preferred over the *unstable* response of Fig. 1.10.

This problem of *stability* of response will be one of our major concerns in this text for obvious reasons. At present, it is sufficient to note that extreme care must be exercised in specifying control systems. In the case considered, the proportional and integral control mechanism described by Eq. (1.9a) will perform satisfactorily if K_R is kept lower than some particular value, as illustrated in Figs. 1.9 and 1.10. However, it is not difficult to construct examples of systems for which the addition of *any* amount of integral control will cause an unstable response. Since integral control usually has the desirable feature of eliminating steady-state error, as it did in Fig. 1.7, it is extremely important that we develop means for predicting the occurrence of unstable response in the design of any control system.

Block Diagram A good overall picture of the relationships among variables in the heated-tank control system may be obtained by preparing a *block diagram*. This diagram, shown in Fig. 1.11, indicates the flow of information around the control system and the function of each part of the system. Much more will be said about block diagrams in Chap. 9, but the reader can undoubtedly form a good intuitive notion about them

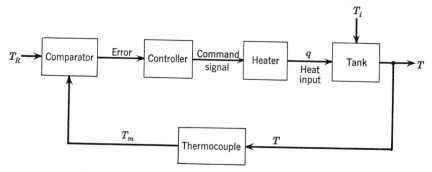

Fig. 1.11 **Block diagram for heated-tank system.**

by comparison of Fig. 1.11 with the physical description of the process given in the previous paragraphs. Particularly significant is the fact that each component of the system is represented by a block, with little regard for the actual physical characteristics of the represented component (e.g., the tank or controller). The major interest is in the relationship between the signals entering the block and those leaving and in the manner in which information flows around the system. For example, T_R and T_m enter the comparator. Their difference, the *error*, leaves the comparator and enters the controller.

Summary   We have had an overall look at a typical control problem and some of its ramifications. At present, the reader has been asked to accept the mathematical results on faith and to concentrate on obtaining a physical understanding of the transient behavior of the heated tank. We shall in the forthcoming chapters develop tools for determining the response of such systems. As this new material is presented, the reader may find it helpful to refer back to this chapter in order to place the material in proper perspective to the overall control problem.

PROBLEMS

1.1   Draw a block diagram for the control system generated when a human being steers an automobile.

part I Laplace Transforms

2 *The Laplace Transform*

Even from our brief look at the control problem of Chap. 1, it is evident that solution of differential equations will be one of our major tasks. The Laplace transform method provides an efficient way to solve linear, ordinary, differential equations with constant coefficients. Because an important class of control problems reduces to the solution of such equations, we devote the next three chapters to a study of Laplace transforms before resuming our investigation of control problems.

Definition of the Transform   The Laplace transform of a function $f(t)$ is *defined* to be $f(s)$ according to the equation

$$f(s) = \int_0^\infty f(t)e^{-st}\,dt \tag{2.1}$$

We often abbreviate this notationally to

$$f(s) = L\{f(t)\}$$

where the operator L is defined by Eq. (2.1).[1]

---

[1] Many texts adopt some notational convention, such as capitalizing the trans-

Example 2.1   Find the Laplace transform of the function

$$f(t) = 1$$

According to Eq. (2.1),

$$f(s) = \int_0^\infty (1)e^{-st}\, dt = -\frac{e^{-st}}{s}\bigg|_{t=0}^{t=\infty} = \frac{1}{s}$$

Thus,

$$L\{1\} = \frac{1}{s}$$

There are several facts worth noting at this point:

1. The Laplace transform $f(s)$ contains no information about the behavior of $f(t)$ for $t < 0$. This is not a limitation for control system study because t will represent the time variable and we shall be interested in the behavior of systems only for positive time. In fact, the variables and systems are usually defined so that $f(t) \equiv 0$ for $t < 0$. This will become clearer as we study specific examples.

2. Since the Laplace transform is defined in Eq. (2.1) by an improper integral, it will not exist for every function $f(t)$. A rigorous definition of the class of functions possessing Laplace transforms is beyond the scope of this text. We merely remark that every function of interest to us does satisfy the requirements for possession of a transform.[1]

3. The Laplace transform is linear. In mathematical notation, this means:

$$L\{af_1(t) + bf_2(t)\} = aL\{f_1(t)\} + bL\{f_2(t)\}$$

where a and b are constants, and f_1 and f_2 are two functions of t.

Proof: Using the definition,

$$
\begin{aligned}
L\{af_1(t) + bf_2(t)\} &= \int_0^\infty [af_1(t) + bf_2(t)]e^{-st}\, dt \\
&= a \int_0^\infty f_1(t)e^{-st}\, dt + b \int_0^\infty f_2(t)e^{-st}\, dt \\
&= aL\{f_1(t)\} + bL\{f_2(t)\}
\end{aligned}
$$

4. The Laplace transform operator transforms a function of the variable t to a function of the variable s. The t variable is eliminated by the integration.

---

formed function as $F(s)$ or putting a bar over it as $\bar{f}(s)$. In general, the appearance of the variable s as the argument or in an equation involving f is sufficient to signify that the function has been transformed, and hence we shall seldom require any such notation in the text.

[1] For details on this and related mathematical topics, see R. V. Churchill, "Operational Mathematics," 2d ed., McGraw-Hill Book Company, New York, 1958.

Transforms of Simple Functions We now proceed to derive the transforms of some simple and useful functions.

1. The step function

$$f(t) = \begin{cases} 0 & t < 0 \\ 1 & t > 0 \end{cases}$$

This important function is known as the unit-step function and will henceforth be denoted by $u(t)$. From Example 2.1, it is clear that

$$L\{u(t)\} = \frac{1}{s}$$

As expected, the behavior of the function for $t < 0$ has no effect on its Laplace transform. Note that as a consequence of linearity, the transform of any constant A, that is, $f(t) = Au(t)$, is just $f(s) = A/s$.

2. The exponential function

$$f(t) = \begin{cases} 0 & t < 0 \\ e^{-at} & t > 0 \end{cases} = u(t)e^{-at}$$

where $u(t)$ is the unit-step function. Again proceeding according to definition,

$$L\{u(t)e^{-at}\} = \int_0^\infty e^{-(s+a)t}\,dt = -\frac{1}{s+a}e^{-(s+a)t}\bigg|_0^\infty = \frac{1}{s+a}$$

provided that $s + a > 0$, that is, $s > -a$. In this case, the convergence of the integral depends upon a suitable choice of s. In case s is a complex number, it may be shown that this condition becomes

$$\text{Re } (s) > -a$$

For problems of interest to us it will always be possible to choose s so that these conditions are satisfied, and the reader uninterested in mathematical niceties can ignore this point.

3. The ramp function

$$f(t) = \begin{cases} 0 & t < 0 \\ t & t > 0 \end{cases} = tu(t)$$

$$L\{tu(t)\} = \int_0^\infty te^{-st}\,dt$$

Integration by parts yields

$$L\{tu(t)\} = -e^{-st}\left(\frac{t}{s} + \frac{1}{s^2}\right)\bigg|_0^\infty = \frac{1}{s^2}$$

Table 2.1

Function	Graph	Transform
$u(t)$		$\dfrac{1}{s}$
$t\,u(t)$	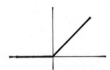	$\dfrac{1}{s^2}$
$t^n u(t)$	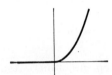	$\dfrac{n!}{s^{n+1}}$
$e^{-at} u(t)$	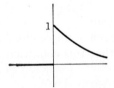	$\dfrac{1}{s+a}$
$t^n e^{-at} u(t)$	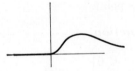	$\dfrac{n!}{(s+a)^{n+1}}$
$\sin kt\; u(t)$	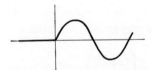	$\dfrac{k}{s^2+k^2}$

Table 2.1 (*Continued*)

Function	Graph	Transform
$\cos kt\, u(t)$		$\dfrac{s}{s^2 + k^2}$
$\sinh kt\, u(t)$		$\dfrac{k}{s^2 - k^2}$
$\cosh kt\, u\,(t)$	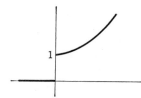	$\dfrac{s}{s^2 - k^2}$
$e^{-at} \sin kt\, u(t)$	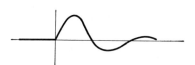	$\dfrac{k}{(s + a)^2 + k^2}$
$e^{-at} \cos kt\, u(t)$	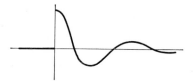	$\dfrac{s + a}{(s + a)^2 + k^2}$
$\delta(t)$, unit impulse	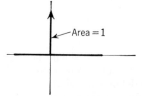	1

1

4. The sine function

$$f(t) = \begin{cases} 0 & t < 0 \\ \sin kt & t > 0 \end{cases} = u(t) \sin kt$$

$$L\{u(t) \sin kt\} = \int_0^\infty \sin kt \, e^{-st} \, dt$$

Integrating by parts,

$$L\{u(t) \sin kt\} = \frac{-e^{-st}}{s^2 + k^2}(s \sin kt + k \cos kt) \Big|_0^\infty$$

$$= \frac{k}{s^2 + k^2}$$

In a like manner, the transforms of other simple functions may be derived. Table 2.1 is a summary of transforms which will be of use to us. Those which we have not derived can be easily established by direct integration, except for the transform of $\delta(t)$, which will be discussed in detail in Chap. 4.

Transforms of Derivatives At this point, the reader may well inquire as to what has been gained by introduction of the Laplace transform. The transform merely changes a function of t into a function of s. The functions of s look no simpler than those of t and, as in the case of $A \rightarrow A/s$, may actually be more complex. In the next few paragraphs, the motivation will become clear. It will be seen that the Laplace transform has the remarkable property of transforming the operation of differentiation with respect to t to that of multiplication by s. Thus, we claim that

$$L\left\{\frac{df(t)}{dt}\right\} = sf(s) - f(0) \tag{2.2}$$

where

$$f(s) = L\{f(t)\}$$

and $f(0)$ is $f(t)$ evaluated at $t = 0$. [It is essential not to interpret $f(0)$ as $f(s)$ with $s = 0$. This will be clear from the proof presented below.][1]

Proof:

$$L\left\{\frac{df(t)}{dt}\right\} = \int_0^\infty \frac{df}{dt} e^{-st} \, dt$$

To integrate this by parts, let

$$u = e^{-st} \qquad dv = \frac{df}{dt} \, dt$$

---

[1] If $f(t)$ is discontinuous at $t = 0$, $f(0)$ should be evaluated at $t = 0^+$, i.e., just to the right of the origin. Since we shall seldom want to differentiate functions that are discontinuous at the origin, this detail is not of great importance. However, the reader is cautioned to watch carefully for situations in which such discontinuities occur.

Then

$$du = -se^{-st}\, dt \qquad v = f(t)$$

Since

$$\int u\, dv = uv - \int v\, du$$

we have

$$\int_0^\infty \frac{df}{dt} e^{-st}\, dt = f(t)e^{-st}\Big|_0^\infty + s\int_0^\infty f(t)e^{-st}\, dt = -f(0) + sf(s) \qquad \text{Q.E.D.}$$

The salient feature of this transformation is that whereas the function of t was to be differentiated with respect to t, the corresponding function of s is merely multiplied by s. We shall find this feature to be extremely useful in the solution of differential equations.

To find the transform of the second derivative we make use of the transform of the first derivative twice, as follows:

$$\begin{aligned}
L\left\{\frac{d^2f}{dt^2}\right\} = L\left\{\frac{d}{dt}\left(\frac{df}{dt}\right)\right\} &= sL\left\{\frac{df}{dt}\right\} - \frac{df(t)}{dt}\Big|_{t=0} \\
&= s[sf(s) - f(0)] - f'(0) \\
&= s^2 f(s) - sf(0) - f'(0)
\end{aligned}$$

where we have abbreviated

$$\frac{df(t)}{dt}\Big|_{t=0} = f'(0)$$

In a similar manner, the reader can easily establish by induction that repeated application of Eq. (2.2) leads to

$$L\left\{\frac{d^n f}{dt^n}\right\} = s^n f(s) - s^{n-1}f(0) - s^{n-2}f^{(1)}(0) - \cdots - sf^{(n-2)}(0) - f^{(n-1)}(0)$$

where $f^{(i)}(0)$ indicates the ith derivative of $f(t)$ with respect to t, evaluated for $t = 0$.

Thus, the Laplace transform may be seen to change the operation of differentiation of the function to that of multiplication of the transform by s, the number of multiplications corresponding to the number of differentiations. In addition, some polynomial terms involving the initial values of $f(t)$ and its first $(n - 1)$ derivatives are involved. In later applications we shall usually define our variables so that these polynomial terms will vanish. Hence, they are of secondary concern here.

Example 2.2 Find the Laplace transform of the function $x(t)$ which satisfies the differential equation and initial conditions

$$\frac{d^3x}{dt^3} + 4\frac{d^2x}{dt^2} + 5\frac{dx}{dt} + 2x = 2$$

$$x(0) = \frac{dx(0)}{dt} = \frac{d^2x(0)}{dt^2} = 0$$

It is permissible mathematically to take the Laplace transforms of both sides of a differential equation and equate them, since equality of functions implies equality of their transforms. Doing this, there is obtained

$$s^3x(s) - s^2x(0) - sx'(0) - x''(0) + 4[s^2x(s) - sx(0) - x'(0)]$$

$$+ 5[sx(s) - x(0)] + 2x(s) = \frac{2}{s}$$

where $x(s) = L\{x(t)\}$. Use has been made of the linearity property and the fact that only positive values of t are of interest. Inserting the initial conditions and solving for $x(s)$,

$$x(s) = \frac{2}{s(s^3 + 4s^2 + 5s + 2)} \tag{2.3}$$

This is the required answer, the Laplace transform of $x(t)$.

Solution of Differential Equations   There are two important points to note regarding this last example. In the first place, application of the transformation resulted in an equation which was solved for the unknown function by *purely algebraic means*. Second, and most important, if the function $x(t)$ which has Laplace transform $2/s(s^3 + 4s^2 + 5s + 2)$ were known, we would have the solution to the differential equation and boundary conditions. This suggests a procedure for solving differential equations which is analogous to that of using logarithms to multiply or divide. To use logarithms, one transforms the pertinent numbers to their logarithms and then adds or subtracts, which is much easier than multiplying or dividing. The result of the addition or subtraction is the logarithm of the desired answer. The answer is found by reference to a table to find the number having this logarithm. In the Laplace transform method for solution of differential equations, the functions are converted to their transforms and the resulting equations are solved for the unknown function *algebraically*. This is much easier than solving a differential equation. However, at the last step the analogy to logarithms is not complete. We obviously cannot hope to construct a table containing the Laplace transform of every function $f(t)$ which possesses a transform. Instead, we shall develop methods for reexpressing complicated transforms, such as $x(s)$ of Example 2.2, in terms of simple transforms which can be found in Table 2.1. For example, it is easily verified that the solution to the differential equation and boundary conditions of Example 2.2 is

$$x(t) = 1 - 2te^{-t} - e^{-2t} \tag{2.4}$$

The Laplace transform of x, using Eq. (2.4) and Table 2.1, is

$$x(s) = \frac{1}{s} - 2\frac{1}{(s + 1)^2} - \frac{1}{s + 2} \tag{2.5}$$

Equation (2.3) is actually the result of placing Eq. (2.5) over a common denominator. Although it is difficult to find $x(t)$ from Eq. (2.3), Eq. (2.5) may be easily inverted to Eq. (2.4) by using Table 2.1. Therefore, what is required is a method for expanding the common-denominator form of Eq. (2.3) to the separated form of Eq. (2.5). This method is provided by the technique of partial fractions, which is developed in Chap. 3.

To summarize, the basis for solving *linear, ordinary, differential equations with constant coefficients* with Laplace transforms has been established.

The procedure is:

1. Take the Laplace transform of both sides of the equation. The initial conditions are incorporated at this step in the transforms of the derivatives.

2. Solve the resulting equation for the Laplace transform of the unknown function, algebraically.

3. Find the function of t which has the Laplace transform obtained in step 2. This function satisfies the differential equation and initial conditions and hence is the desired solution. This is frequently the most difficult or tedious step and will be developed further in the next chapter. It is called inversion of the transform. Although there are other techniques available for inversion, the one which we shall develop and make consistent use of is that of partial-fraction expansion.

A simple example will serve to illustrate steps 1 and 2, and a trivial case of step 3.

Example 2.3   Solve

$$\frac{dx}{dt} + 3x = 0$$

$$x(0) = 2$$

We number our steps according to the discussion in the preceding paragraphs:

1. $sx(s) - 2 + 3x(s) = 0$

2. $x(s) = \dfrac{2}{s+3} = 2\,\dfrac{1}{s+3}$

3. $x(t) = 2e^{-3t}$

3 *Inversion by Partial Fractions*

Our study of the application of Laplace transforms to linear differential equations with constant coefficients has enabled us to rapidly establish the Laplace transform of the solution. We now wish to develop methods for inverting the transforms to obtain the solution in the time domain. The first part of this chapter will be a series of examples which illustrate the partial-fraction technique. After a generalization of these techniques, we proceed to a discussion of the qualitative information which can be obtained from the transform of the solution without inverting it.

The equations to be solved are all of the general form

$$a_n \frac{d^n x}{dt^n} + a_{n-1} \frac{d^{n-1} x}{dt^{n-1}} + \cdots + a_1 \frac{dx}{dt} + a_0 x = f(t)$$

The unknown function of time is $x(t)$, and a_n, a_{n-1}, . . . , a_1, a_0 are constants. The given function $f(t)$ is called the *forcing function*. In addition, for all problems of interest in control system analysis, the initial conditions are given. In other words, values of x, dx/dt, . . . , $d^{n-1}x/dt^{n-1}$ are specified at time zero. The problem is to determine $x(t)$ for all $t \geq 0$.

It is evident that Eq. (1.5), which must be solved to determine the behavior of the heated-tank control system, is included in this class. The unknown function is T, and the forcing function is a sum of terms involving T_i, T_R, and q_s.

Partial Fractions In the series of examples which follow, we set forth the technique of partial-fraction inversion for solution of this class of differential equations.

Example 3.1 Solve:

$$\frac{dx}{dt} + x = 1$$

$$x(0) = 0$$

Application of the Laplace transform yields

$$sx(s) + x(s) = \frac{1}{s}$$

or

$$x(s) = \frac{1}{s(s+1)}$$

The theory of partial fractions enables us to write this as

$$x(s) = \frac{1}{s(s+1)} = \frac{A}{s} + \frac{B}{s+1} \tag{3.1}$$

where A and B are constants. Hence, using Table 2.1, it follows that

$$x(t) = A + Be^{-t} \tag{3.2}$$

Therefore, if A and B were known, we would have the solution. The conditions on A and B are that they must be chosen to make Eq. (3.1) an identity in s.

To determine A, multiply both sides of Eq. (3.1) by s.

$$\frac{1}{s+1} = A + \frac{Bs}{s+1} \tag{3.3}$$

Since this must hold for all s, it must hold for $s = 0$. Putting $s = 0$ in Eq. (3.3) yields

$$A = 1$$

To find B, multiply both sides of Eq. (3.1) by $(s+1)$.

$$\frac{1}{s} = \frac{A}{s}(s+1) + B \tag{3.4}$$

Since this must hold for all s, it must hold for $s = -1$. This yields

$$B = -1$$

Hence,

$$\frac{1}{s(s+1)} = \frac{1}{s} - \frac{1}{s+1} \tag{3.5}$$

and therefore,

$$x(t) = 1 - e^{-t} \tag{3.6}$$

Equation (3.5) may be checked by putting the right side over a common denominator, and Eq. (3.6) by substitution into the original differential equation and initial condition.

Example 3.2   Solve:

$$\frac{d^3x}{dt^3} + 2\frac{d^2x}{dt^2} - \frac{dx}{dt} - 2x = 4 + e^{2t}$$

$$x(0) = 1 \qquad x'(0) = 0 \qquad x''(0) = -1$$

Taking the Laplace transform of both sides,

$$[s^3x(s) - s^2 + 1] + 2[s^2x(s) - s] - [sx(s) - 1] - 2x(s) = \frac{4}{s} + \frac{1}{s - 2}$$

Solving algebraically for $x(s)$,

$$x(s) = \frac{s^4 - 6s^2 + 9s - 8}{s(s - 2)(s^3 + 2s^2 - s - 2)}$$

The cubic in the denominator may be factored, and $x(s)$ expanded in partial fractions

$$x(s) = \frac{s^4 - 6s^2 + 9s - 8}{s(s - 2)(s + 1)(s + 2)(s - 1)} = \frac{A}{s} + \frac{B}{s - 2} + \frac{C}{s + 1} + \frac{D}{s + 2} + \frac{E}{s - 1} \quad (3.7)$$

To find A, multiply both sides of Eq. (3.7) by s and then set $s = 0$; the result is

$$A = \frac{-8}{(-2)(1)(2)(-1)} = -2$$

The other constants are determined in the same way.   The procedure and results are summarized in the following table:

To determine	multiply (3.7) by	and set s to	Result
B	$s - 2$	2	$B = \frac{1}{12}$
C	$s + 1$	-1	$C = 1\frac{1}{3}$
D	$s + 2$	-2	$D = -1\frac{7}{12}$
E	$s - 1$	1	$E = \frac{2}{3}$

Accordingly, the solution to the problem is

$$x(t) = -2 + \frac{1}{12}e^{2t} + 1\frac{1}{3}e^{-t} - 1\frac{7}{12}e^{-2t} + \frac{2}{3}e^{t}$$

A comparison between this method and the classical method, as applied to Example 3.2, may be profitable.   In the classical method for solution of differential equations we first write down the characteristic function of the homogeneous equation:

$$s^3 + 2s^2 - s - 2 = 0$$

This must be factored, as was also required in the Laplace transform method, to obtain the roots -1, -2, and $+1$.   Thus, the complementary solution is

$$x_c(t) = C_1e^{-t} + C_2e^{-2t} + C_3e^{t}$$

Furthermore, by inspection of the forcing function, we know that the particular solution has the form

$$x_p(t) = A + Be^{2t}$$

The constants A and B are determined by substitution into the differential equation and, as expected, are found to be -2 and $\frac{1}{12}$, respectively.   Then

$$x(t) = -2 + \frac{1}{12}e^{2t} + C_1e^{-t} + C_2e^{-2t} + C_3e^t$$

and the constants C_1, C_2, and C_3 are determined by the three initial conditions.   The Laplace transform method has systematized the evaluation of these constants, avoiding the solution of three simultaneous equations. Four points are worth noting:

1. In both methods, one must find the roots of the characteristic equation.   The roots give rise to terms in the solution *whose form is independent of the forcing function*.   These terms make up the *complementary solution*.

2. The forcing function gives rise to terms in the solution *whose form depends on the form of the forcing function and is independent of the left side of the equation*.   These terms comprise the *particular solution*.

3. The only interaction between these sets of terms, i.e., between the right side and left side of the differential equation, occurs in the evaluation of the constants involved.

4. The only effect of the initial conditions is in the evaluation of the constants.   This is because the initial conditions affect only the numerator of $x(s)$, as may be seen from the solution of this example.

In the two examples we have discussed, the denominator of $x(s)$ factored into real factors only.   In the next example, we consider the complications which arise when the denominator of $x(s)$ has complex factors.

Example 3.3   Solve:

$$\frac{d^2x}{dt^2} + 2\frac{dx}{dt} + 2x = 2$$

$$x(0) = x'(0) = 0$$

Application of the Laplace transform yields

$$x(s) = \frac{2}{s(s^2 + 2s + 2)}$$

The quadratic term in the denominator may be factored by use of the quadratic formula.   The roots are found to be $(-1 - j)$ and $(-1 + j)$.   This gives the partial-fraction expansion

$$x(s) = \frac{2}{s(s + 1 + j)(s + 1 - j)} = \frac{A}{s} + \frac{B}{(s + 1 + j)} + \frac{C}{(s + 1 - j)} \tag{3.8}$$

where A, B, and C are constants to be evaluated, so that this relation is an identity in s. The presence of complex factors does not alter the procedure at all. However, the computations may be slightly more tedious.

To obtain A, multiply Eq. (3.8) by s and set $s = 0$:

$$A = \frac{2}{(1+j)(1-j)} = 1$$

To obtain B, multiply Eq. (3.8) by $(s + 1 + j)$ and set $s = (-1 - j)$:

$$B = \frac{2}{(-1-j)(-2j)} = \frac{-1-j}{2}$$

To obtain C, multiply Eq. (3.8) by $(s + 1 - j)$ and set $s = (-1 + j)$:

$$C = \frac{2}{(-1+j)(2j)} = \frac{-1+j}{2}$$

Therefore,

$$x(s) = \frac{1}{s} + \frac{-1-j}{2}\frac{1}{s+1+j} + \frac{-1+j}{2}\frac{1}{s+1-j}$$

This is the desired result. To invert $x(s)$, we may now use the fact that $1/(s + a)$ is the transform of e^{-at}. The fact that a is complex does not invalidate this result, as can be seen by returning to the derivation of the transform of e^{-at}. The result is

$$x(t) = 1 + \frac{-1-j}{2}e^{-(1+j)t} + \frac{-1+j}{2}e^{-(1-j)t}$$

Using the identity

$$e^{(a+jb)t} = e^{at}(\cos bt + j \sin bt)$$

this can be converted to

$$x(t) = 1 - e^{-t}(\cos t + \sin t)$$

The details of this conversion are recommended as an exercise for the reader.

A more general discussion of this case is warranted. It was seen in Example 3.3 that the complex conjugate roots of the denominator of $x(s)$ gave rise to a pair of complex terms in the partial-fraction expansion. The constants in these terms, B and C, proved to be complex conjugates $(-1 - j)/2$ and $(-1 + j)/2$. When these terms were combined through a trigonometric identity, it was found that the complex terms canceled, leaving a real result for $x(t)$. Of course, it is necessary that $x(t)$ be real, since the original differential equation and initial conditions are real.

This information may be utilized as follows: The general case of complex conjugate roots arises in the form

$$x(s) = \frac{F(s)}{(s + k_1 + jk_2)(s + k_1 - jk_2)} \tag{3.9}$$

where $F(s)$ is some real function of s.

For example, in Example 3.3, we had

$$F(s) = \frac{2}{s} \qquad k_1 = 1 \qquad k_2 = 1$$

Expanding (3.9) in partial fractions,

$$\frac{F(s)}{(s + k_1 + jk_2)(s + k_1 - jk_2)} = F_1(s)$$
$$+ \left(\frac{a_1 + jb_1}{s + k_1 + jk_2} + \frac{a_2 + jb_2}{s + k_1 - jk_2} \right) \quad (3.10)$$

where a_1, a_2, b_1, b_2 are the constants to be evaluated in the partial fraction expansion and $F_1(s)$ is a series of fractions arising from $F(s)$.

Again, in Example 3.3,

$$a_1 = -\frac{1}{2} \qquad a_2 = -\frac{1}{2} \qquad b_1 = -\frac{1}{2} \qquad b_2 = \frac{1}{2} \qquad F_1(s) = \frac{1}{s}$$

Now, since the left side of Eq. (3.10) is real for all real s, the right side must also be real for all real s. Since two complex numbers will add to form a real number if they are complex conjugates, it is seen that the right side will be real *for all real s* if and only if the two terms are complex conjugates. Since the denominators of the terms are conjugates, this means that the numerators must also be conjugates, or

$$a_2 = a_1$$
$$b_2 = -b_1$$

This is exactly the result obtained in the specific case of Example 3.3. With this information, Eq. (3.10) becomes

$$\frac{F(s)}{(s + k_1 + jk_2)(s + k_1 - jk_2)} = F_1(s)$$
$$+ \left(\frac{a_1 + jb_1}{s + k_1 + jk_2} + \frac{a_1 - jb_1}{s + k_1 - jk_2} \right) \quad (3.11)$$

Hence, it has been established that terms in the inverse transform arising from the complex conjugate roots may be written in the form

$$(a_1 + jb_1)e^{(-k_1 - jk_2)t} + (a_1 - jb_1)e^{(-k_1 + jk_2)t}$$

Again, using the identity

$$e^{(C_1 + jC_2)t} = e^{C_1 t}(\cos C_2 t + j \sin C_2 t)$$

this reduces to

$$2e^{-k_1 t}(a_1 \cos k_2 t + b_1 \sin k_2 t) \tag{3.12}$$

Let us now rework Example 3.3 using Eq. (3.12).   We return to the point where we arrived by our usual techniques at the conclusion that

$$B = \frac{-1 - j}{2}$$

Comparison of Eqs. (3.8) and (3.11) and the result for B show that we have two possible ways to assign a_1, b_1, k_1, and k_2 so that we match the form of Eq. (3.11).   They are

$$
\begin{array}{ll}
a_1 = -\frac{1}{2} & a_1 = -\frac{1}{2} \\
b_1 = -\frac{1}{2} & b_1 = \frac{1}{2} \\
k_1 = 1 & k_1 = 1 \\
k_2 = 1 & k_2 = -1
\end{array}
$$

or

The first way corresponds to matching the term involving B with the first term of the conjugates of Eq. (3.11), and the second to matching it with the second term.   *In either case*, substitution of these constants into Eq. (3.12) yields

$$-e^{-t}(\cos t + \sin t)$$

which is, as we have seen, the correct term in $x(t)$.

What this means is that one can proceed directly from the evaluation of one of the partial-fraction constants, in this case B, to the complete term in the inverse transform, in this case $-e^{-t}(\cos t + \sin t)$.   It is not necessary to perform all the algebra, since it has been done in the general case to arrive at Eq. (3.12).

Another example will serve to emphasize the application of this technique.

Example 3.4   Solve:

$$\frac{d^2x}{dt^2} + 4x = 2e^{-t}$$
$$x(0) = x'(0) = 0$$

The Laplace transform method yields

$$x(s) = \frac{2}{(s^2 + 4)(s + 1)}$$

Factoring and expanding into partial fractions,

$$\frac{2}{(s + 1)(s + 2j)(s - 2j)} = \frac{A}{s + 1} + \frac{B}{s + 2j} + \frac{C}{s - 2j} \qquad (3.13)$$

Multiplying Eq. (3.13) by $(s + 1)$ and setting $s = -1$ yield

$$A = \frac{2}{(-1 + 2j)(-1 - 2j)} = \frac{2}{5}$$

Multiplying Eq. (3.13) by $(s + 2j)$ and setting $s = -2j$ yield

$$B = \frac{2}{(-2j + 1)(-4j)} = \frac{-2 + j}{10}$$

Matching the term

$$\frac{(-2 + j)/10}{s + 2j}$$

with the first term of the conjugates of Eq. (3.11) requires that

$$a_1 = -\tfrac{2}{10} = -\tfrac{1}{5}$$
$$b_1 = \tfrac{1}{10}$$
$$k_1 = 0$$
$$k_2 = 2$$

Substituting in (3.12) results in

$$-\tfrac{2}{5} \cos 2t + \tfrac{1}{5} \sin 2t$$

Hence the complete answer is

$$x(t) = \tfrac{2}{5}e^{-t} - \tfrac{2}{5} \cos 2t + \tfrac{1}{5} \sin 2t$$

The reader should verify that this answer satisfies the differential equation and boundary conditions. In addition, he should show that it can also be obtained by matching the term with the second term of the conjugates of Eq. (3.11) or by determining C instead of B.

In the next example, an exceptional case is considered; the denominator of $x(s)$ has repeated roots. The procedure in this case will vary slightly from that of the previous cases.

Example 3.5   Solve:

$$\frac{d^3x}{dt^3} + \frac{3d^2x}{dt^2} + \frac{3dx}{dt} + x = 1$$
$$x(0) = x'(0) = x''(0) = 0$$

Application of the Laplace transform yields

$$x(s) = \frac{1}{s(s^3 + 3s^2 + 3s + 1)}$$

Factoring and expanding in partial fractions,

$$x(s) = \frac{1}{s(s + 1)^3} = \frac{A}{s} + \frac{B}{(s + 1)^3} + \frac{C}{(s + 1)^2} + \frac{D}{s + 1} \tag{3.14}$$

As in the previous cases, to determine A, multiply both sides by s and then set s to zero. This yields

$$A = 1$$

Multiplication of both sides of Eq. (3.14) by $(s + 1)^3$ results in

$$\frac{1}{s} = \frac{A(s + 1)^3}{s} + B + C(s + 1) + D(s + 1)^2 \tag{3.15}$$

Setting $s = -1$ in Eq. (3.15) gives

$$B = -1$$

However, if we attempt to determine C by multiplying Eq. (3.14) by $(s + 1)^2$ and then setting $s = -1$, the term involving B becomes infinite. Similar remarks apply to determination of D. Hence, an alternate procedure is required to find C and D.

If Eq. (3.15) is differentiated with respect to s, the result is

$$\frac{-1}{s^2} = \frac{A(s + 1)^2(2s - 1)}{s^2} + C + 2D(s + 1) \tag{3.16}$$

The result of this differentiation is advantageous, because it eliminates B and frees C of the factor $(s + 1)$ while leaving A and D with this factor. Now, setting $s = -1$ in Eq. (3.16) yields

$$C = -1$$

To obtain D, differentiate again with respect to s.

$$\frac{2}{s^3} = 2A \frac{(s + 1)(s^2 - s + 1)}{s^3} + 2D \tag{3.17}$$

Now we have eliminated C, freed D of the $(s + 1)$, and still have A multiplied by $(s + 1)$. Setting $s = -1$ in Eq. (3.17) gives

$$D = -1$$

The final result is then

$$x(s) = \frac{1}{s} - \frac{1}{(s + 1)^3} - \frac{1}{(s + 1)^2} - \frac{1}{s + 1} \tag{3.18}$$

Referring to Table 2.1, this can be inverted to

$$x(t) = 1 - e^{-t}\left(\frac{t^2}{2} + t + 1\right) \tag{3.19}$$

The reader should verify that Eq. (3.18) placed over a common denominator results in the original form

$$x(s) = \frac{1}{s(s + 1)^3}$$

and that Eq. (3.19) satisfies the differential equation and initial conditions.

The result of Example 3.5 may be generalized. The appearance of the factor $(s + a)^n$ in the denominator of $x(s)$ leads to n terms in the partial-fraction expansion:

$$\frac{C_1}{(s + a)^n}, \frac{C_2}{(s + a)^{n-1}}, \cdots, \frac{C_n}{s + a}$$

The constant C_1 can be determined as usual by multiplying the expansion by $(s + a)^n$ and setting $s = -a$. The other constants are determined by successive differentiations of the result of the multiplication. These terms, according to Table 2.1, lead to the following expression as the inverse transform:

$$\left[\frac{C_1}{(n-1)!} t^{n-1} + \frac{C_2}{(n-2)!} t^{n-2} + \cdots + C_{n-1}t + C_n \right] e^{-at} \qquad (3.20)$$

It is interesting to recall that, in the classical method for solving these equations, one treats repeated roots of the characteristic equation by postulating the form of Eq. (3.20) and selecting the constants to fit the initial conditions.

Qualitative Nature of Solutions If we are interested *only in the form* of the solution $x(t)$, which is often the case in our work, *this information may be obtained directly from the roots of the denominator of $x(s)$.* As an illustration of this "qualitative" approach to differential equations consider Example 3.3 in which

$$x(s) = \frac{2}{s(s^2 + 2s + 2)} = \frac{A}{s} + \frac{B}{s+1+j} + \frac{C}{s+1-j}$$

is the transformed solution of

$$\frac{d^2x}{dt^2} + \frac{2dx}{dt} + 2x = 2$$

It is evident by inspection of the partial-fraction expansion, *without* evaluation of the constants, that the s in the denominator of $x(s)$ will give rise to a constant in $x(t)$. Also, since the roots of the quadratic term are $-1 \pm j$, it is known that $x(t)$ must contain terms of the form $e^{-t}(C_1 \cos t + C_2 \sin t)$. This may be sufficient information for our purposes. Alternatively, we may be interested in the behavior of $x(t)$ as $t \to \infty$. It is clear that the terms involving sin and cos vanish because of the factor e^{-t}. Therefore, $x(t)$ ultimately approaches the constant, which by inspection must be unity.

The qualitative nature of the solution $x(t)$ can be related to the location of the roots of the denominator of $x(s)$ in the complex plane. These roots are the roots of the characteristic equation and the roots of the denominator of the transformed forcing function. Consider Fig. 3.1, a drawing of the complex plane, in which several typical roots are located and labeled with their coordinates. Table 3.1 gives the form of the terms

Table 3.1

Roots	Terms in $x(t)$ for $t > 0$
s_1	$C_1 e^{-a_1 t}$
$s_2,\ s_2^*$	$e^{-a_2 t}(C_1 \cos b_2 t + C_2 \sin b_2 t)$
$s_3,\ s_3^*$	$C_1 \cos b_3 t + C_2 \sin b_3 t$
$s_4,\ s_4^*$	$e^{a_4 t}(C_1 \cos b_4 t + C_2 \sin b_4 t)$
s_5	$C_1 e^{a_5 t}$
s_6	C_1

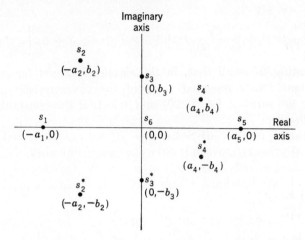

Fig. 3.1 **Location of typical roots of characteristic equation.**

in the equation for $x(t)$, corresponding to these roots.   Note that all constants, a_1, a_2, . . . , b_1, b_2, . . . , are taken as positive.   The constants C_1 and C_2 are arbitrary and can be determined by the partial-fraction expansion techniques.   As discussed above, this determination is often not necessary for our work.

If any of these roots are repeated, the term given in Table 3.1 is multiplied by a power series in t,

$$K_1 + K_2 t + K_3 t^2 + \cdots K_r t^{r-1}$$

where r is the number of repetitions of the root and the constants K_1, K_2, . . . , K_r can be evaluated by partial-fraction expansion.

It is thus evident that the imaginary axis divides the root locations into distinct areas, with regard to the behavior of the corresponding terms in $x(t)$ as t becomes large.   Terms corresponding to roots to the left of the imaginary axis vanish exponentially in time, while those corresponding to roots to the right of the imaginary axis increase exponentially in time. Terms corresponding to roots at the origin behave as power series in time, a constant being considered as a degenerate power series.   Terms corresponding to roots located elsewhere on the imaginary axis oscillate with constant amplitude in time unless they are multiple roots in which case the amplitude of oscillation increases as a power series in time.   Much use will be made of this information in later sections of the text.

Summary   The reader now has at his disposal the basic tools for use of Laplace transforms to solve differential equations.   In addition, it is now possible to obtain considerable information about the qualitative nature of the solution with a minimum of labor.   It should be pointed out that it is always necessary to factor the denominator of $x(s)$ in order to obtain any information about $x(t)$. If this denominator is a polynomial of order three or more, this may be far from a trivial problem. Chapters 15 to 17 are largely devoted to a solution of this problem within the context of control applications.

The next chapter is a grouping of several Laplace transform theorems which will find later application. In addition, a discussion of the impulse function $\delta(t)$ is presented there. Unavoidably, this chapter is rather dry. It may be desirable for the reader to skip directly to Chap. 5, where our control studies begin. At each point where a theorem of Chap. 4 is applied, reference to the appropriate section of Chap. 4 can be made.

PROBLEMS

3.1 Solve the following using Laplace transforms:

a. $\dfrac{d^2x}{dt^2} + \dfrac{dx}{dt} + x = 1$ $\qquad x(0) = x'(0) = 0$

b. $\dfrac{d^2x}{dt^2} + \dfrac{2\,dx}{dt} + x = 1$ $\qquad x(0) = x'(0) = 0$

c. $\dfrac{d^2x}{dt^2} + \dfrac{3\,dx}{dt} + x = 1$ $\qquad x(0) = x'(0) = 0$

Sketch the behavior of these solutions on a single graph. What is the effect of the coefficient of dx/dt?

3.2 Solve the following differential equations by Laplace transforms:

a. $\dfrac{d^4x}{dt^4} + \dfrac{d^3x}{dt^3} = \cos t$ $\qquad x(0) = x'(0) = x'''(0) = 0$ $\qquad x''(0) = 1$

b. $\dfrac{d^2q}{dt^2} + \dfrac{dq}{dt} = t^2 + 2t$ $\qquad q(0) = 4$ $\qquad q'(0) = -2$

3.3 Invert the following transforms:

a. $\dfrac{3s}{(s^2 + 1)(s^2 + 4)}$

b. $\dfrac{1}{s(s^2 - 2s + 5)}$

c. $\dfrac{3s^3 - s^2 - 3s + 2}{s^2(s - 1)^2}$

4 *Further Properties of Transforms*

This chapter is a collection of theorems and results relative to the Laplace transformation. The theorems are selected because of their applicability to problems in control theory. Other theorems and properties of the Laplace transformation are available in standard texts.[1] In later chapters, the theorems presented here will be used as needed.

Final-value Theorem If $f(s)$ is the Laplace transform of $f(t)$, then

$$\lim_{t \to \infty} [f(t)] = \lim_{s \to 0} [sf(s)]$$

provided that $sf(s)$ does not become infinite for any value of s satisfying Re $(s) \geq 0$. If this condition does not hold, $f(t)$ does not approach a limit as $t \to \infty$.

---

[1] See, for example, R. V. Churchill, "Operational Mathematics," 2d ed., chaps. 2 and 3, McGraw-Hill Book Company, New York, 1958.

Proof:

From the Laplace transform of a derivative, we have

$$\int_0^\infty \frac{df}{dt} e^{-st}\, dt = sf(s) - f(0)$$

Hence,

$$\lim_{s \to 0} \int_0^\infty \frac{df}{dt} e^{-st}\, dt = \lim_{s \to 0} [sf(s)] - f(0)$$

It can be shown that the order of the integration and limit operation on the left side of this equation can be interchanged if the conditions of the theorem hold. Doing this gives

$$\int_0^\infty \frac{df}{dt}\, dt = \lim_{s \to 0} [sf(s)] - f(0)$$

Evaluating the integral,

$$\lim_{t \to \infty} [f(t)] - f(0) = \lim_{s \to 0} [sf(s)] - f(0)$$

which immediately yields the desired result.

Example 4.1 Find the final value of the function $x(t)$ for which the Laplace transform is

$$x(s) = \frac{1}{s(s^3 + 3s^2 + 3s + 1)}$$

Direct application of the final-value theorem yields

$$\lim_{t \to \infty} [x(t)] = \lim_{s \to 0} \frac{1}{s^3 + 3s^2 + 3s + 1} = 1$$

As a check, note that this transform was inverted in Example 3.5 to give

$$x(t) = 1 - e^{-t}\left(\frac{t^2}{2} + t + 1\right)$$

which approaches unity as t approaches infinity. Note that, since the denominator of $sx(s)$ can be factored to $(s + 1)^3$, the conditions of the theorem are satisfied; that is, $(s + 1)^3 \neq 0$ unless $s = -1$.

Example 4.2 Find the final value of the function $x(t)$ for which the Laplace transform is

$$x(s) = \frac{s^4 - 6s^2 + 9s - 8}{s(s - 2)(s^3 + 2s^2 - s - 2)}$$

In this case, the function $sx(s)$ can be written

$$sx(s) = \frac{s^4 - 6s^2 + 9s - 8}{(s + 1)(s + 2)(s - 1)(s - 2)}$$

Since this becomes infinite for $s = 1$ and $s = 2$, the conditions of the theorem are not satisfied.   Note that we inverted this transform in Example 3.2, where it was found that

$$x(t) = -2 + \tfrac{1}{12} e^{2t} + 11\tfrac{1}{3} e^{-t} - 17\tfrac{1}{12} e^{-2t} + \tfrac{2}{3} e^{t}$$

This function continues to grow exponentially with t and, as expected, does not approach a limit.

The proof of the next theorem closely parallels the proof of the last one and is left as an exercise for the reader.

Initial-value Theorem

$$\lim_{t \to 0} [f(t)] = \lim_{s \to \infty} [sf(s)]$$

The conditions on this theorem are not so stringent as those upon the last because for functions of interest to us the order of integration and limiting process need not be interchanged to establish the result.

Example 4.3   Find the initial value $x(0)$ of the function which has the following transform:

$$x(s) = \frac{s^4 - 6s^2 + 9s - 8}{s(s - 2)(s^3 + 2s^2 - s - 2)}$$

The function $sx(s)$ is written in the form

$$sx(s) = \frac{s^4 - 6s^2 + 9s - 8}{s^4 - 5s^2 + 4}$$

Upon performing the indicated long division, this becomes

$$sx(s) = 1 - \frac{s^2 - 9s + 12}{s^4 - 5s^2 + 4}$$

which clearly goes to unity as s becomes infinite.   Hence

$$x(0) = 1$$

which again checks Example 3.2.

Translation of Transform   If $L\{f(t)\} = f(s)$, then

$$L\{e^{-at}f(t)\} = f(s + a)$$

In other words, the variable in the transform s is translated by a.

Proof:

$$L\{e^{-at}f(t)\} = \int_0^\infty f(t)e^{-(s+a)t} \, dt = f(s + a)$$

Example 4.4   Find $L\{e^{-at} \cos kt\}$.   Since

$$L\{\cos kt\} = \frac{s}{s^2 + k^2}$$

then by the previous theorem,

$$L\{e^{-at} \cos kt\} = \frac{s+a}{(s+a)^2 + k^2}$$

which checks Table 2.1.

A primary use for this theorem is in the inversion of transforms. For example, using this theorem the transform

$$x(s) = \frac{1}{(s+a)^2}$$

can be immediately inverted to

$$x(t) = te^{-at}$$

Translation of Function

If $L\{f(t)\} = f(s)$, then

$$L\{f(t - t_0)\} = e^{-st_0}f(s)$$

provided that

$$f(t) = 0 \qquad \text{for } t < 0$$

(which will always be true for functions we use).

Before proving this theorem, it may be desirable to clarify the relationship between $f(t - t_0)$ and $f(t)$. This is done for an arbitrary function $f(t)$ in Fig. 4.1, where it can be seen that $f(t - t_0)$ is simply translated horizontally from $f(t)$ through a distance t_0.

Proof:

$$L\{f(t - t_0)\} = \int_0^\infty f(t - t_0)e^{-st}\,dt$$
$$= e^{-st_0}\int_{-t_0}^\infty f(t - t_0)e^{-s(t-t_0)}\,d(t - t_0)$$

But since $f(t) = 0$ for $t < 0$, the lower limit of this integral may be replaced by zero. Since $(t - t_0)$ is now the dummy variable of integration, the integral may be recognized as the Laplace transform of $f(t)$; thus, the theorem is proved.

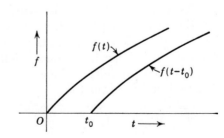

Fig. 4.1   Illustration of $f(t - t_0)$ as related to $f(t)$.

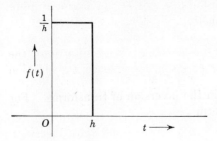

Fig. 4.2 Pulse function of Example 4.5.

This result is also useful in inverting transforms. Thus, it follows that, if $f(t)$ is the inverse transform of $f(s)$, then the inverse transform of

$$e^{-st_0}f(s)$$

is the function

$$g(t) = \begin{cases} 0 & t < t_0 \\ f(t - t_0) & t > t_0 \end{cases}$$

Example 4.5   Find the Laplace transform of

$$f(t) = \begin{cases} 0 & t < 0 \\ \dfrac{1}{h} & 0 < t < h \\ 0 & t > h \end{cases}$$

This function is pictured in Fig. 4.2. It is clear that $f(t)$ may be represented by the difference of two functions,

$$f(t) = \frac{1}{h}[u(t) - u(t - h)]$$

where $u(t - h)$ is the unit-step function translated h units to the right. We may now use the linearity of the transform and the previous theorem to write immediately

$$f(s) = \frac{1}{h}\frac{1 - e^{-hs}}{s}$$

This result is of considerable value in establishing the transform of the unit-impulse function, as will be described in the next section.

Transform of the Unit-impulse Function   Consider again the function of Example 4.5. If we allow h to shrink to zero, we obtain a new function which is zero everywhere except at the origin, where it is infinite. However, it is important to note that the area under this function always remains equal to unity. We call this new function $\delta(t)$, and the fact that its area is unity means that

$$\int_{-\infty}^{\infty} \delta(t)\, dt = 1$$

The graph of $\delta(t)$ appears as a line of infinite height at the origin, as indicated in Table 2.1. $\delta(t)$ is called the unit-impulse function or, alternatively, the delta function.

We remark here that, in the strict mathematical sense of a limit, the function $f(t)$ does not possess a limit as h goes to zero. Hence, the function $\delta(t)$ does not fit the strict mathematical definition of a function. To assign a mathematically precise meaning to the unit-impulse function requires use of the theory of distributions, which is clearly beyond the scope of this text. However, for our work in automatic control, we shall be able to obtain useful results by formal manipulation of the delta function, and hence we ignore these mathematical difficulties.

We have derived in Example 4.5 the Laplace transform of $f(t)$ as

$$L\{f(t)\} = \frac{1 - e^{-hs}}{hs}$$

Formally, then, the Laplace transform of $\delta(t)$ can be obtained by letting h go to zero in $L\{f(t)\}$. Applying L'Hôpital's rule,

$$L\{\delta(t)\} = \lim_{h \to 0} \frac{1 - e^{-hs}}{hs} = \lim_{h \to 0} \frac{se^{-hs}}{s} = 1 \tag{4.1}$$

This "verifies" the entry in Table 2.1.

It is interesting to note that, since we rewrote $f(t)$ in Example 4.5 as

$$f(t) = \frac{1}{h}[u(t) - u(t - h)]$$

then $\delta(t)$ can be written as

$$\delta(t) = \lim_{h \to 0} \frac{u(t) - u(t - h)}{h}$$

In this form, the delta function appears as the derivative of the unit-step function. The reader may find it interesting to ponder this statement in relation to the graphs of $\delta(t)$ and $u(t)$ and in relation to the integral of $\delta(t)$ discussed above.

The unit-impulse function finds use as an idealized disturbance in control systems analysis and design.

Transform of an Integral If $L\{f(t)\} = f(s)$, then

$$L\left\{\int_0^t f(t) \, dt\right\} = \frac{f(s)}{s}$$

This important theorem is closely related to the theorem on differentiation. Since the operations of differentiation and integration are inverses of each other when applied to the time functions, i.e.,

$$\frac{d}{dt} \int_0^t f(t) \, dt = \int_0^t \frac{df}{dt} \, dt = f(t) \tag{4.2}$$

it is to be expected that these operations when applied to the transforms will also be inverses. Thus assuming the theorem to be valid, Eq. (4.2) in the transformed variable s becomes

$$s \frac{f(s)}{s} = \frac{1}{s} \, sf(s) = f(s)$$

In other words multiplication of $f(s)$ by s corresponds to differentiation of $f(t)$ with respect to t, and division of $f(s)$ by s corresponds to integration of $f(t)$ with respect to t.

The proof follows from a straightforward integration by parts.

$$f(s) = \int_0^\infty f(t) e^{-st} \, dt$$

Let

$$u = e^{-st} \qquad dv = f(t) \, dt$$

Then

$$du = -se^{-st} \, dt \qquad v = \int_0^t f(t) \, dt$$

Hence,

$$f(s) = e^{-st} \int_0^t f(t) \, dt \, \bigg|_0^\infty + s \int_0^\infty \left[\int_0^t f(t) \, dt \right] e^{-st} \, dt$$

Since $f(t)$ must satisfy the requirements for possession of a transform, it can be shown that the infinite integral appearing in the upper limit of the first term on the right side is finite. This means that the upper limit vanishes because of the factor e^{-st}. Furthermore, the lower limit clearly vanishes, and hence, there is no contribution from the first term. The second term may be recognized as $sL \left\{ \int_0^t f(t) \, dt \right\}$, and the theorem follows immediately.

Example 4.6   Solve the following equation for $x(t)$:

$$\frac{dx}{dt} = \int_0^t x(t) \, dt - t$$
$$x(0) = 3$$

Taking the Laplace transform of both sides, and making use of the previous theorem

$$sx(s) - 3 = \frac{x(s)}{s} - \frac{1}{s^2}$$

Solving for $x(s)$,

$$x(s) = \frac{3s^2 - 1}{s(s^2 - 1)} = \frac{3s^2 - 1}{s(s + 1)(s - 1)}$$

This may be expanded into partial fractions according to the usual procedure to give

$$x(s) = \frac{1}{s} + \frac{1}{s + 1} + \frac{1}{s - 1}$$

Hence,

$$x(t) = 1 + e^{-t} + e^t$$

The reader should verify that this function satisfies the original equation.

PROBLEMS

4.1 If a forcing function $f(t)$ has the Laplace transform

$$f(s) = \frac{1}{s} + \frac{e^{-s} - e^{-2s}}{s^2} - \frac{e^{-3s}}{s}$$

graph the function $f(t)$.

4.2 Solve the following equation for $y(t)$:

$$\int_0^t y(\tau)\, d\tau = \frac{dy\,(t)}{dt} \qquad y(0) = 1$$

part **II** **Linear Open-loop Systems**

5 Response of First-order Systems

Before discussing a complete control system, it is necessary to become familiar with the responses of some of the simple basic systems which often form the building blocks of a control system. In this chapter and the three to follow, we shall describe in detail the behavior of several basic systems and show that a great variety of physical systems can be represented by a combination of these basic systems. Some of the terms and conventions which have become well established in the field of automatic control will also be introduced.

TRANSFER FUNCTION

Mercury Thermometer We shall develop the *transfer function* for a *first-order system* by considering the unsteady-state behavior of an ordinary mercury-in-glass thermometer. A cross-sectional view of the bulb is shown in Fig. 5.1.

Consider the thermometer to be located in a flowing stream of fluid for which the temperature x varies with time. Our problem is to calculate

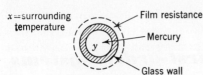

$x =$ surrounding
temperature

Film resistance

Mercury

Glass wall

Fig. 5.1   **Cross-sectional view of thermometer.**

the *response* or the time variation of the thermometer reading y for a particular change in x.[1]

The following assumptions[2] will be used in this analysis:

1. All the resistance to heat transfer resides in the film surrounding
the bulb (i.e., the resistance offered by the glass and mercury is neglected).

2. All the thermal capacity is in the mercury.   Furthermore, at any
instant the mercury assumes a uniform temperature throughout.

3. The glass wall containing the mercury does not expand or contract
during the transient response.   (In an actual thermometer, the expansion
of the wall has an additional effect on the response of the thermometer
reading.[3])

It is assumed that the thermometer is initially at steady state.   This
means that, before time zero, there is no change in temperature with time.
At time zero the thermometer will be subjected to some change in the
surrounding temperature $x(t)$.

By applying the unsteady-state energy balance

Input rate $-$ output rate = rate of accumulation

we get the result

$$hA(x - y) - 0 = mC\frac{dy}{dt} \tag{5.1}$$

where A = surface area of bulb for heat transfer, ft²
       C = heat capacity of mercury, Btu/(lb$_m$)(°F)
       m = mass of mercury in bulb, lb$_m$
       t = time, hr
       h = film coefficient of heat transfer, Btu/(hr)(ft²)(°F)
For illustrative purposes, typical engineering units have been used.

[1] In order that the result of the analysis of the thermometer be general and therefore
applicable to other first-order systems, the symbols x and y have been selected to represent surrounding temperature and thermometer reading.

[2] Making the first two assumptions is often referred to as *lumping of parameters*
because all the resistance is "lumped" into one location and all the capacitance into
another.   As shown in the analysis, these assumptions make it possible to represent the
dynamics of the system by an ordinary differential equation.   If such assumptions were
not made, the analysis would lead to a partial differential equation, and the representation would be referred to as a *distributed-parameter system*.   In Chap. 25, distributed-
parameter systems will be considered in detail.

[3] See Iinoya Koichi and R. J. Altepeter, Inverse Response in Process Control, IEC,
54(7): 39–43 (July, 1962).

Equation (5.1) states that the rate of flow of heat through the film resistance surrounding the bulb causes the internal energy of the mercury to increase at the same rate. The increase in internal energy is manifested by a change in temperature and a corresponding expansion of mercury which causes the mercury column, or "reading" of the thermometer, to rise.

The coefficient h will depend on the flow rate and properties of the surrounding fluid and the dimensions of the bulb. We shall assume that h is constant for a particular installation of the thermometer.

Our analysis has resulted in Eq. (5.1), which is a first-order differential equation. Before solving this equation by means of the Laplace transform we shall introduce *deviation variables* into Eq. (5.1). The reason for these new variables will soon become apparent. Prior to the change in x, the thermometer is at steady state and the derivative dy/dt is zero. For the steady-state condition, Eq. (5.1) may be written

$$hA(x_s - y_s) = 0 \qquad t < 0 \tag{5.2}$$

The subscript s is used to indicate that the variable is the steady-state value. Equation (5.2) simply states that $y_s = x_s$, or the thermometer reads the true bath temperature. Subtracting Eq. (5.2) from Eq. (5.1) gives

$$hA[(x - x_s) - (y - y_s)] = mC \frac{d(y - y_s)}{dt} \tag{5.3}$$

Notice that $d(y - y_s)/dt = dy/dt$ because y_s is a constant.

If we define the deviation variables to be the differences between the variables and their steady-state values

$$X = x - x_s$$
$$Y = y - y_s$$

Eq. (5.3) becomes

$$hA(X - Y) = mC \frac{dY}{dt} \tag{5.4}$$

If we let $mC/hA = \tau$, Eq. (5.4) becomes

$$X - Y = \tau \frac{dY}{dt} \tag{5.5}$$

Taking the Laplace transform of Eq. (5.5) gives

$$X(s) - Y(s) = \tau s Y(s) \tag{5.6}$$

Rearranging Eq. (5.6) as a ratio of $Y(s)$ to $X(s)$ gives

$$\frac{Y(s)}{X(s)} = \frac{1}{\tau s + 1} \tag{5.7}$$

The parameter τ is called the *time constant* of the system and has the units of time.

The expression on the right side of Eq. (5.7) is called the *transfer function* of the system. It is the ratio of the Laplace transform of the deviation in thermometer reading to the Laplace transform of the deviation in the surrounding temperature. In examining other physical systems, we shall usually attempt to obtain a transfer function.

Any physical system for which the relation between Laplace transforms of input and output deviation variables is of the form given by Eq. (5.7) is called a *first-order system*. Synonyms for first-order system are first-order lag and single exponential stage. All these terms are motivated by the fact that Eq. (5.7) results from a first-order, linear differential equation, Eq. (5.5). In Chap. 6, we shall discuss a number of other physical systems which are first-order.

By reviewing the steps leading to Eq. (5.7), it can be seen that the introduction of deviation variables prior to taking the Laplace transform of the differential equation results in a transfer function that is free of initial conditions because the initial values of X and Y are zero. In control system engineering, we are primarily concerned with the deviations of system variables from their steady-state values. The use of deviation variables is, therefore, natural as well as convenient.

Properties of Transfer Functions   In general, a transfer function relates two variables in a physical process; one of these is the cause (forcing function or input variable) and the other is the effect (response or output variable). In terms of the example of the mercury thermometer, the surrounding temperature was the cause or input, whereas the thermometer reading was the effect or output. We may write

$$\text{Transfer function} = G(s) = \frac{Y(s)}{X(s)}$$

where $G(s)$ = symbol for transfer function
   $X(s)$ = transform of forcing function or input, in deviation form
   $Y(s)$ = transform of response or output, in deviation form

The transfer function completely describes the dynamic characteristics of the system. If we select a particular input variation $X(t)$ for which the transform is $X(s)$, the response of the system is simply

$$Y(s) = G(s)X(s) \tag{5.8}$$

By taking the inverse of $Y(s)$, we get $Y(t)$, the response of the system.

The transfer function results from a linear differential equation; therefore, the principle of superposition is applicable. This means that the transformed response of a system with transfer function $G(s)$ to a forcing

Fig. 5.2 **Block diagram.**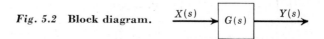

function

$$X(s) = a_1 X_1(s) + a_2 X_2(s)$$

where X_1 and X_2 are particular forcing functions and a_1 and a_2 are constants, is

$$\begin{aligned} Y(s) &= G(s)X(s) \\ &= a_1 G(s) X_1(s) + a_2 G(s) X_2(s) \\ &= a_1 Y_1(s) + a_2 Y_2(s) \end{aligned}$$

$Y_1(s)$ and $Y_2(s)$ are the responses to X_1 and X_2 alone, respectively. For example, the response of the mercury thermometer to a sudden change in surrounding temperature of 10°F is simply twice the response to a sudden change of 5°F in surrounding temperature.

The functional relationship contained in a transfer function is often expressed by a *block-diagram* representation, as shown in Fig. 5.2. The arrow entering the box is the forcing function or input variable, and the arrow leaving the box is the response or output variable. Inside the box is placed the transfer function. We state that the transfer function $G(s)$ in the box "operates" on the input function $X(s)$ to produce an output function $Y(s)$. The usefulness of the block diagram will be appreciated in Chap. 9, when a complete control system containing several blocks is analyzed.

TRANSIENT RESPONSE

Now that the transfer function of a first-order system has been established, we can easily obtain its transient response to *any* forcing function. Since this type of system occurs so frequently in practice, it is worthwhile to study its response to several common forcing functions: step, impulse, and sinusoidal. These forcing functions have been found to be most useful in theoretical and experimental aspects of process control. They will be used extensively in our studies of this subject, and hence we discuss each one further before studying the transient response of the first-order system to these forcing functions.

Forcing Functions *Step function* Mathematically, the step function of magnitude A can be expressed as

$$X(t) = Au(t)$$

where $u(t)$ is the unit-step function defined in Chap. 2. A graphical representation is shown in Fig. 5.3.

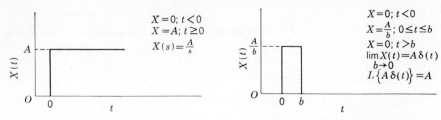

$$X = 0;\ t < 0$$
$$X = A;\ t \geq 0$$
$$X(s) = \frac{A}{s}$$

$$X = 0;\ t < 0$$
$$X = \frac{A}{b};\ 0 \leq t \leq b$$
$$X = 0;\ t > b$$
$$\lim_{b \to 0} X(t) = A\delta(t)$$
$$L.\{A\delta(t)\} = A$$

Fig. 5.3 Step input.   Fig. 5.4 Impulse function.

The transform of this function is $X(s) = A/s$. A step function can be approximated very closely in practice. For example, a step change in flow rate can be obtained by the sudden opening of a valve.

Impulse function Mathematically, the impulse function of magnitude A is defined as

$$X(t) = A\delta(t)$$

where $\delta(t)$ is the unit-impulse function defined and discussed in Chap. 4. A graphical representation of this function, before the limit is taken, is shown in Fig. 5.4.

The true impulse function, obtained by letting $b \to 0$ in Fig. 5.4, has Laplace transform A. It is used more frequently as a mathematical aid than as an actual input to a physical system. For some systems it is difficult even to approximate an impulse forcing function. For this reason the representation of Fig. 5.4 is valuable, since this form can usually be approximated physically by application and removal of a step function. If the time duration b is sufficiently small, we shall see in Chap. 6 that the forcing function of Fig. 5.4 gives a response which closely resembles the response to a true impulse. In this sense, we often justify the use of A as the Laplace transform of the physically realizable forcing function of Fig. 5.4.

Sinusoidal input This function is represented mathematically by the equations

$$X = 0 \qquad\qquad t < 0$$
$$X = A \sin \omega t \qquad t \geq 0$$

where A is the amplitude and ω is the radian frequency. The radian frequency ω is related to the frequency f in cycles per unit time by $\omega = 2\pi f$. Figure 5.5 shows the graphical representation of this function. The transform is $X(s) = A\omega/(s^2 + \omega^2)$. This forcing function forms the basis of an important branch of control theory known as *frequency response*. Historically, a large segment of the development of control theory was based on frequency-response methods, which will be presented in Chaps. 18 to 21. Physically, it is more difficult to obtain a sinusoidal forcing function in most process variables than to obtain a step function.

This completes the discussion of some of the common forcing functions. We shall now devote our attention to the transient response of the first-order system to each of the forcing functions just discussed.

Step Response   If a step change of magnitude A is introduced into a first-order system, the transform of $X(t)$ is

$$X(s) = \frac{A}{s} \tag{5.9}$$

The transfer function, which is given by Eq. (5.7), is

$$\frac{Y(s)}{X(s)} = \frac{1}{\tau s + 1} \tag{5.7}$$

Combining Eqs. (5.7) and (5.9) gives

$$Y(s) = \frac{A}{s} \frac{1}{\tau s + 1} \tag{5.10}$$

This can be expanded by partial fractions to give

$$Y(s) = \frac{A/\tau}{(s)(s + 1/\tau)} = \frac{C_1}{s} + \frac{C_2}{s + 1/\tau} \tag{5.11}$$

Solving for the constants C_1 and C_2 by the techniques covered in Chap. 3 gives $C_1 = A$ and $C_2 = -A$. Inserting these constants into Eq. (5.11) and taking the inverse transform give the time response for Y:

$$\begin{aligned} Y(t) &= 0 & t &< 0 \\ Y(t) &= A(1 - e^{-t/\tau}) & t &\geq 0 \end{aligned} \tag{5.12}$$

Hereafter, for the sake of brevity, it will be understood that, as in Eq. (5.12), the response is zero before $t = 0$. Equation (5.12) is plotted in Fig. 5.6 in terms of the dimensionless quantities $Y(t)/A$ and t/τ.

Having obtained the step response, Eq. (5.12), from a purely mathematical approach, we should consider whether or not the result seems to be correct from physical principles. Immediately after the thermometer is placed in the new environment, the temperature difference between the

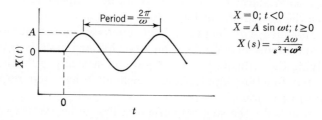

Fig. 5.5   Sinusoidal input.

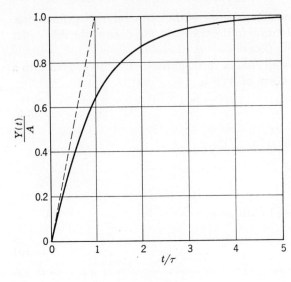

Fig. 5.6 Response of a first-order system to a step input.

mercury in the bulb and the bath temperature is at its maximum value. With our simple lumped-parameter model, we should expect the flow of heat to commence immediately, with the result that the mercury temperature rises, causing a corresponding rise in the column of mercury. As the mercury temperature rises, the driving force causing heat to flow into the mercury will diminish, with the result that the mercury temperature changes at a slower rate as time proceeds. We see that this description of the response based on physical grounds does agree with the response given by Eq. (5.12) and shown graphically in Fig. 5.6.

Several features of this response, worth remembering, are

1. The value of $Y(t)$ reaches 63.2 percent of its ultimate value when the time elapsed is equal to one time constant τ. When the time elapsed is 2τ, 3τ, and 4τ, the percent response is 86.5, 95, and 98, respectively. From these facts, one can consider the response essentially completed in three to four time constants.

2. One can show from Eq. (5.12) that the slope of the response curve at the origin in Fig. 5.6 is 1. This means that, if the initial rate of change of $Y(t)$ were maintained, the response would be complete in one time constant. (See the dotted line in Fig. 5.6.)

3. A consequence of the principle of superposition is that the response to a step input of any magnitude A may be obtained directly from Fig. 5.6 by multiplying the ordinate by A. Figure 5.6 actually gives the response to a unit-step function input, from which all other step responses are derived by superposition.

These results for the step response of a first-order system will now be applied to the following example.

Example 5.1   A thermometer having a time constant[1] of 0.1 min is at a steady-state temperature of 90°F.   At time $t = 0$, the thermometer is placed in a temperature bath maintained at 100°F.   Determine the time needed for the thermometer to read 98°.

In terms of symbols used in this chapter, we have

$$\tau = 0.1 \text{ min} \qquad x_s = 90° \qquad A = 10°$$

The ultimate thermometer reading will, of course, be 100°, and the ultimate value of the deviation variable $Y(\infty)$ is 10°.   When the thermometer reads 98°, $Y(t) = 8°$.

Substituting into Eq. (5.12) the appropriate values of Y, A, and τ gives

$$8 = 10(1 - e^{-t/0.1})$$

Solving this equation for t yields

$$t = 0.161 \text{ min}$$

The same result can also be obtained by referring to Fig. 5.6, where it is seen that $Y/A = 0.8$ at $t/\tau = 1.6$.

Impulse Response   The impulse response of a first-order system will now be developed.   Anticipating the use of superposition, we consider a unit impulse for which the Laplace transform is

$$X(s) = 1 \tag{5.13}$$

Combining this with the transfer function for a first-order system, which is given by Eq. (5.7), results in

$$Y(s) = \frac{1}{\tau s + 1} \tag{5.14}$$

This may be rearranged to

$$Y(s) = \frac{1/\tau}{s + 1/\tau} \tag{5.15}$$

The inverse of $Y(s)$ can be found directly from the table of transforms and can be written in the form

$$\tau Y(t) = e^{-t/\tau} \tag{5.16}$$

A plot of this response is shown in Fig. 5.7 in terms of the variables t/τ and $\tau Y(t)$.   The response to an impulse of magnitude A is obtained, as usual, by multiplying $\tau Y(t)$ from Fig. 5.7 by A/τ.

Notice that the response rises immediately to 1.0 and then decays exponentially.   Such an abrupt rise is, of course, physically impossible, but as we shall see in Chap. 6, it is approached by the response to a finite pulse of narrow width, such as that of Fig. 5.4.

[1] The time constant given in this problem applies to the thermometer when it is located in the temperature bath.   The time constant for the thermometer in air will be considerably different from that given because of the lower heat-transfer coefficient in air.

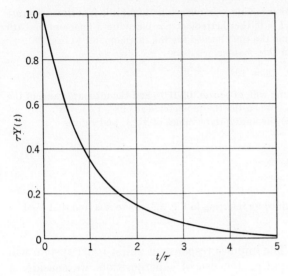

Fig. 5.7   **Unit-impulse response of a first-order system.**

Sinusoidal Response   To investigate the response of a first-order system to a sinusoidal forcing function, the example of the mercury thermometer will be considered again.   Consider a thermometer to be in equilibrium with a temperature bath at temperature x_s.   At some time $t = 0$, the bath temperature begins to vary according to the relationship

$$x = x_s + A \sin \omega t \qquad t > 0 \tag{5.17}$$

where x = temperature of bath
$\quad x_s$ = temperature of bath before sinusoidal disturbance is applied
$\quad A$ = amplitude of variation in temperature
$\quad \omega$ = radian frequency, rad/time
  In anticipation of a simple result we shall introduce a deviation variable X which is defined as

$$X = x - x_s \tag{5.18}$$

Using this new variable in Eq. (5.17) gives

$$X = A \sin \omega t \tag{5.19}$$

By referring to a table of transforms, the transform of Eq. (5.19) is

$$X(s) = \frac{A\omega}{s^2 + \omega^2} \tag{5.20}$$

Combining Eqs. (5.7) and (5.20) to eliminate $X(s)$ yields

$$Y(s) = \frac{A\omega}{s^2 + \omega^2} \frac{1/\tau}{s + 1/\tau} \tag{5.21}$$

This equation can be solved for $Y(t)$ by means of a partial-fraction expansion, as described in Chap. 3. The result is

$$Y(t) = \frac{A\omega\tau e^{-t/\tau}}{\tau^2\omega^2 + 1} - \frac{A\omega\tau}{\tau^2\omega^2 + 1} \cos \omega t + \frac{A}{\tau^2\omega^2 + 1} \sin \omega t \tag{5.22}$$

Equation (5.22) can be written in another form by using the trigonometric identity

$$p \cos A + q \sin A = r \sin (A + \theta) \tag{5.23}$$

where

$$r = \sqrt{p^2 + q^2} \qquad \tan \theta = \frac{p}{q}$$

Applying the identity of Eq. (5.23) to (5.22) gives

$$Y(t) = \frac{A\omega\tau}{\tau^2\omega^2 + 1} e^{-t/\tau} + \frac{A}{\sqrt{\tau^2\omega^2 + 1}} \sin (\omega t + \phi) \tag{5.24}$$

where

$$\phi = \tan^{-1} (-\omega\tau)$$

As $t \to \infty$, the first term on the right-hand side of Eq. (5.24) vanishes and leaves only the ultimate periodic solution which is sometimes called the steady-state solution

$$Y(t) \bigg|_s = \frac{A}{\sqrt{\tau^2\omega^2 + 1}} \sin (\omega t + \phi) \tag{5.25}$$

By comparing Eq. (5.19) for the input forcing function with Eq. (5.25) for the ultimate periodic response, we see that

1. The output is a sine wave with a frequency ω equal to that of the input signal.

2. The ratio of output amplitude to input amplitude is $1/\sqrt{\tau^2\omega^2 + 1}$. This is always smaller than 1. We often state this by saying that the signal is attenuated.

3. The output lags behind the input by an angle $|\phi|$. It is clear that lag occurs, for the sign of ϕ is always negative.[1]

For a particular system for which the time constant τ is a fixed quantity, it is seen from Eq. (5.25) that the attenuation of amplitude and the

[1] By convention, the output sinusoid lags the input sinusoid if ϕ in Eq. (5.25) is negative. In terms of a recording of input and output, this means that the input peak occurs before the output peak. If ϕ is positive in Eq. (5.25), the system exhibits phase *lead*, or the output leads the input. In this book we shall always use the term phase angle (ϕ) and interpret whether there is lag or lead by the convention

$\phi < 0$ phase lag
$\phi > 0$ phase lead

phase angle ϕ depend only upon the frequency ω. The attenuation and phase lag increase with frequency, but the phase lag can never exceed 90° and approaches this value asymptotically.

The sinusoidal response is interpreted in terms of the mercury thermometer by the following example:

Example 5.2   A mercury thermometer having a time constant of 0.1 min is placed in a temperature bath at 100°F and allowed to come to equilibrium with the bath.   At time $t = 0$, the temperature of the bath begins to vary sinusoidally about its average temperature of 100°F with an amplitude of 2°F.   If the frequency of oscillation is $10/\pi$ cycles/min, plot the ultimate response of the thermometer reading as a function of time. What is the phase lag?

In terms of the symbols used in this chapter

$$\tau = 0.1$$
$$x_s = 100°F$$
$$A = 2°F$$
$$f = \frac{10}{\pi} \text{ cycles/min}$$
$$\omega = 2\pi f = 2\pi \frac{10}{\pi} = 20 \text{ rad/min}$$

From Eq. (5.25), the amplitude of the response and the phase angle are calculated; thus

$$\frac{A}{\sqrt{\tau^2\omega^2 + 1}} = \frac{2}{\sqrt{4 + 1}} = 0.896°F$$
$$\phi = -\tan^{-1} 2 = -63.5°$$

or

Phase lag $= 63.5°$

The response of the thermometer is therefore

$$Y(t) = 0.896 \sin (20t - 63.5°)$$

or

$$y(t) = 100 + 0.896 \sin (20t - 63.5°)$$

To obtain the lag in terms of time rather than angle, we proceed as follows: A frequency of $10/\pi$ cycles/min means that a complete cycle (peak to peak) occurs in $(10/\pi)^{-1}$ min.   Since 1 cycle is equivalent to 360° and the lag is 63.5°, the time corresponding to this lag is

$$\frac{63.5}{360} \times (\text{time for 1 cycle})$$

or

$$\text{Lag} = \frac{63.5}{360} \frac{\pi}{10} = 0.0555 \text{ min}$$

In general, the lag in units of time is given by

$$\text{Lag} = \frac{|\phi|}{360f}$$

when ϕ is expressed in degrees.

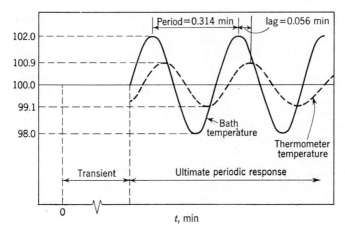

Fig. 5.8 **Response of thermometer in Example 5.2.**

The response of the thermometer reading and the variation in bath temperature are shown in Fig. 5.8. It should be noted that the response shown in this figure holds only after sufficient time has elapsed for the nonperiodic term of Eq. (5.24) to become negligible. For all practical purposes this term becomes negligible after a time equal to about 3τ. If the response were desired beginning from the time the bath temperature begins to oscillate, it would be necessary to plot the complete response as given by Eq. (5.24).

SUMMARY

In this chapter several basic concepts and definitions of control theory have been introduced. These include input variable, output variable, deviation variable, transfer function, response, time constant, first-order system, block diagram, attenuation, and phase lag. Each of these ideas arose naturally in the study of the dynamics of the first-order system, which was the basic subject matter of the chapter. As might be expected, the concepts will find frequent use in succeeding chapters.

In addition to introducing new concepts, we have listed the response of the first-order system to forcing functions of major interest. This information on the dynamic behavior of the first-order system will be of significant value in the remainder of our studies.

PROBLEMS

5.1 A thermometer having a time constant of 0.2 min is placed in a temperature bath, and after the thermometer comes to equilibrium with the bath, the temperature of the bath is increased linearly with time at a rate of 1°/min. What is the difference between the indicated temperature and the bath temperature (**a**) 0.1 min., (**b**) 1.0 min. after the change in temperature begins?

   c. What is the maximum deviation between indicated temperature and bath temperature, and when does it occur?

   d. Plot the forcing function and response on the same graph. After a long enough time, by how many minutes does the response lag the input?

   5.2   A mercury thermometer bulb is $\frac{1}{2}$ in. long by $\frac{1}{8}$ in. diameter. The glass envelope is very thin. Calculate the time constant in water flowing at 10 ft/sec at a temperature of 100°F. In your solution, give a summary which includes

   a. Assumptions used
   b. Source of data
   c. Results

   5.3   Given a system with the transfer function $Y(s)/X(s) = (T_1s + 1)/(T_2s + 1)$. Find $Y(t)$ if $X(t)$ is a unit-step function. If $T_1/T_2 = 5$, sketch $Y(t)$ versus t/T_2. Show the numerical values of minimum, maximum, and ultimate values which may occur during the transient. Check these using the initial-value and final-value theorems of Chap. 4.

   5.4   A thermometer having first-order dynamics with a time constant of 1 min is placed in a temperature bath at 100°F. After the thermometer reaches steady state, it is suddenly placed in a bath at 110°F at $t = 0$ and left there for 1 min, after which it is immediately returned to the bath at 100°F.

   a. Draw a sketch showing the variation of the thermometer reading with time.
   b. Calculate the thermometer reading at $t = 0.5$ min and at $t = 2.0$ min.

   5.5   Repeat Prob. 5.4 if the thermometer is in the 110°F bath for only 10 sec.

   5.6   A mercury thermometer, which has been on a table for some time, is registering the room temperature, 75°F. Suddenly, it is placed in a 400°F oil bath. The following data are obtained for the response of the thermometer.

Time, sec	Thermometer reading, °F
0	75
1	107
2.5	140
5	205
8	244
10	282
15	328
30	385

Give two independent estimates of the thermometer time constant.

   5.7   Rewrite the sinusoidal response of a first-order system [Eq. (5.24)] in terms of a cosine wave. Reexpress the forcing function [Eq. (5.19)] as a cosine wave, and compute the phase difference between input and output cosine waves.

   5.8   The mercury thermometer of Prob. 5.6 is again allowed to come to equilibrium in the room air at 75°F. Then it is placed in the 400°F oil bath for a length of time less than 1 sec, and quickly removed from the bath and reexposed to the 75°F ambient conditions. It may be estimated that the heat-transfer coefficient to the thermometer in air is one-fifth that in the oil bath. If 10 sec after the thermometer is removed from the bath it reads 98°F, estimate the length of time that the thermometer was in the bath.

6 *Physical Examples of First-order Systems*

In the first part of this chapter, we shall consider several physical systems which can be represented by a first-order transfer function. In the second part, a method for approximating the dynamic response of a nonlinear system by a linear response will be presented. This approximation is called linearization.

EXAMPLES OF FIRST-ORDER SYSTEMS

Liquid Level Consider the system shown in Fig. 6.1 which consists of a tank of uniform cross-sectional area A to which is attached a flow resistance R such as a valve, a pipe, or a weir. Assume that q_o, the volumetric flow rate (volume/time) through the resistance, is related to the head h by the linear relationship

$$q_o = \frac{h}{R} \tag{6.1}$$

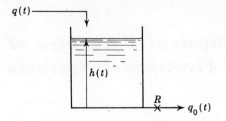

$q(t)$

$h(t)$

R

$q_0(t)$

Fig. 6.1   **Liquid-level system.**

A resistance which has this linear relationship between flow and head is referred to as a *linear resistance.*[1]   A time-varying volumetric flow q of liquid of constant density ρ enters the tank.   Determine the transfer function which relates head to flow.

We can analyze this system by writing a transient mass balance around the tank:

Mass flow in − mass flow out =
rate of accumulation of mass in the tank

In terms of the variables used in this analysis, the mass balance becomes

$$\rho q(t) - \rho q_o(t) = \frac{d(\rho A h)}{dt}$$

$$q(t) - q_o(t) = A\frac{dh}{dt} \tag{6.2}$$

Combining Eqs. (6.1) and (6.2) to eliminate $q_o(t)$ gives the following linear differential equation:

$$q - \frac{h}{R} = A\frac{dh}{dt} \tag{6.3}$$

We shall introduce deviation variables into the analysis before proceeding to the transfer function.   Initially, the process is operating at steady state, which means that $dh/dt = 0$ and we can write Eq. (6.3) as

$$q_s - \frac{h_s}{R} = 0 \tag{6.4}$$

---

[1] A pipe is a linear resistance if the flow is in the laminar range.   A specially contoured weir, called a Sutro weir, produces a linear head-flow relationship.   Formulas used to prepare the shape of such a weir have been reported in the literature: See E. W. Rettger, *Eng. News,* (26):1409–1410 (June 25, 1914); E. A. Pratt, *Eng. News,* **72**(9): 462–463 (Aug. 27, 1914); or E. Soucek, H. E. Howe, and F. T. Mavis, *Eng. News-Record,* Nov. 12, 1936, pp. 679–680.   Turbulent flow through pipes and valves is generally proportional to $\sqrt{h}$.   Flow through weirs having simple geometric shapes can be expressed as Kh^n, where K and n are positive constants.   For example, the flow through a rectangular-shaped weir is proportional to $h^{3/2}$.

where the subscript s has been used to indicate the steady-state value of the variable.

Subtracting Eq. (6.4) from Eq. (6.3) gives

$$(q - q_s) = \frac{1}{R}(h - h_s) + A\frac{d(h - h_s)}{dt} \tag{6.5}$$

If we define the deviation variables as

$$Q = q - q_s$$
$$H = h - h_s$$

Eq. (6.5) can be written

$$Q = \frac{1}{R}H + A\frac{dH}{dt} \tag{6.6}$$

Taking the transform of Eq. (6.6) gives

$$Q(s) = \frac{1}{R}H(s) + AsH(s) \tag{6.7}$$

Notice that $H(0)$ is zero and therefore the transform of dH/dt is simply $sH(s)$.

Equation (6.7) can be rearranged into the standard form of the first-order lag to give

$$\frac{H(s)}{Q(s)} = \frac{R}{\tau s + 1} \tag{6.8}$$

where $\tau = AR$.

In comparing the transfer function of the tank given by Eq. (6.8) with the transfer function for the thermometer given by Eq. (5.7), we see that Eq. (6.8) contains the factor R. The term R is simply the conversion factor which relates $h(t)$ to $q(t)$ when the system is at steady state. For this reason, a factor K in the transfer function $K/(\tau s + 1)$ is often called the steady-state gain. We can readily show this name to be appropriate by applying the final-value theorem of Chap. 4 to the determination of the steady-state value of H when the flow rate $Q(t)$ changes according to a unit-step change; thus

$$Q(t) = u(t)$$

where $u(t)$ is the symbol for the unit-step change. The transform of $Q(t)$ is

$$Q(s) = \frac{1}{s}$$

Combining this forcing function with Eq. (6.8) gives

$$H(s) = \frac{1}{s} \frac{R}{\tau s + 1}$$

Applying the final-value theorem, proved in Chap. 4, to $H(s)$ gives

$$H(t) \Big|_{t \to \infty} = \lim_{s \to 0} [sH(s)] = \lim_{s \to 0} \frac{R}{\tau s + 1} = R$$

This shows that the ultimate change in $H(t)$ for a unit change in $Q(t)$ is simply R.

If the transfer function relating the inlet flow $q(t)$ to the outlet flow is desired, note that we have from Eq. (6.1)

$$q_{o_s} = \frac{h_s}{R} \tag{6.9}$$

Subtracting Eq. (6.9) from Eq. (6.1) and using the deviation variable $Q_o = q_o - q_{o_s}$ gives

$$Q_o = \frac{H}{R} \tag{6.10}$$

Taking the transform of Eq. (6.10) gives

$$Q_o(s) = \frac{H(s)}{R} \tag{6.11}$$

Combining Eqs. (6.11) and (6.8) to eliminate $H(s)$ gives

$$\frac{Q_o(s)}{Q(s)} = \frac{1}{\tau s + 1} \tag{6.12}$$

Notice that the steady-state gain for this transfer function is dimensionless, which is to be expected because the input variable $q(t)$ and the output variable $q_o(t)$ have the same units (volume/time).

The possibility of approximating an impulse forcing function in the flow rate to the liquid-level system is quite real. Recalling that the unit-impulse function is defined as a pulse of unit area as the duration of the pulse approaches zero, the impulse function can be approximated by suddenly increasing the flow to a large value for a very short time; i.e. we may pour very quickly a volume of liquid into the tank. The nature of the impulse response for a liquid-level system will be described by the following example.

Example 6.1   A tank having a time constant of 1 min and a resistance of $\frac{1}{9}$ ft/cfm is operating at steady state with an inlet flow of 10 ft³/min. At time $t = 0$, the flow is suddenly increased to 100 ft³/min for 0.1 min by adding an additional 9 ft³ of water to the

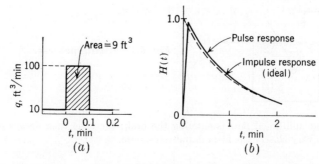

Fig. 6.2 Approximation of an impulse function in a liquid-level system (Example 6.1). (*a*) Pulse input; (*b*) response of tank level.

tank uniformly over a period of 0.1 min. (See Fig. 6.2 for this input disturbance.) Plot the response in tank level and compare with the impulse response.

Before proceeding with the details of the computation, we should observe that, as the time interval over which the 9 ft³ of water is added to the tank is shortened, the input approaches an impulse function having a magnitude of 9.

From the data given in this example, the transfer function of the process is

$$\frac{H(s)}{Q(s)} = \frac{1}{9} \frac{1}{s+1}$$

The input may be expressed as the difference in step functions, as was done in Example 4.5.

$$Q(t) = 90[u(t) - u(t - 0.1)]$$

The transform of this is

$$Q(s) = \frac{90}{s} (1 - e^{-0.1s})$$

Combining this and the transfer function of the process, we obtain

$$H(s) = 10 \left(\frac{1}{s(s+1)} - \frac{e^{-0.1s}}{s(s+1)} \right)$$

The inverse of $H(s)$ is

$$
\begin{aligned}
H(t) &= 10(1 - e^{-t}) & t &< 0.1 \\
H(t) &= 10\{(1 - e^{-t}) - [1 - e^{-(t-0.1)}]\} & t &> 0.1
\end{aligned}
\tag{6.13}
$$

This result is obtained by using the theorem on translation of functions presented in Chap. 4. Simplifying the expression for $H(t)$ for $t > 0.1$ gives

$$H(t) = 1.052e^{-t} \qquad t > 0.1$$

From Eq. (5.16), the response of the system to an impulse of magnitude 9 is given by

$$H(t)\Big|_{\text{impulse}} = (9)\tfrac{1}{9}e^{-t} = e^{-t}$$

In Fig. 6.2, the pulse response of the liquid-level system and the ideal impulse response are shown for comparison. Notice that the level rises very rapidly during the

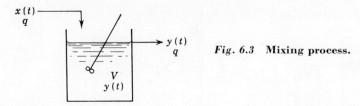

Fig. 6.3   **Mixing process.**

0.1 min that additional flow is entering the tank; the level then decays exponentially and follows very closely the ideal impulse response.

The responses to step and sinusoidal forcing functions are the same for the liquid-level system as for the mercury thermometer of Chap. 5. Hence, they need not be rederived. This is the advantage of characterizing all first-order systems by the same transfer function. The next example of a first-order system is a mixing process.

Mixing Process   Consider the mixing process shown in Fig. 6.3 in which a stream of solution containing dissolved salt flows at a constant volumetric flow rate q into a tank of constant holdup volume V. The concentration of the salt in the entering stream, x(mass of salt/volume), varies with time. It is desired to determine the transfer function relating the outlet concentration y to the inlet concentration x.

Assuming the density of the solution to be constant, the flow rate in must equal the flow rate out, since the holdup volume is fixed. We may analyze this system by writing a transient mass balance for the salt; thus

Flow rate of salt in − flow rate of salt out

$$= \text{rate of accumulation of salt in the tank}$$

Expressing this mass balance in terms of symbols gives

$$qx - qy = \frac{d(Vy)}{dt} \tag{6.14}$$

We shall again introduce deviation variables as we have in the previous examples. At steady state, Eq. (6.14) may be written

$$qx_s - qy_s = 0 \tag{6.15}$$

Subtracting Eq. (6.15) from Eq. (6.14) and introducing the deviation variables

$$X = x - x_s$$
$$Y = y - y_s$$

give

$$qX - qY = V\frac{dY}{dt}$$

Taking the Laplace transform of this expression and rearranging the result give

$$\frac{Y(s)}{X(s)} = \frac{1}{\tau s + 1} \tag{6.16}$$

where $\tau = V/q$.

This mixing process is, therefore, another first-order process, for which the dynamics are now well known. We next bring in an example from DC circuit theory.

RC Circuit Consider the simple RC circuit shown in Fig. 6.4 in which a voltage source $v(t)$ is applied to a series combination of a resistance R and a capacitance C. For $t < 0$, $v(t) = v_s$. Determine the transfer function relating $e_c(t)$ to $v(t)$, where $e_c(t)$ is the voltage across the capacitor.

Applying Kirchhoff's law, which states that in any loop the sum of voltage rises [$v(t)$ in this example] must equal the sum of the voltage drops, gives

$$v(t) = Ri(t) + \frac{1}{C} \int i \, dt \tag{6.17}$$

Recalling that the current is the rate of change of charge with respect to time (coulombs per second), we may replace i by dq/dt in Eq. (6.17) to obtain

$$v(t) = R\frac{dq(t)}{dt} + \frac{1}{C} q(t) \tag{6.18}$$

Since the voltage across the capacitance is given by the relationship

$$e_c = \frac{q}{C} \tag{6.19}$$

the initial charge on the capacitor is simply

$$q_s = Ce_{c_s}$$

Initially, when the circuit is at steady state and the capacitor is fully charged, the voltage across the capacitor is equal to the source voltage v_s; therefore, Eq. (6.18) can be written for these steady-state conditions as

$$v_s = \frac{1}{C} q_s = e_{c_s} \tag{6.20}$$

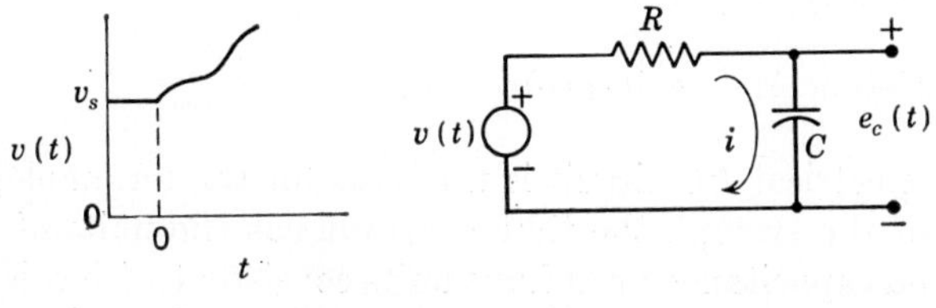

Fig. 6.4 RC circuit.

Subtracting Eq. (6.20) from Eq. (6.18) and introducing the deviation variables

$$V = v - v_s$$
$$Q = q - q_s$$
$$E_c = e_c - e_{c_s} = \frac{Q}{C} \tag{6.21}$$

we obtain the result

$$V = R\frac{dQ}{dt} + \frac{Q}{C} \tag{6.22}$$

or

$$V = RC\frac{dE_c}{dt} + E_c \tag{6.23}$$

Taking the transform of Eq. (6.23) and rearranging the result give

$$\frac{E_c(s)}{V(s)} = \frac{1}{\tau s + 1} \tag{6.24}$$

where $\tau = RC$. Again we obtain a first-order transfer function.

The three examples which have been presented in this section are intended to show that the dynamic characteristics of many physical systems can be represented by a first-order transfer function. In the remainder of the book, more examples of first-order systems will appear as we discuss a variety of control systems.

Summary In each example of a first-order system, the time constant has been expressed in terms of system parameters; thus

$$\tau = \frac{mC}{hA} \quad \text{for thermometer, Eq. (5.5)}$$
$$\tau = AR \quad \text{for liquid-level process, Eq. (6.8)}$$
$$\tau = \frac{V}{q} \quad \text{for mixing process, Eq. (6.16)}$$
$$\tau = RC \quad \text{for } RC \text{ circuit, Eq. (6.24)}$$

In each case we may consider the time constant to be the product of a resistance and a capacitance; thus

$$\tau = (\text{resistance})(\text{capacitance})$$

In fact, the electrical RC circuit is the basis for the terminology used here. Regardless of the system, the time constant has the units of time and the resistance and capacitance must have units consistent with this requirement.

In general, capacitance is defined as the change in storage of a quantity

Table 6.1

System	Stored quantity	Driving force	Flow	Resistance	Capacitance
			Term		
Thermometer	Internal energy, Btu	Temperature, °F	Heat flow, Btu/min	$\dfrac{1}{hA}$	mC
Liquid-level	Volume, ft^3	Head, ft	Flow, cfm	R	$\dfrac{\text{Tank volume}}{\text{Head}} = \dfrac{hA}{h} = A$
RC circuit	Electric charge, coul	Voltage, volt	Current, coul/sec	R	C
Mixing process	Mass of solute, lb	Concentration, lb/ft^3	Solute flow, lb/min	$\dfrac{y}{qy} = \dfrac{1}{q}$	$\dfrac{\left(\begin{array}{c}\text{Mass of solute}\\\text{in tank}\end{array}\right)}{\text{Concentration}} = \dfrac{Vy}{y} = V$

(such as coulombs or Btu) per unit change in driving force or potential; thus

$$\text{Capacitance} = \frac{\Delta \text{ storage}}{\Delta \text{ driving force}}$$

The driving force or potential is said to create the flow of the quantity under consideration. For example, in a liquid-level system, the head of fluid (feet) is the driving force which produces the flow (cubic feet per minute), and the capacitance is the increase in volume (cubic feet) for a unit change in head.

The resistance is the ratio of the change in driving force to the change in flow:

$$\text{Resistance} = \frac{\Delta \text{ driving force}}{\Delta \text{ flow}}$$

In terms of these general concepts, we can identify a resistance and a capacitance for each of the systems studied so far. In Table 6.1, the driving force, flow, resistance, and capacitance are listed for each physical system in terms of the names, symbols, and dimensions commonly associated with the particular system.

LINEARIZATION

Thus far, all the examples of physical systems, including the liquid-level system of Fig. 6.1, have been linear. Actually, most physical systems of practical importance are nonlinear.

Characterization of a dynamic system by a transfer function can be done only for linear systems (those described by linear differential equations). The convenience of using transfer functions for dynamic analysis, which we have already seen in applications, provides significant motivation for approximating nonlinear systems by linear ones. A very important technique for such approximation is illustrated by the following discussion of the liquid-level system of Fig. 6.1.

We now assume that the resistance follows the square-root relationship

$$q_o = Ch^{\frac{1}{2}} \tag{6.25}$$

where C is a constant.

For a liquid of constant density and a tank of uniform cross-sectional area A, a material balance around the tank gives

$$q(t) - q_o(t) = A\,\frac{dh}{dt} \tag{6.26}$$

Combining Eqs. (6.25) and (6.26) gives the nonlinear differential equation

$$q - Ch^{\frac{1}{2}} = A\,\frac{dh}{dt} \tag{6.27}$$

At this point, we cannot proceed as before and take the Laplace transform. This is owing to the presence of the nonlinear term $h^{\frac{1}{2}}$, for which there is no simple transform. This difficulty can be circumvented as follows:

By means of a Taylor-series expansion, the function $q_o(h)$ may be expanded around the steady-state value h_s; thus

$$q_o = q_o(h_s) + q_o'(h_s)(h - h_s) + \frac{q_o''(h_s)(h - h_s)^2}{2!} + \cdots$$

where $q_o'(h_s)$ is the first derivative of q_o evaluated at h_s, $q_o''(h_s)$ the second derivative, etc. If we keep only the linear term, the result is

$$q_o \cong q_o(h_s) + q_o'(h_s)(h - h_s) \tag{6.28}$$

Introducing Eq. (6.25) into Eq. (6.28) gives

$$q_o = q_{o_s} + \frac{1}{R_1}\,(h - h_s) \tag{6.29}$$

where $q_{o_s} = q_o(h_s)$

$$(R_1)^{-1} = \tfrac{1}{2}Ch_s^{-\frac{1}{2}}$$

Substituting Eq. (6.29) into (6.26) gives

$$q - q_{o_s} - \frac{h - h_s}{R_1} = A\,\frac{dh}{dt} \tag{6.30}$$

At steady state the flow entering the tank equals the flow leaving the tank; thus

$$q_s = q_{o_s} \tag{6.31}$$

Introducing this last equation into Eq. (6.30) gives

$$A \frac{dh}{dt} + \frac{h - h_s}{R_1} = q - q_s \tag{6.32}$$

Introducing deviation variables $Q = q - q_s$ and $H = h - h_s$ into Eq. (6.32) and transforming give

$$\frac{H(s)}{Q(s)} = \frac{R_1}{\tau s + 1} \tag{6.33}$$

where

$$R_1 = 2h_s^{\frac{1}{2}}/C$$
$$\tau = R_1 A$$

We see that a transfer function is obtained which is identical in form with that of the linear system, Eq. (6.8). However, in this case, the resistance R_1 depends on the steady-state conditions around which the process operates. Graphically, the resistance R_1 is the reciprocal of the slope of the tangent line passing through the point (q_{o_s}, h_s) as shown in Fig. 6.5. Furthermore, the linear approximation given by Eq. (6.28) is the equation of the tangent line itself. From the graphical representation, it should be clear that the linear approximation improves as the deviation in h becomes smaller. If one does not have an analytic expression such as $h^{\frac{1}{2}}$ for the nonlinear function, but only a graph of the function, the technique can still be applied by representing the function by the tangent line passing through the point of operation.

Whether or not the linearized result is a valid representation depends on the operation of the system. If the level is being maintained by a controller at or close to a fixed level h_s, then by the very nature of the control imposed on the system, deviations in level should be small (for good control)

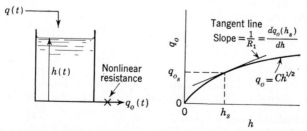

Fig. 6.5 **Liquid-level system with nonlinear resistance**

and the linearized equation is adequate. On the other hand, if the level should change over a wide range, the linear approximation may be very poor and the system may deviate significantly from the prediction of the linear transfer function. In such cases, it may be necessary to use the more difficult methods of nonlinear analysis, some of which are discussed in Chaps. 26 through 29. We shall extend the discussion of linearization to more complex systems in Chap. 25.

In summary, we have characterized, in an approximate sense, a nonlinear system by a linear transfer function. In general, this technique may be applied to any nonlinearity which can be expressed in a Taylor series (or, equivalently, has a unique slope at the operating point). Since this includes most nonlinearities arising in process control, we have ample justification for studying linear systems in considerable detail.

PROBLEMS

6.1 Derive the transfer function $H(s)/Q(s)$, for the liquid-level system of Fig. P6.1 when

  a. The tank level operates about the steady-state value of $h_s = 1$ ft.
  b. The tank level operates about the steady-state value of $h_s = 3$ ft.

The pump removes water at a constant rate of 10 cfm; this rate is independent of head. The cross-sectional area of the tank is 1.0 ft² and the resistance R is 0.5 ft/cfm.

q, ft³/min

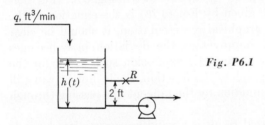

Fig. P6.1

6.2 A liquid-level system, such as the one shown in Fig. 6.1, has a cross-sectional area of 3.0 ft². The valve characteristics are

$$q = 8 \sqrt{h}$$

where q = flow rate, cfm
  h = level above the valve, ft
Calculate the time constant for this system if the average operating level is

  a. 3 ft
  b. 9 ft

6.3 A tank having a cross-sectional area of 2 ft² is operating at steady state with an inlet flow rate of 2.0 cfm. The flow-head characteristics are shown in Fig. P6.3.

  a. Find the transfer function $H(s)/Q(s)$.
  b. If the flow to the tank increases from 2.0 to 2.2 cfm according to a step change, calculate the level h two minutes after the change occurs.

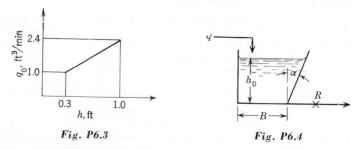

Fig. P6.3   Fig. P6.4

6.4 Develop a formula for finding the time constant of the liquid-level system shown in Fig. P6.4 when the average operating level is h_o. The resistance R is linear. The tank has three vertical walls and one which slopes at an angle α from the vertical as shown. The distance separating the parallel walls is 1.

6.5 Consider the stirred-tank reactor shown in Fig. P6.5. The reaction occurring is

$$A \rightarrow B$$

and it proceeds at a rate

$$r = kC_o$$

where r = moles A reacting/(volume)(time)
$\quad k$ = reaction velocity constant
$\quad C_o(t)$ = concentration of A in reactor, moles/volume
$\quad V$ = volume of mixture in reactor
Further let F = constant feed rate, volume/time
$\quad C_i(t)$ = concentration of A in feed stream
Assuming constant density and constant V, derive the transfer function relating the concentration in the reactor to the feed-stream concentration. Prepare a block diagram

Fig. P6.5

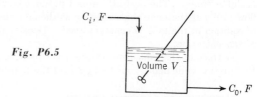

for the reactor. Sketch the response of the reactor to a unit-step change in C_i. Represent the time constant as the product of a resistance and capacitance as in Table 6.1.

6.6 A thermocouple junction of area A, mass m, heat capacity C, and emissivity e is located in a furnace which normally is at $T_{i_s}°$C. At these temperatures convective and conductive heat transfer to the junction are negligible compared with radiative heat transfer. Determine the linearized transfer function between the furnace temperature T_i and the junction temperature T_o. For the case

$\quad m = 0.1$ g
$\quad C = 0.12$ cal/(g)(°C)
$\quad e = 0.7$
$\quad A = 0.1$ cm^2
$\quad T_{i_r} = 1100$°C

plot the response of the thermocouple to a 10°C step change in furnace temperature.
Compare this with the true response obtained by integration of the differential equation.

6.7   A liquid-level system has the following properties:
Tank dimensions: 10 ft high by 5 ft diameter
Steady-state operating characteristics:

Inflow, gal/hr	Steady-state level, ft
0	0
5,000	.7
10,000	1.1
15,000	2.3
20,000	3.9
25,000	6.3
30,000	8.8

a. Plot the level response of the tank under the following circumstances: The
   inlet flow rate is held at 300 gal/min for 1 hr and then suddenly raised to 400
   gal/min.
b. How accurate is the steady-state level calculated from the dynamic response
   in part *a* when compared with the value given by the table above?
c. The tank is now connected in series with a second tank which has identical
   operating characteristics, but which has dimensions 8 ft high by 4 ft diameter.   Plot the response of the original tank (which is upstream of the new
   tank) to the change described in part *a* when the connection is such that the
   tanks are (1) interacting, (2) noninteracting.   (See Chap. 7.)

6.8   A mixing process may be described as follows: A stream with solute concentration C_i (pounds/volume) is fed to a perfectly stirred tank at a constant flow rate of q
(volume/time).   The perfectly mixed product is withdrawn from the tank, also at the
flow rate q at the same concentration as the material in the tank, C_o.   The total volume
of solution in the tank is constant at V.   Density may be considered to be independent
of concentration.

A trace of the tank concentration versus time appears as shown in Fig. P6.8.   *(a)*
Plot on this same figure your best guess of the *quantitative* behavior of the inlet concentra-

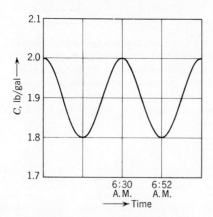

Fig. P6.8

tion versus time. Be sure to label the graph with *quantitative* information regarding times and magnitudes and any other data which will demonstrate your understanding of the situation. (*b*) Write an equation for C_i as a function of time.

Data: Tank dimensions: 8 ft high by 5 ft diameter
      Tank volume V: 700 gal
      Flow rate q: 100 gal/min
      Average density: 70 lb/ft^3

7 *Response of First-order Systems in Series*

Introductory Remarks   Very often a physical system can be represented by several first-order processes connected in series.   To illustrate this type of system, consider the liquid-level systems shown in Fig. 7.1 in which two tanks are arranged so that the outlet flow from the first tank is the inlet flow to the second tank.

Two possible piping arrangements are shown in Fig. 7.1.   In Fig. 7.1a the outlet flow from tank 1 discharges directly into the atmosphere before spilling into tank 2 and the flow through R_1 depends only on h_1.   The variation in h_2 in tank 2 does not affect the transient response occurring in tank 1.   This type of system is referred to as a *noninteracting* system. In contrast to this, the system shown in Fig. 7.1b is said to be *interacting* because the flow through R_1 now depends on the difference between h_1 and h_2.   We shall consider first the noninteracting system of Fig. 7.1a.

Noninteracting System   As in the previous liquid-level example, we shall assume the liquid to be of constant density, the tanks to have uniform

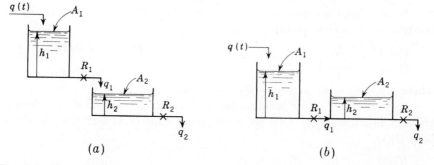

Fig. 7.1 **Two-tank liquid-level system.** (*a*) **Noninteracting;** (*b*) **interacting.**

cross-sectional area, and the flow resistances to be linear. Our problem is to find a transfer function which relates h_2 to q, that is, $H_2(s)/Q(s)$. The approach will be to obtain a transfer function for each tank, $Q_1(s)/Q(s)$ and $H_2(s)/Q_1(s)$, by writing a transient mass balance around each tank; these transfer functions will then be combined to eliminate the intermediate flow $Q_1(s)$ and produce the desired transfer function.

A balance on tank 1 gives

$$q - q_1 = A_1 \frac{dh_1}{dt} \tag{7.1}$$

A balance on tank 2 gives

$$q_1 - q_2 = A_2 \frac{dh_2}{dt} \tag{7.2}$$

The flow-head relationships for the two linear resistances are given by the expressions

$$q_1 = \frac{h_1}{R_1} \tag{7.3}$$

$$q_2 = \frac{h_2}{R_2} \tag{7.4}$$

Combining Eqs. (7.1) and (7.3) in exactly the same manner as was done in Chap. 6 and introducing deviation variables give the transfer function for tank 1; thus

$$\frac{Q_1(s)}{Q(s)} = \frac{1}{\tau_1 s + 1} \tag{7.5}$$

where $Q_1 = q_1 - q_{1_s}$, $Q = q - q_s$, and $\tau_1 = R_1 A_1$.

In the same manner, we can combine Eqs. (7.2) and (7.4) to obtain the transfer function for tank 2; thus

$$\frac{H_2(s)}{Q_1(s)} = \frac{R_2}{\tau_2 s + 1} \tag{7.6}$$

where $H_2 = h_2 - h_{2_s}$ and $\tau_2 = R_2 A_2$.

Having the transfer function for each tank, we can obtain the overall transfer function $H_2(s)/Q(s)$ by multiplying Eqs. (7.5) and (7.6) to eliminate $Q_1(s)$:

$$\frac{H_2(s)}{Q(s)} = \frac{1}{\tau_1 s + 1}\frac{R_2}{\tau_2 s + 1} \tag{7.7}$$

Notice that the overall transfer function of Eq. (7.7) is the product of two first-order transfer functions, each one of which is the transfer function of a single tank operating independently of the other. In the case of the interacting system of Fig. 7.1b, the overall transfer function *cannot* be found by simply multiplying together the separate transfer functions; this will become apparent when the interacting system is analyzed later.

Example 7.1 Two noninteracting tanks are connected in series as shown in Fig. 7.1a. The time constants are $\tau_2 = 1$ and $\tau_1 = 0.5$; $R_2 = 1$. Sketch the response of the level in tank 2 if a unit-step change is made in the inlet flow rate to tank 1.

The transfer function for this system is found directly from Eq. (7.7); thus

$$\frac{H_2(s)}{Q(s)} = \frac{R_2}{(\tau_1 s + 1)(\tau_2 s + 1)} \tag{7.8}$$

For a unit-step change in Q, we obtain

$$H_2(s) = \frac{1}{s}\frac{R_2}{(\tau_1 s + 1)(\tau_2 s + 1)} \tag{7.9}$$

Inversion by means of partial-fraction expansion gives

$$H_2(t) = R_2\left[1 - \frac{\tau_1 \tau_2}{\tau_1 - \tau_2}\left(\frac{1}{\tau_2} e^{-t/\tau_1} - \frac{1}{\tau_1} e^{-t/\tau_2}\right)\right] \tag{7.10}$$

Substituting in the values of τ_1, τ_2, and R_2 gives

$$H_2(t) = 1 - (2e^{-t} - e^{-2t}) \tag{7.11}$$

A plot of this response is shown in Fig. 7.2. Notice that the response is S-shaped and the

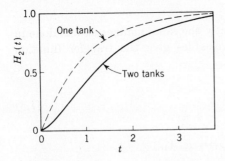

Fig. 7.2 **Transient response of liquid-level system (Example 7.1).**

slope dH_2/dt at the origin is zero. If the change in flow rate were introduced into the second tank, the response would be first-order and is shown for comparison in Fig. 7.2 by the dotted curve.

Generalization for Several Noninteracting Systems in Series

Having observed that the overall transfer function for two noninteracting first-order systems connected in series is simply the product of the individual transfer functions, we may now generalize by considering n noninteracting first-order systems as represented by the block diagram of Fig. 7.3.

The block diagram is equivalent to the relationships

$$\frac{X_1(s)}{X_0(s)} = \frac{k_1}{\tau_1 s + 1}$$

$$\frac{X_2(s)}{X_1(s)} = \frac{k_2}{\tau_2 s + 1}$$

$$\cdots \cdots \cdots \cdots$$

$$\frac{X_n(s)}{X_{n-1}(s)} = \frac{k_n}{\tau_n s + 1}$$

To obtain the overall transfer function, we simply multiply together the individual transfer functions; thus

$$\frac{X_n(s)}{X_0(s)} = \prod_{i=1}^{n} \frac{k_i}{\tau_i s + 1} \tag{7.12}$$

From Example 7.1, notice that the step response of a system consisting of two first-order systems is S-shaped and that the response changes very slowly just after introduction of the step input. This sluggishness or delay is sometimes called *transfer lag* and is always present when two or more first-order systems are connected in series. For a single first-order system, there is no transfer lag; i.e., the response begins immediately after the step change is applied, and the rate of change of the response (slope of response curve) is maximal at $t = 0$.

In order to show how the transfer lag is increased as the number of stages increases, Fig. 7.4 gives the unit-step response curves for several systems containing one or more first-order stages in series.

Interacting System　To illustrate an interacting system, we shall derive the transfer function for the system shown in Fig. 7.1b. The analysis is started by writing mass balances on the tanks as was done for the noninteracting case. The balances on tanks 1 and 2 are the same as before and

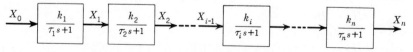

Fig. 7.3 **Noninteracting first-order systems.**

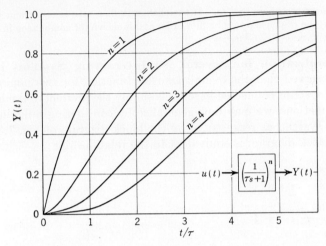

Fig. 7.4 Step response of noninteracting first-order systems.

are given by Eqs. (7.1) and (7.2). However, the flow-head relationship for tank 1 is now

$$q_1 = \frac{1}{R_1} (h_1 - h_2) \tag{7.13}$$

The flow-head relationship for R_2 is the same as before and is expressed by Eq. (7.4). A simple way to combine Eqs. (7.1), (7.2), (7.4), and (7.13) is to first express them in terms of deviation variables, transform the resulting equations, and then combine the transformed equations to eliminate the unwanted variables.

At steady state, Eqs. (7.1) and (7.2) can be written

$$q_s - q_{1_s} = 0 \tag{7.14}$$
$$q_{1_s} - q_{2_s} = 0 \tag{7.15}$$

Subtracting Eq. (7.14) from Eq. (7.1) and Eq. (7.15) from Eq. (7.2) and introducing deviation variables give

$$Q - Q_1 = A_1 \frac{dH_1}{dt} \tag{7.16}$$

$$Q_1 - Q_2 = A_2 \frac{dH_2}{dt} \tag{7.17}$$

Expressing Eqs. (7.13) and (7.4) in terms of deviation variables gives

$$Q_1 = \frac{H_1 - H_2}{R_1} \tag{7.18}$$

$$Q_2 = \frac{H_2}{R_2} \tag{7.19}$$

Transforming Eqs. (7.16) through (7.19) gives

$$Q(s) - Q_1(s) = A_1 s H_1(s) \tag{7.20}$$
$$Q_1(s) - Q_2(s) = A_2 s H_2(s) \tag{7.21}$$
$$R_1 Q_1(s) = H_1(s) - H_2(s) \tag{7.22}$$
$$R_2 Q_2(s) = H_2(s) \tag{7.23}$$

The analysis has produced four algebraic equations containing five unknowns: (Q, Q_1, Q_2, H_1, and H_2). These equations may be combined to eliminate Q_1, Q_2, and H_1 and arrive at the desired transfer function:

$$\frac{H_2(s)}{Q(s)} = \frac{R_2}{\tau_1 \tau_2 s^2 + (\tau_1 + \tau_2 + A_1 R_2)s + 1} \tag{7.24}$$

Notice that the product of the transfer functions for the tanks operating separately, Eqs. (7.5) and (7.6), does not produce the correct result for the interacting system. The difference between the transfer function for the noninteracting system, Eq. (7.7), and the interacting system, Eq. (7.24), is the presence of the term $A_1 R_2$ in the coefficient of s.

The term *interacting* is often referred to as *loading*. The second tank of Fig. 7.1b is said to *load* the first tank.

To understand the effect of interaction on the transient response of a system, consider a two-tank system for which the time constants are equal ($\tau_1 = \tau_2 = \tau$). If the tanks are noninteracting, the transfer function relating inlet flow to outlet flow is

$$\frac{Q_2(s)}{Q(s)} = \left(\frac{1}{\tau s + 1}\right)^2 \tag{7.25}$$

The unit-step response for this transfer function can be obtained by the usual procedure to give

$$Q_2(t) = 1 - e^{-t/\tau} - \frac{t}{\tau} e^{-t/\tau} \tag{7.26}$$

If the tanks are interacting, the overall transfer function, according to Eq. (7.24), is (assuming further that $A_1 = A_2$)

$$\frac{Q_2(s)}{Q(s)} = \frac{1}{\tau^2 s^2 + 3\tau s + 1} \tag{7.27}$$

By application of the quadratic formula, the denominator of this transfer function can be written as

$$\frac{Q_2(s)}{Q(s)} = \frac{1}{(0.38\tau s + 1)(2.62\tau s + 1)} \tag{7.28}$$

For this example, we see that the effect of interaction has been to change the effective time constants of the interacting system. One time

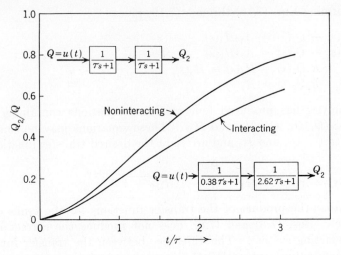

Fig. 7.5   **Effect of interaction on step response of two-tank system.**

constant has become considerably larger and the other smaller than the time constant τ of either tank in the noninteracting system. The response of $Q_2(t)$ to a unit-step change in $Q(t)$ for the interacting case [Eq. (7.28)] is

$$Q_2(t) = 1 + 0.17e^{-t/0.38\tau} - 1.17e^{-t/2.62\tau} \tag{7.29}$$

In Fig. 7.5, the unit-step responses [Eqs. (7.26) and (7.29)] for the two cases are plotted to show the effect of interaction. From this figure, it can be seen that interaction slows up the response. This result can be understood on physical grounds in the following way: If the same size step change is introduced into the two systems of Fig. 7.1, the flow from tank 1 (q_1) for the noninteracting case will not be reduced by the increase in level in tank 2. However, for the interacting case, the flow q_1 will be reduced by the build-up of level in tank 2. At any time t_1 following the introduction of the step input, q_1 for the interacting case will be less than for the noninteracting case with the result that h_2 (or q_2) will increase at a slower rate.

In general, the effect of interaction on a system containing two first-order lags is to change the ratio of effective time constants in the interacting system. In terms of the transient response, this means that the interacting system is more sluggish than the noninteracting system.

This chapter concludes our specific discussion of first-order systems. We shall make continued use of the material developed here in the succeeding chapters.

PROBLEMS

7.1   Determine the transfer function $H(s)/Q(s)$ for the liquid-level system shown in Fig. P7.1. Resistances R_1 and R_2 are linear. The flow rate from tank 3 is maintained

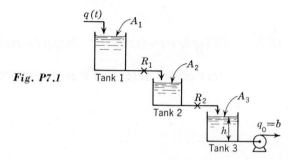

Fig. P7.1

constant at b by means of a pump; i.e., the flow rate from tank 3 is independent of head h. The tanks are noninteracting.

7.2 The mercury thermometer in Chap. 5 was considered to have all its resistance in the convective film surrounding the bulb and all its capacitance in the mercury. A more detailed analysis would consider both the convective resistance surrounding the bulb and that between the bulb and mercury. In addition, the capacitance of the glass bulb would be included. Let

A_i = inside area of bulb, for heat transfer to mercury
A_o = outside area of bulb, for heat transfer from surrounding fluid
m = mass of mercury in bulb
m_b = mass of glass bulb
C = heat capacity of mercury
C_b = heat capacity of glass bulb
h_i = convective coefficient between bulb and mercury
h_o = convective coefficient between bulb and surrounding fluid
T = temperature of mercury
T_b = temperature of glass bulb
T_f = temperature of surrounding fluid

Determine the transfer function between T_f and T. What is the effect of the bulb resistance and capacitance on the thermometer response? Note that the inclusion of the bulb results in a pair of interacting systems, which give an overall transfer function somewhat different from that of Eq. (7.24).

7.3 There are N storage tanks of volume V arranged so that when water is fed into the first tank, an equal volume of liquid overflows from the first tank into the second tank, and so on. Each tank initially contains component A at some concentration C_o and is equipped with a perfect stirrer. At time zero, a stream of zero concentration is fed into the first tank at a volumetric rate q. Find the resulting concentration in each tank as a function of time.

8 *Higher-order Systems: Second-order and Transportation Lag*

SECOND - ORDER SYSTEM

Transfer Function   In this section, we shall introduce a basic system called a *second-order system* or a *quadratic lag*.   A second-order transfer function will be developed by considering a classical example from mechanics.   This is the damped vibrator, which is shown in Fig. 8.1.

A block of mass W resting on a horizontal frictionless table is attached to a linear spring.   A viscous damper (dashpot) is also attached to the block. Assume that the system is free to oscillate horizontally under the influence of a forcing function $F(t)$.   The origin of the coordinate system is taken as the right edge of the block when the spring is in the relaxed or unstretched condition.   At time zero, the block is assumed to be at rest at this origin.[1]

---

[1] In effect, this assumption makes the displacement variable $Y(t)$ a deviation variable.   Also, the assumption that the block is initially at rest permits derivation of the second-order transfer function in its standard form.   An initial velocity has the same effect as a forcing function.   Hence, this assumption is in no way restrictive.

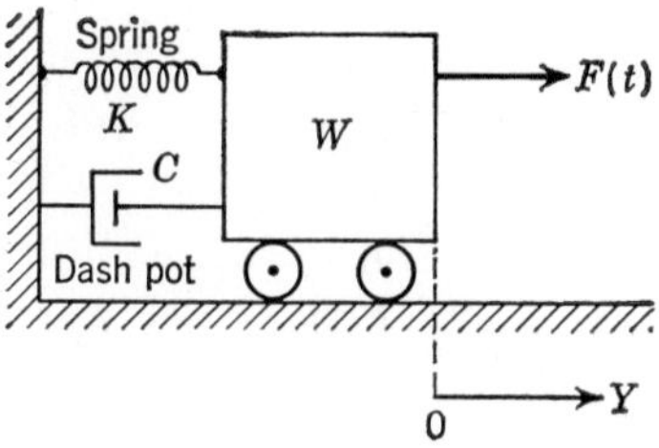

Fig. 8.1   Damped vibrator.

Positive directions for force and displacement are indicated by the arrows in Fig. 8.1.

Consider the block at some instant when it is to the right of $Y = 0$ and when it is moving toward the right (positive direction).   Under these conditions, the position Y and the velocity dY/dt are both positive.   At this particular instant, the following forces are acting on the block:

1. The force exerted by the spring (toward the left) of $-KY$ where K is a positive constant, called Hooke's constant
2. The viscous friction force (acting to the left) of $-C \, dY/dt$, where C is a positive constant called the damping coefficient
3. The external force $F(t)$ (acting toward the right)

Newton's law of motion, which states that the sum of all forces acting on the mass is equal to the rate of change of momentum (mass $\times$ acceleration), takes the form

$$\frac{W}{g_c} \frac{d^2Y}{dt^2} = -KY - C \frac{dY}{dt} + F(t) \tag{8.1}$$

Rearrangement gives

$$\frac{W}{g_c} \frac{d^2Y}{dt^2} + C \frac{dY}{dt} + KY = F(t) \tag{8.2}$$

where W = mass of block, lb_m

g_c = 32.2$(lb_m)(ft)/(lb_f)(sec^2)$

C = viscous damping coefficient, $lb_f/(ft/sec)$

K = Hooke's constant, lb_f/ft

$F(t)$ = driving force, a function of time, lb_f

Dividing Eq. (8.2) by K gives

$$\frac{W}{g_c K} \frac{d^2Y}{dt^2} + \frac{C}{K} \frac{dY}{dt} + Y = \frac{F(t)}{K} \tag{8.3}$$

For convenience, this is written as

$$\tau^2 \frac{d^2Y}{dt^2} + 2\zeta\tau \frac{dY}{dt} + Y = X(t) \tag{8.4}$$

where

$$\tau^2 = \frac{W}{g_c K} \tag{8.5}$$

$$2\zeta\tau = \frac{C}{K} \tag{8.6}$$

$$X(t) = \frac{F(t)}{K} \tag{8.7}$$

Solving for τ and ζ from Eqs. (8.5) and (8.6) gives

$$\tau = \sqrt{\frac{W}{g_c K}} \quad \text{sec} \tag{8.8}$$

$$\zeta = \sqrt{\frac{g_c C^2}{4WK}} \quad \text{dimensionless} \tag{8.9}$$

By definition, both τ and ζ must be positive.   The reason for introducing τ and ζ in the particular form shown in Eq. (8.4) will become clear when we discuss the solution of Eq. (8.4) for particular forcing functions $X(t)$.

Equation (8.4) is written in a standard form which is widely used in control theory.   Notice that, because of superposition, $X(t)$ can be considered as a forcing function because it is proportional to the force $F(t)$.

If the block is motionless $(dY/dt = 0)$ and located at its rest position $(Y = 0)$ before the forcing function is applied, the Laplace transform of Eq. (8.4) becomes

$$\tau^2 s^2 Y(s) + 2\zeta\tau s Y(s) + Y(s) = X(s) \tag{8.10}$$

From this, the transfer function follows:

$$\frac{Y(s)}{X(s)} = \frac{1}{\tau^2 s^2 + 2\zeta\tau s + 1} \tag{8.11}$$

The transfer function given by Eq. (8.11) is written in standard form, and we shall show later that other physical systems can be represented by a transfer function having the denominator $\tau^2 s^2 + 2\zeta\tau s + 1$.   All such systems are defined as second-order.   Note that it requires two parameters, τ and ζ, to characterize the dynamics of a second-order system in contrast to only one parameter for a first-order system.   For the time being, the variables and parameters of Eq. (8.11) can be interpreted in terms of the damped vibrator.   We shall now discuss the response of a second-order system to some of the common forcing functions, namely, step, impulse, and sinusoidal.

Step Response   If the forcing function is a unit-step function, we have

$$X(s) = \frac{1}{s} \tag{8.12}$$

In terms of the damped vibrator shown in Fig. 8.1 this is equivalent to suddenly applying a force of magnitude K directed toward the right at time $t = 0$. This follows from the fact that X is defined by the relationship $X(t) = F(t)/K$. Superposition will enable us to determine easily the response to a step function of any other magnitude.

Combining Eq. (8.12) with the transfer function of Eq. (8.11) gives

$$Y(s) = \frac{1}{s} \frac{1}{\tau^2 s^2 + 2\zeta\tau s + 1} \tag{8.13}$$

The quadratic term in this equation may be factored into two linear terms that contain the roots

$$s_1 = -\frac{\zeta}{\tau} + \frac{\sqrt{\zeta^2 - 1}}{\tau} \tag{8.14}$$

$$s_2 = -\frac{\zeta}{\tau} - \frac{\sqrt{\zeta^2 - 1}}{\tau} \tag{8.15}$$

Equation (8.13) can now be written

$$Y(s) = \frac{1/\tau^2}{(s)(s - s_1)(s - s_2)} \tag{8.16}$$

The response of the system $Y(t)$ can be found by inverting Eq. (8.16). The roots s_1 and s_2 will be real or complex depending on the parameter ζ. The nature of the roots will, in turn, affect the form of $Y(t)$. The problem may be divided into the three cases shown in Table 8.1. Each case is discussed below.

Case I **Step response for $\zeta < 1$** For this case, the inversion of Eq. (8.16) yields the result

$$Y(t) = 1 - \frac{1}{\sqrt{1 - \zeta^2}} e^{-\zeta t/\tau} \sin\left(\sqrt{1 - \zeta^2} \frac{t}{\tau} + \tan^{-1} \frac{\sqrt{1 - \zeta^2}}{\zeta}\right) \tag{8.17}$$

To derive Eq. (8.17), use is made of the techniques of Chap. 3. Since $\zeta < 1$, Eqs. (8.14) to (8.16) indicate a pair of complex conjugate roots in the left-half plane and a root at the origin. In terms of the symbols of Fig. 3.1, the complex roots correspond to s_2 and s_2^* and the root at the origin to s_6.

Table 8.1

Case	ζ	Nature of roots	Description of response
I	<1	Complex	Underdamped or oscillatory
II	=1	Real and equal	Critically damped
III	>1	Real	Overdamped or nonoscillatory

By referring to Table 3.1, we see that $Y(t)$ has the form

$$Y(t) = C_1 + e^{-\zeta t/\tau}\left(C_2 \cos \sqrt{1 - \zeta^2}\,\frac{t}{\tau} + C_3 \sin \sqrt{1 - \zeta^2}\,\frac{t}{\tau}\right) \qquad (8.18)$$

The constants C_1, C_2, and C_3 are found by partial fractions. The resulting equation is then put in the form of Eq. (8.17) by applying the trigonometric identity used in Chap. 5, Eq. (5.23). The details are left as an exercise for the reader. It is evident from Eq. (8.17) that $Y(t) \to 1$ as $t \to \infty$.

The nature of the response can be understood most clearly by plotting Eq. (8.17) as shown in Fig. 8.2, where $Y(t)$ is plotted against the dimensionless variable t/τ for several values of ζ, including those above unity which will be considered in the next section. Note that, for $\zeta < 1$, all the response curves are oscillatory in nature and become less oscillatory as ζ is increased. The slope at the origin in Fig. 8.2 is zero for all values of ζ. The response of a second-order system for $\zeta < 1$ is said to be *underdamped*.

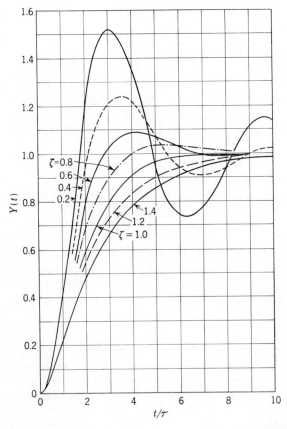

Fig. 8.2   **Response of a second-order system to a unit-step forcing function.**

Case II   Step response for ζ = 1   For this case, the response is given by the expression

$$Y(t) = 1 - \left(1 + \frac{t}{\tau}\right)e^{-t/\tau} \tag{8.19}$$

This is derived as follows: Equations (8.14) and (8.15) show that the roots s_1 and s_2 are real and equal.   Referring to Fig. 3.1 and Table 3.1, it is seen that Eq. (8.19) is in the correct form.   The constants are obtained, as usual, by partial fractions.

The response, which is plotted in Fig. 8.2, is nonoscillatory.   This condition, ζ = 1, is called *critical damping* and allows most rapid approach of the response to $Y = 1$ without oscillation.

Case III   Step response for ζ > 1   For this case, the inversion of Eq. (8.16) gives the result

$$Y(t) = 1 - e^{-\zeta t/\tau}\left(\cosh \sqrt{\zeta^2 - 1}\,\frac{t}{\tau} + \frac{\zeta}{\sqrt{\zeta^2 - 1}}\sinh \sqrt{\zeta^2 - 1}\,\frac{t}{\tau}\right) \tag{8.20}$$

where the hyperbolic functions are defined as

$$\sinh a = \frac{e^a - e^{-a}}{2}$$

$$\cosh a = \frac{e^a + e^{-a}}{2}$$

The procedure for obtaining Eq. (8.20) is parallel to that used in the previous cases.

The response has been plotted in Fig. 8.2 for several values of ζ. Notice that the response is nonoscillatory and becomes more "sluggish" as ζ increases.   This is known as an *overdamped* response.   As in previous cases, all curves eventually approach the line $Y = 1$.

Actually, the response for ζ > 1 is not new.   We met it previously in the discussion of the step response of a system containing two first-order systems in series, for which the transfer function is

$$\frac{Y(s)}{X(s)} = \frac{1}{(\tau_1 s + 1)(\tau_2 s + 1)} \tag{8.21}$$

This is true for ζ > 1 because the roots s_1 and s_2 are real, and the denominator of Eq. (8.11) may be factored into two real linear factors.   Therefore, Eq. (8.11) is equivalent to Eq. (8.21) in this case.   By comparing the linear factors of the denominator of Eq. (8.11) with those of Eq. (8.21), it follows that

$$\tau_1 = (\zeta + \sqrt{\zeta^2 - 1})\tau \tag{8.22}$$
$$\tau_2 = (\zeta - \sqrt{\zeta^2 - 1})\tau \tag{8.23}$$

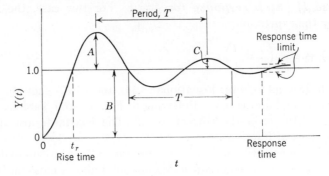

Fig. 8.3   **Terms used to describe an underdamped second-order response.**

Note that, if $\tau_1 = \tau_2$, then $\tau = \tau_1 = \tau_2$ and $\zeta = 1$. The reader should verify these results.

Terms Used to Describe an Underdamped System   Of these three cases, the underdamped response occurs most frequently in control systems.   Hence a number of terms are used to describe the underdamped response quantitatively.   Equations for some of these terms are listed below for future reference.   In general, the terms depend on ζ and/or τ. All these equations can be derived from the time response as given by Eq. (8.17); however, the mathematical derivations are left to the reader as exercises.

1. Overshoot.   Overshoot is a measure of how much the response exceeds the ultimate value following a step change and is expressed as the ratio A/B in Fig. 8.3.

The overshoot for a unit step is related to ζ by the expression

$$\text{Overshoot} = \exp\left(-\pi\zeta/\sqrt{1 - \zeta^2}\right) \tag{8.24}$$

This relation is plotted in Fig. 8.4.   The overshoot increases for decreasing ζ.

2. Decay ratio.   The decay ratio is defined as the ratio of the sizes of successive peaks and is given by C/A in Fig. 8.3.   The decay ratio is related to ζ by the expression

$$\text{Decay ratio} = \exp\left(-2\pi\zeta/\sqrt{1 - \zeta^2}\right) = (\text{overshoot})^2 \tag{8.25}$$

which is plotted in Fig. 8.4.   Notice that larger ζ means greater damping, hence greater decay.

3. Rise time.   This is the time required for the response to first reach its ultimate value and is labeled t_r in Fig. 8.3.   The reader can verify from Fig. 8.2 that t_r increases with increasing ζ.

4. Response time.   This is the time required for the response to come within ± 5 percent of its ultimate value and remain there.   The response time is indicated in Fig. 8.3.   The limits ± 5 percent are arbitrary, and other limits have been used in other texts for defining a response time.

5. Period of oscillation. From Eq. (8.17), the radian frequency (radians/time) is the coefficient of t in the sine term; thus,

$$\omega, \text{ radian frequency} = \frac{\sqrt{1 - \zeta^2}}{\tau} \tag{8.26}$$

Since the radian frequency ω is related to the cyclical frequency f by $\omega = 2\pi f$, it follows that

$$f = \frac{1}{T} = \frac{1}{2\pi} \frac{\sqrt{1 - \zeta^2}}{\tau} \tag{8.27}$$

where T is the period of oscillation (time/cycle). In terms of Fig. 8.3, T is the time elapsed between peaks. It is also the time elapsed between alternate crossings of the line $Y = 1$.

6. Natural period of oscillation. If the damping is eliminated [$C = 0$ in Eq. (8.1), or $\zeta = 0$], the system oscillates continuously without attenuation in amplitude. Under these "natural" or undamped conditions, the radian frequency is $1/\tau$, as shown by Eq. (8.26) when $\zeta = 0$. This frequency is referred to as the natural frequency ω_n:

$$\omega_n = \frac{1}{\tau} \tag{8.28}$$

The corresponding natural cyclical frequency f_n and period T_n are related by the expression

$$f_n = \frac{1}{T_n} = \frac{1}{2\pi\tau} \tag{8.29}$$

Thus, τ has the significance of the undamped period.

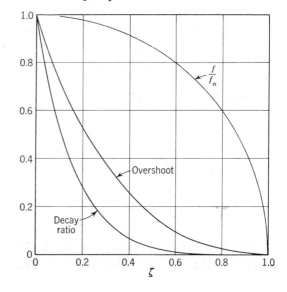

Fig. 8.4   Characteristics of a step response of an underdamped second-order system.

From Eqs. (8.27) and (8.29), the natural frequency is related to the actual frequency by the expression

$$\frac{f}{f_n} = \sqrt{1 - \zeta^2}$$

which is plotted in Fig. 8.4. Notice that, for $\zeta < 0.5$, the natural frequency is nearly the same as the actual frequency.

In summary, it is evident that ζ is a measure of the degree of damping, or the oscillatory character, and τ is a measure of the period, or speed, of the response of a second-order system.

Impulse Response If a unit impulse $\delta(t)$ is applied to the second-order system, then from Eqs. (8.11) and (4.1) the transform of the response is

$$Y(s) = \frac{1}{\tau^2 s^2 + 2\zeta\tau s + 1} \tag{8.30}$$

As in the case of the step input, the nature of the response to a unit impulse will depend on whether the roots of the denominator of Eq. (8.30) are real or complex. The problem is again divided into the three cases shown in Table 8.1, and each is discussed below.

Case I Impulse response for $\zeta < 1$ The inversion of Eq. (8.30) for $\zeta < 1$ yields the result

$$Y(t) = \frac{1}{\sqrt{1 - \zeta^2}\,\tau}\, e^{-\zeta t/\tau} \sin \sqrt{1 - \zeta^2}\,\frac{t}{\tau} \tag{8.31}$$

which is plotted in Fig. 8.5. The slope at the origin in Fig. 8.5 is 1.0 for all values of ζ.

A simple way to obtain Eq. (8.31) from the step response of Eq. (8.17) is to take the derivative of Eq. (8.17). Comparison of Eqs. (8.13) and (8.30) shows that

$$Y(s)\Big|_{\text{impulse}} = sY(s)\Big|_{\text{step}} \tag{8.32}$$

The presence of s on the right-hand side of Eq. (8.32) implies differentiation with respect to t in the time response. In other words, the inverse transform of Eq. (8.32) is

$$Y(t)\Big|_{\text{impulse}} = \frac{d}{dt}\left(Y(t)\Big|_{\text{step}}\right) \tag{8.33}$$

Application of Eq. (8.33) to Eq. (8.17) yields Eq. (8.31). This principle also yields the results for the next two cases.

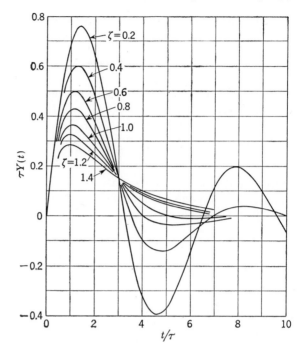

Fig. 8.5 **Response of a second-order system to a unit-impulse forcing function.**

Case II **Impulse response for** $\zeta = 1$ For the critically damped case, the response is given by

$$Y(t) = \frac{1}{\tau^2} te^{-t/\tau} \tag{8.34}$$

which is plotted in Fig. 8.5.

Case III **Impulse response for** $\zeta > 1$ For the overdamped case, the response is given by

$$Y(t) = \frac{1}{\tau} \frac{1}{\sqrt{\zeta^2 - 1}} e^{-\zeta t/\tau} \sinh \sqrt{\zeta^2 - 1} \frac{t}{\tau} \tag{8.35}$$

which is plotted in Fig. 8.5.

To summarize, the impulse-response curves of Fig. 8.5 show the same general behavior as the step-response curves of Fig. 8.2. However, the impulse response always returns to zero. Terms such as decay ratio, period of oscillation, etc., may also be used to describe the impulse response. Many control systems exhibit transient responses such as those of Fig. 8.5. This is illustrated by Fig. 1.7 for the stirred-tank heat exchanger.

Sinusoidal Response   If the forcing function applied to the second-order system is sinusoidal,

$$X(t) = A \sin \omega t$$

then it follows from Eqs. (8.11) and (5.20) that

$$Y(s) = \frac{A\omega}{(s^2 + \omega^2)(\tau^2 s^2 + 2\zeta\tau s + 1)} \tag{8.36}$$

The inversion of Eq. (8.36) may be accomplished by first factoring the two quadratic terms to give

$$Y(s) = \frac{A\omega/\tau^2}{(s - j\omega)(s + j\omega)(s - s_1)(s - s_2)} \tag{8.37}$$

Here s_1 and s_2 are the roots of the denominator of the transfer function and are given by Eqs. (8.14) and (8.15). For the case of an underdamped system ($\zeta < 1$), the roots of the denominator of Eq. (8.37) are a pair of pure imaginary roots $(+j\omega, -j\omega)$ contributed by the forcing function and a pair of complex roots $(-\zeta/\tau + j\sqrt{1 - \zeta^2}/\tau, -\zeta/\tau - j\sqrt{1 - \zeta^2}/\tau)$. We may write the form of the response $Y(t)$ by referring to Fig. 3.1 and Table 3.1; thus

$$Y(t) = C_1 \cos \omega t + C_2 \sin \omega t$$
$$+ e^{-\zeta t/\tau}\left(C_3 \cos \sqrt{1 - \zeta^2}\,\frac{t}{\tau} + C_4 \sin \sqrt{1 - \zeta^2}\,\frac{t}{\tau}\right) \tag{8.38}$$

The constants are evaluated by partial fractions. Notice in Eq. (8.38) that, as $t \to \infty$, only the first two terms do not become zero. These remaining terms are the ultimate periodic solution; thus

$$Y(t)\Big|_{t \to \infty} = C_1 \cos \omega t + C_2 \sin \omega t \tag{8.39}$$

The reader should verify that Eq. (8.39) is also true for $\zeta \geq 1$. From this little effort, we see already that the response of the second-order system to a sinusoidal driving function is ultimately sinusoidal and has the same frequency as the driving function. If the constants C_1 and C_2 are evaluated, we get from Eqs. (5.23) and (8.39)

$$Y(t) = \frac{A}{\sqrt{[1 - (\omega\tau)^2]^2 + (2\zeta\omega\tau)^2}} \sin(\omega t + \phi) \tag{8.40}$$

where

$$\phi = -\tan^{-1}\frac{2\zeta\omega\tau}{1 - (\omega\tau)^2}$$

By comparing Eq. (8.40) with the forcing function

$$X(t) = A \sin \omega t$$

it is seen that:

1. The ratio of the output amplitude to the input amplitude is

$$\frac{1}{\sqrt{[1 - (\omega\tau)^2]^2 + (2\zeta\omega\tau)^2}}$$

It will be shown in Chap. 18 that this may be greater or less than 1, depending on ζ and $\omega\tau$. This is in direct contrast to the sinusoidal response of the first-order system, where the ratio of the output amplitude to the input amplitude is always *less than* 1.

2. The output lags the input by phase angle $|\phi|$. It can be seen from Eq. (8.40), and will be shown in Chap. 18, that $|\phi|$ approaches 180° asymptotically as ω increases. The phase lag of the first-order system, on the other hand, can never exceed 90°. Discussion of other characteristics of the sinusoidal response will be deferred until Chap. 18.

We now have at our disposal considerable information about the dynamic behavior of the second-order system. It happens that many control systems which are not truly second-order exhibit step responses very similar to those of Fig. 8.2. Such systems are often characterized by second-order equations for approximate mathematical analysis. Hence, the second-order system is quite important in control theory, and frequent use will be made of the material in this chapter.

TRANSPORTATION LAG

A phenomenon which is often present in flow systems is the *transportation lag*. Synonyms for this term are *dead time* and *distance velocity lag*. As an example, consider the system shown in Fig. 8.6, in which a liquid flows through an insulated tube of uniform cross-sectional area A and length L at a constant volumetric flow rate q. The density ρ and the heat capacity C are constant. The tube wall has negligible heat capacity, and the velocity profile is flat (plug flow).

The temperature x of the entering fluid varies with time, and it is desired to find the response of the outlet temperature $y(t)$ in terms of a transfer function.

As usual, it is assumed that the system is initially at steady state; for this system, it is obvious that the inlet temperature equals the outlet

Fig. 8.6 **System with transportation lag.**

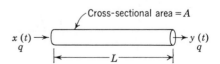

temperature; i.e.,

$$x_s = y_s \tag{8.41}$$

If a step change were made in $x(t)$ at $t = 0$, the change would not be detected at the end of the tube until τ sec later, where τ is the time required for the entering fluid to pass through the tube. This simple step response is shown in Fig. 8.7a.

If the variation in $x(t)$ were some arbitrary function, as shown in Fig. 8.7b, the response $y(t)$ at the end of the pipe would be identical with $x(t)$ but again delayed by τ units of time. The transportation lag parameter τ is simply the time needed for a particle of fluid to flow from the entrance of the pipe to the exit, and it can be calculated from the expression

$$\tau = \frac{\text{volume of pipe}}{\text{volumetric flow rate}}$$

or

$$\tau = \frac{AL}{q} \tag{8.42}$$

It can be seen from Fig. 8.7 that the relationship between $y(t)$ and $x(t)$ is

$$y(t) = x(t - \tau) \tag{8.43}$$

Subtracting Eq. (8.41) from (8.43) and introducing the deviation variables $X = x - x_s$ and $Y = y - y_s$ give

$$Y(t) = X(t - \tau) \tag{8.44}$$

If the Laplace transform of $X(t)$ is $X(s)$, the Laplace transform of $X(t - \tau)$ is $e^{-s\tau}X(s)$. This result follows from the theorem on translation of a function which was discussed in Chap. 4. Equation (8.44) becomes

$$Y(s) = e^{-s\tau}X(s)$$

or

$$\frac{Y(s)}{X(s)} = e^{-s\tau} \tag{8.45}$$

Therefore, the transfer function of a transportation lag is $e^{-s\tau}$.

The transportation lag is quite common in the chemical process industries where a fluid is transported through a pipe. We shall see in a

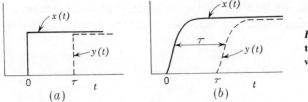

Fig. 8.7 Response of transportation lag to various inputs.

later chapter that the presence of a transportation lag in a control system can make it much more difficult to control. In general, such lags should be avoided if possible by placing equipment close together. They can seldom be entirely eliminated.

PROBLEMS

8.1 A step change of magnitude 4 is introduced into a system having the transfer function

$$\frac{Y(s)}{X(s)} = \frac{10}{s^2 + 1.6s + 4}$$

Determine

    a. Percent overshoot
    b. Rise time
    c. Maximum value of $Y(t)$
    d. Ultimate value of $Y(t)$
    e. Period of oscillation

8.2 The two-tank system shown in Fig. P8.2 is operating at steady state. At time $t = 0$, 10 ft³ of water are quickly added to the first tank. Using appropriate figures and

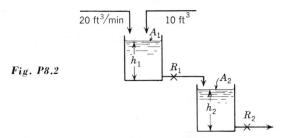

Fig. P8.2

equations in the text, determine the maximum deviation in level (feet) in both tanks from the ultimate steady-state values and the time at which each maximum occurs.
Data:

    $A_1 = A_2 = 10$ ft²
    $R_1 = 0.1$ ft/cfm
    $R_2 = 0.35$ ft/cfm

8.3 The two-tank liquid-level system shown in Fig. P8.3 is operating at steady state when a step change is made in the flow rate to tank 1. The transient response is critically

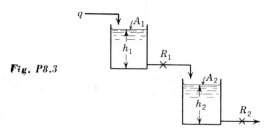

Fig. P8.3

damped, and it takes 1.0 min for the change in level of the second tank to reach 50 percent of the total change.

If the ratio of the cross-sectional areas of the tanks is $A_1/A_2 = 2$, calculate the ratio R_1/R_2. Calculate the time constant for each tank. How long does it take for the change in level of the first tank to reach 90 percent of the total change?

8.4 A mercury manometer is depicted in Fig. P8.4. Assuming the flow in the manometer to be laminar and the steady-state friction law for drag force in laminar flow

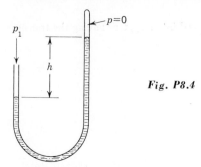

Fig. P8.4

to apply at each instant, determine a transfer function between the applied pressure p_1 and the manometer reading h. It will simplify the calculations if, for inertial terms, the velocity profile is assumed to be flat. From your transfer function, written in standard second-order form, list (*a*) the steady-state gain, (*b*) τ, and (*c*) ζ. Comment on these parameters as they are related to the physical nature of the problem.

8.5 Design a mercury manometer which will measure pressures up to 2 atm absolute and will give responses which are slightly underdamped (that is, $\zeta \approx 0.7$).

8.6 *a.* A system with second-order dynamics is subjected to an input

$$X = at$$

After *sufficient time has elapsed* for the response to move with constant velocity, find the relation between the response Y and input X. In particular give the lag and error, where lag is defined as the time elapsed between crossing of a particular value by the input and output and error is the difference at any time between input and output.

b. Consider an idealized spring-mass-viscous friction system with an applied force $F(t) = at$. $a = 0.15$ lb$_f$/sec. The spring constant $K = 20$ lb$_f$/ft. The friction force is proportional to the velocity of the block. The proportionality constant is $C = 1.8$ lb$_f$ sec/ft. The block weighs 5 lb$_m$.

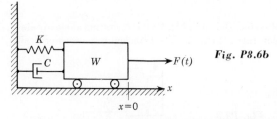

Fig. P8.6b

Determine the response of the system after sufficient time has elapsed for the velocity of the block to become constant and plot this response and $F(t)$ exactly (i.e., position versus time). Also sketch your best estimate of the response from initial application of the force up to the "sufficient time."

8.7   Verify Eqs. (8.17), (8.19), and (8.20).

8.8   Verify Eqs. (8.24) and (8.25).

8.9   Verify Eq. (8.40).

8.10   A second-order system is observed to exhibit a step response similar to the oscillatory curves (for $\zeta < 0.7$) of Fig. 8.2. (**a**) Suggest various methods for estimating the parameters ζ and τ from the step response for this system. (**b**) Suggest methods for evaluating these parameters from the response to a sinusoidal input.

8.11   If a second-order system is overdamped, it is more difficult to determine the parameters ζ and τ experimentally. One method for determining the parameters from a step response has been suggested by Oldenbourg and Sartorius (The Dynamics of Automatic Controls, *Trans. ASME*, p. 78, 1948), as described below.

    a. Show that the unit-step response for the overdamped case may be written in the form

$$S(t) = 1 - \frac{r_1 e^{r_2 t} - r_2 e^{r_1 t}}{r_1 - r_2}$$

    where r_1 and r_2 are the (real and negative) roots of

$$\tau^2 s^2 + 2\zeta\tau s + 1 = 0$$

    b. Show that $S(t)$ has an inflection point at

$$t_i = \frac{\ln (r_2/r_1)}{r_1 - r_2}$$

    c. Show that the slope of the step response at the inflection point

$$\frac{dS(t)}{dt}\bigg|_{t=t_i} = S'(t_i)$$

    has the value

$$S'(t_i) = -r_1 e^{r_1 t_i} = -r_2 e^{r_2 t_i}$$
$$= -r_1 \left(\frac{r_2}{r_1}\right)^{r_1/(r_1-r_2)}$$

    d. Show that the value of the step response at the inflection point is

$$S(t_i) = 1 + \frac{r_1 + r_2}{r_1 r_2} S'(t_i)$$

    and that hence

$$\frac{1 - S(t_i)}{S'(t_i)} = -\frac{1}{r_1} - \frac{1}{r_2}$$

    e. On a typical sketch of a unit-step response, show distances equal to

$$\frac{1}{S'(t_i)} \quad \text{and} \quad \frac{1 - S(t_i)}{S'(t_i)}$$

    and hence present two simultaneous equations resulting from a graphical method for determination of r_1 and r_2.

    f. Relate ζ and τ to r_1 and r_2.

8.12   **a.** If the two roots r_1 and r_2 are almost equal, show that the method proposed in Prob. 8.11 becomes inaccurate.

    b. Show that for $r_1 = r_2$, the unit-step response is

$$S(t) = 1 - (1 - r_1 t)e^{r_1 t}$$

c. Show that for this case

$$S(t_i) = 1 - \frac{2}{e} = 0.265$$

$$\frac{1 - S(t_i)}{S'(t_i)} = \frac{-2}{r_1}$$

d. Show that, for $r_1 \neq r_2$,

$$S(t_i) < 0.265$$

(*Hint:* Consider the limiting case for a first-order system.)

e. Show that if

$$S(t_i) > 0.265$$

the system is underdamped.

f. Show that if r_2 is very large compared with r_1,

$$S(t_i) \approx 0$$

and

$$\frac{1 - S(t_i)}{S'(t_i)} \approx \frac{-1}{r_1}$$

g. On the basis of the results derived in parts **a** through **f**, present a comprehensive graphical technique for analysis of a nonoscillatory response of a second-order system to a unit-step input.

part III Linear Closed-loop Systems

9 The Control System

INTRODUCTION

In the previous chapters, the dynamic behavior of several basic systems was examined. With this background, we can extend the discussion to a complete control system and introduce the fundamental concept of feedback. In order to work with a familiar system, the treatment will be based on the illustrative example of Chap. 1, which is concerned with a stirred-tank heater.

Figure 9.1 is a sketch of the apparatus. To orient the reader, the physical description of this control system will be reviewed. A liquid stream at a temperature T_i enters an insulated, well-stirred tank at a constant flow rate w (mass/time). It is desired to maintain (or control) the temperature in the tank at T_R by means of the controller. If the indicated tank temperature T_m differs from the desired temperature T_R, the controller senses the difference or *error*, $\epsilon = T_R - T_m$, and changes the heat input in such a way as to reduce the magnitude of ϵ. If the controller changes the heat input to the tank by an amount which is proportional to ϵ, we have

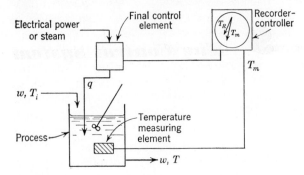

Fig. 9.1   **Control system for a stirred-tank heater.**

proportional control.   The reader should review the description of proportional control in Chap. 1.

In Fig. 9.1, it is indicated that the source of heat input q may be electricity or steam.   If an electrical source were used, the final control element might be a variable transformer which is used to adjust current to a resistance heating element; if steam were used, the final control element would be a control valve which adjusts the flow of steam.   In either case, the output signal from the controller should adjust q in such a way as to maintain control of the temperature in the tank.

Components of a Control System   The system shown in Fig. 9.1 may be divided into the following components:

1. Process (stirred-tank heater)
2. Measuring element (thermometer)
3. Controller
4. Final control element (variable transformer or control valve)

Each of these components can be readily identified as a separate physical item in the process.   In general, these four components will constitute most of the control systems which we shall consider in this text; however, the reader should realize that more complex control systems exist in which more components are used.   For example, there are some processes which require a cascade control system in which two controllers and two measuring elements are used.

Block Diagram   For computational purposes, it is convenient to represent the control system of Fig. 9.1 by means of the block diagram shown in Fig. 9.2.   Such a diagram makes it much easier to visualize the relationships among the various signals.   New terms which appear in Fig. 9.2 are *set point* and *load*.   The set point is a synonym for the desired value of the controlled variable.   The load refers to a change in any variable which may cause the controlled variable of the process to change.   In this example, the inlet temperature T_i is a load variable.   Other possible loads for this system are changes in flow rate and heat loss from the tank.   (These loads are not shown on the diagram.)

The control system shown in Fig. 9.2 is called a *closed-loop* system or a feedback system because the measured value of the controlled variable is returned or "fed back" to a device called the *comparator*. In the comparator, the controlled variable is compared with the desired value or *set point*. If there is any difference between the measured variable and the set point, an error is generated. This error enters a *controller*, which in turn adjusts the *final control element* in order to return the controlled variable to the set point.

Negative Feedback versus Positive Feedback   Several terms have been used which may need further clarification. The feedback principle, which is illustrated by Fig. 9.2, involves the use of the controlled variable T to maintain itself at a desired value T_R. The arrangement of the apparatus of Fig. 9.2 is often described as *negative feedback* to contrast with another arrangement called positive feedback. Negative feedback ensures that the difference between T_R and T_m is used to adjust the control element so that the tendency is to reduce the error. For example, assume that the system is at steady state and that $T = T_m = T_R$. If the load T_i should increase, T and T_m would start to increase, which would cause the error ϵ to become negative. With proportional control, the decrease in error would cause the controller and final control element to *decrease* the flow of heat to the system with the result that the flow of heat would eventually be reduced to a value such that T approaches T_R. A verbal description of the operation of a feedback control system, such as the one just given, is admittedly inadequate, for this description necessarily is given as a sequence of events. Actually all the components operate simultaneously, and the only adequate description of what is occurring is a set of simultaneous differential equations. This more accurate description is the primary subject matter of the present and succeeding chapters.

If the signal to the comparator were obtained by adding T_R and T_m, we would have a *positive feedback* system, which is inherently unstable. To see that this is true, again assume that the system is at steady state and that $T = T_m = T_R$. If T_i were to increase, T and T_m would increase, which would cause the signal from the comparator (ϵ in Fig. 9.2) to increase,

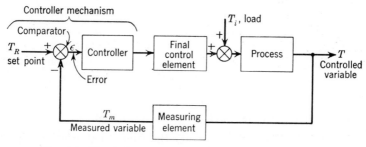

Fig. 9.2   **Block diagram of a simple control system.**

with the result that the heat to the system would increase. However, this action, which is just the opposite of that needed, would cause T to increase further. It should be clear that this situation would cause T to "run away" and control would not be achieved. For this reason, positive feedback would never be used intentionally in the system of Fig. 9.2. However, in more complex systems it may arise naturally. An example of this is discussed in Chap. 25.

Servo Problem versus Regulator Problem   The control system of Fig. 9.2 can be considered from the point of view of its ability to handle either of two types of situations. In the first situation, which is called the servomechanism-type (or servo) problem, we assume that there is no change in load T_i and that we are interested in changing the bath temperature according to some prescribed function of time. For this problem, the set point T_R would be changed in accordance with the desired variation in bath temperature. If the variation is sufficiently slow, the bath temperature may be expected to follow the variation in T_R very closely. There are occasions when a control system in the chemical industry will be operated in this manner. For example, one may be interested in varying the temperature of a reactor according to a prescribed time-temperature pattern, in which case a mechanism such as a time-driven cam would be arranged to move the set point in the desired fashion. However, the majority of problems which may be described as the servo type come from fields other than the chemical industry. The tracking of missiles and aircraft and the automatic machining of intricate parts from a master pattern are well-known examples of the servo-type problem. The other situation will be referred to as the regulator problem. In this case, the desired value T_R is to remain fixed and the purpose of the control system is to maintain the controlled variable at T_R in spite of changes in load T_i. This problem is very common in the chemical industry, and a complicated industrial process will often have many self-contained control systems, each of which maintains a particular process variable at a desired value. These control systems are of the regulator type.

In considering control systems in the following chapters, we shall frequently discuss the response of a linear control system to a change in set point (servo problem) separately from the response to a change in load (regulator problem). However, it should be realized that this is done only for convenience. The basic approach to obtaining the response of either type is essentially the same, and the two responses may be superimposed to obtain the response to any linear combination of set-point and load changes.

DEVELOPMENT OF BLOCK DIAGRAM

Each block in Fig. 9.2 represents the functional relationship existing between the input and output of a particular component. In the previous chapters, such input-output relations were developed in the form of transfer functions. In block-diagram representations of control systems, the variables selected are *deviation variables,* and inside each block is placed the transfer function relating the input-output pair of variables. Finally, the blocks are combined to give the overall block diagram. This is the procedure to be followed in developing Fig. 9.2.

Process Consider first the block for the process. This block will be seen to differ somewhat from those presented in previous chapters in that two input variables are present; however, the procedure for developing the transfer function remains the same.

An unsteady-state energy balance[1] around the heating tank gives

$$q + wC(T_i - T_o) - wC(T - T_o) = \rho C V \frac{dT}{dt} \tag{9.1}$$

where T_o is the reference temperature.

At steady state, dT/dt is zero, and Eq. (9.1) can be written

$$q_s + wC(T_{i_s} - T_o) - wC(T_s - T_o) = 0 \tag{9.2}$$

where the subscript s has been used to indicate steady state.

Subtracting Eq. (9.2) from Eq. (9.1) gives

$$q - q_s + wC[(T_i - T_{i_s}) - (T - T_s)] = \rho C V \frac{d(T - T_s)}{dt} \tag{9.3}$$

Notice that the reference temperature T_o cancels in the subtraction. If we introduce the deviation variables

$$T'_i = T_i - T_{i_s} \tag{9.4}$$
$$Q = q - q_s \tag{9.5}$$
$$T' = T - T_s \tag{9.6}$$

Eq. (9.3) becomes

$$Q + wC(T'_i - T') = \rho C V \frac{dT'}{dt} \tag{9.7}$$

Taking the Laplace transform of Eq. (9.7) gives

$$Q(s) + wC[T'_i(s) - T'(s)] = \rho C V s T'(s) \tag{9.8}$$

or

$$T'(s) \left(\frac{\rho V}{w} s + 1 \right) = \frac{Q(s)}{wC} + T'_i(s) \tag{9.9}$$

[1] In this analysis, it is assumed that the flow rate of heat q is instantaneously available and independent of the temperature in the tank. In some stirred-tank heaters, such as a jacketed kettle, q depends on both the temperature of the fluid in the jacket and the temperature of the fluid in the kettle. In this introductory chapter, systems (electrically heated tank or direct steam-heated tank) are selected for which this complication can be ignored. In Chap. 25, the analysis of a steam-jacketed kettle is given in which the effect of kettle temperature on q is taken into account.

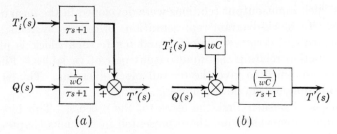

Fig. 9.3 **Block diagram for process.**

This last expression can be written

$$T'(s) = \frac{1/wC}{\tau s + 1} Q(s) + \frac{1}{\tau s + 1} T'_i(s) \tag{9.10}$$

where

$$\tau = \frac{\rho V}{w}$$

If there is a change in $Q(t)$ only, then $T'_i(t) = 0$ and the transfer function relating T' to Q is

$$\frac{T'(s)}{Q(s)} = \frac{1/wC}{\tau s + 1} \tag{9.11}$$

If there is a change in $T'_i(t)$ only, then $Q(t) = 0$ and the transfer function relating T' to T'_i is

$$\frac{T'(s)}{T'_i(s)} = \frac{1}{\tau s + 1} \tag{9.12}$$

Equation (9.10) is represented by the block diagram shown in Fig. 9.3a. This diagram is simply an alternate way to express Eq. (9.10) in terms of the transfer functions of Eqs. (9.11) and (9.12). Superposition makes this representation possible. Notice that, in Fig. 9.3, we have indicated summation of signals by the symbol shown in Fig. 9.4, which is called a *summing junction*. Subtraction can also be indicated with this symbol by placing a minus sign at the appropriate input. The summing junction was used previously as the symbol for the comparator of the controller (see Fig. 9.2). This symbol, which is standard in the control literature, may have several inputs but only one output.

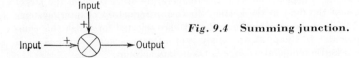

Fig. 9.4 **Summing junction.**

A block diagram which is equivalent to Fig. 9.3a is shown in Fig. 9.3b. That this diagram is correct can be seen by rearranging Eq. (9.10); thus

$$T'(s) = [Q(s) + wCT'_i(s)] \frac{1/wC}{\tau s + 1} \tag{9.13}$$

In Fig. 9.3b, the input variables $Q(s)$ and $wCT'_i(s)$ are summed before being operated on by the transfer function $1/wC/(\tau s + 1)$.

The physical situation that exists for the control system (Fig. 9.1) if steam heating is used requires more careful analysis to show that Fig. 9.3 is an equivalent block diagram. Assume that a supply of steam at constant conditions is available for heating the tank. One method for introducing heat to the system is to let the steam flow through a control valve and discharge directly into the water in the tank, where it will condense completely and become part of the stream leaving the tank (see Fig. 9.5).

If the flow of steam, f (pounds/time), is small compared with the inlet flow w, the total outlet flow is approximately equal to w. When the system is at steady state, the heat balance may be written

$$wC(T_{i_s} - T_o) - wC(T_s - T_o) + f_s(H_g - H_{l_s}) = 0 \tag{9.14}$$

where T_o = reference temperature used to evaluate enthalpy of all streams entering and leaving tank
  H_g = specific enthalpy of the steam supplied, a constant
  H_{l_s} = specific enthalpy of the condensed steam flowing out at T_s, as part of the total stream

The term H_{l_s} may be written in terms of heat capacity and temperature; thus

$$H_{l_s} = C(T_s - T_o) \tag{9.15}$$

From this, we see that, if the steady-state temperature changes, H_{l_s} changes. In Eq. (9.14), $f_s(H_g - H_{l_s})$ is equivalent to the steady-state input q_s used previously, as can be seen by comparing Eq. (9.2) with (9.14).

Now consider an *unsteady-state* operation in which f is much less than w and the temperature T of the bath does not deviate significantly from the steady-state temperature T_s. For these conditions, we may write the unsteady-state balance approximately; thus

$$wC(T_i - T_o) - wC(T - T_o) + f(H_g - H_{l_s}) = \rho CV \frac{dT}{dt} \tag{9.16}$$

In a practical situation for steam, H_g will be about 1000 Btu/lb$_m$. If the temperature of the bath, T, never deviates from T_s by more than 10°, the error in using the term $f(H_g - H_{l_s})$ instead of $f(H_g - H_l)$ will be no more than 1 percent. Under these conditions, Eq. (9.16) represents the system closely, and by comparing Eq. (9.16) with Eq.

Fig. 9.5 Supplying heat by steam.

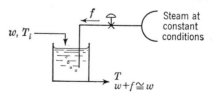

(9.1), it is clear that

$$q = f(H_g - H_{l_s}) \tag{9.17}$$

Therefore, q is proportional to the flow of steam f, which may be varied by means of a control valve. It should be emphasized that the analysis presented here is only approximate. Both f and the deviation in T must be small. The smaller they become, the more closely Eq. (9.16) represents the actual physical system. An exact analysis of the problem leads to a differential equation with time-varying coefficients, and the transfer-function approach does not apply. The problem becomes considerably more difficult. A better approximation will be discussed in Chap. 25, where linearization techniques are used.

Measuring Element The temperature-measuring element, which senses the bath temperature T and transmits a signal T_m to the controller, may exhibit some dynamic lag. From the discussion of the mercury thermometer in Chap. 5, we observed this lag to be first-order. In this example, we shall assume that the temperature-measuring element is a first-order system, for which the transfer function is

$$\frac{T'_m(s)}{T'(s)} = \frac{1}{\tau_m s + 1} \tag{9.18}$$

where the input-output variables T' and T'_m are deviation variables, defined as

$$T' = T - T_s$$
$$T'_m = T_m - T_{m_s}$$

Note that, when the control system is at steady state, $T_s = T_{m_s}$, which means that the temperature-measuring element reads the true bath temperature. The transfer function for the measuring element may be represented by the block diagram shown in Fig. 9.6.

Controller and Final Control Element For convenience, the blocks representing the controller and the final control element are combined into one block. In this way, we need be concerned only with the overall response between the error and the heat input to the tank. Also, it is assumed that the controller is a proportional controller. (In the next chapter, the response of other controllers, which are commonly used in control systems, will be described.) The relationship for a proportional

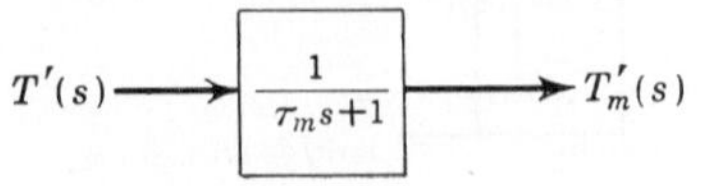

$$T'(s) \longrightarrow \boxed{\dfrac{1}{\tau_m s + 1}} \longrightarrow T'_m(s)$$

Fig. 9.6 Block diagram of measuring element.

controller is

$$q = K_c\epsilon + A \qquad (9.19)$$

where $\epsilon = T'_R - T'_m$
  $T'_R = T_R - T_{R_s}$
  T_{R_s} = "normal" set-point temperature
  K_c = proportional sensitivity or controller gain
  A = heat input when $\epsilon = 0$
Use of the deviation variable T'_R allows us to consider the effects of set-point changes. It is clear that the system should be designed so that $T_{R_s} = T_s$. The value of K_c can be varied by turning a knob in the controller. The mechanism for accomplishing this is discussed more fully in Chap. 22.

At steady state, it is assumed that the error ϵ is zero, so that

$$T'_R = T'_m = T' = 0$$

and therefore

$$A = q_s \qquad (9.20)$$

Subtracting Eq. (9.20) from Eq. (9.19) and using the deviation variable $Q = q - q_s$ give[1]

$$Q = K_c\epsilon \qquad (9.21)$$

The transform of Eq. (9.21) is simply

$$Q(s) = K_c\epsilon(s) \qquad (9.22)$$

This transfer function is represented by the block diagram shown in Fig. 9.7.

We have now completed the development of the separate blocks. If these are combined according to Fig. 9.2, there is obtained the block diagram for the complete control system shown in Fig. 9.8. The reader should verify this figure.

---

[1] Notice that assuming $\epsilon = 0$ at steady state implies that ϵ itself is a deviation variable; thus
  Deviation in $\epsilon = \epsilon(t) - \epsilon_s$
    $= \epsilon(t) - 0$
    $= \epsilon(t)$

Fig. 9.7   Block diagram of proportional controller.

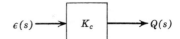

$\epsilon(s) \longrightarrow \boxed{K_c} \longrightarrow Q(s)$

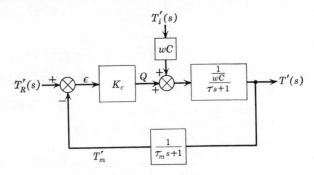

Fig. 9.8   **Block diagram of control system.**

SUMMARY

It has been shown that a control system can be translated into a block diagram which includes the transfer functions of the various components. It should be emphasized that a block diagram is simply a systematic way of writing the simultaneous differential and algebraic equations which describe the dynamic behavior of the components. In the present case, these were Eqs. (9.10), (9.18), and (9.22) and the definition of ϵ. The block diagram clarifies the relationships among the variables of these simultaneous equations. Another advantage of the block-diagram representation is that it clearly shows the feedback relationship between measured variable and desired variable and how the difference in these two signals (the error ϵ) is used to maintain control. A set of equations generally does not clearly indicate the relationships shown by the block diagram.

In the next several chapters, tools will be developed that will enable us to reduce a block diagram such as the one in Fig. 9.8 to a single block that relates $T'(s)$ to T'_i or T'_R. We shall then obtain the transient response of the control system shown in Fig. 9.8 to some specific changes in T'_R and T'_i. However, we shall first pause in Chap. 10 to look more carefully at the controller and control element blocks, which have been skimmed over in the present chapter.

10 *Controllers and Final Control Elements*

In the previous chapter, the block-diagram representation of a simple control system (Fig. 9.2) was developed. In this chapter, we shall focus attention on the controller and final control element and discuss the dynamic characteristics of some of these components which are in common use. As shown in Fig. 9.2, the input signal to the controller is the error and the output signal of the controller is fed to the final control element. In many process control systems, this output signal is an air pressure and the final control element is a pneumatic valve which opens and closes as air pressure on the diaphragm changes.

For the mathematical analysis of control systems, it is sufficient to regard the controller as a simple computer. For example, a proportional controller may be thought of as a device which receives the error signal and puts out a signal proportional to it. Similarly, the final control element may be regarded as a device which produces corrective action on the process. The corrective action is regarded as mathematically related to the output signal from the controller. However, it is desirable to have some appreciation of the actual physical mechanisms used to accomplish this. For this reason, we begin this chapter with a physical description of a pneumatic control valve and a simplified, rudimentary description of a pneumatic proportional controller. The pneumatic system is chosen because, at the time of this writing, it is by far the most widely used type of control action in the process industries. However, in recent years, there has been an increased use of electronic controllers, especially in new plants. The transfer functions which are presented in this chapter apply to either type of controller, and the discussion is in no way restrictive.

After the introductory mechanistic discussion, transfer functions will be presented for simplified or idealized versions of these pneumatic devices. These transfer functions, for practical purposes, will adequately represent the dynamic behavior of control elements and controllers. Hence, they will be used in subsequent chapters for mathematical analysis and design of control systems. In Chap. 22, a more complete discussion of the physical characteristics of pneumatic controllers will be given. At that point, the reader will have sufficient background and perspective to appreciate the more complete treatment.

MECHANISMS

Control Valve The control valve shown in Fig. 10.1 contains a pneumatic device (valve motor) which moves the valve stem as the pressure on a spring-loaded diaphragm changes. The stem positions a plug in the orifice of the valve body. As the pressure increases, the plug moves downward and restricts the flow of fluid through the valve. This action is referred to as air-to-close. The valve may also be constructed to have air-to-open action. Valve motors are often constructed so that the valve stem position is proportional to the valve-top pressure. Most commercial valves move from fully open to fully closed as the valve-top pressure changes from 3 to 15 psig.

In general, the flow rate of fluid through the valve depends upon the upstream and downstream fluid pressures and the size of the opening through

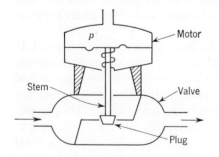

Fig. 10.1 **Pneumatic control valve (air-to-close).**

the valve. When the plug and seat (or orifice) are shaped, various relationships between stem position and size of opening (hence flow rate) can be obtained.[1] In our example, we shall assume for simplicity that at *steady state* the flow (for fixed upstream and downstream fluid pressures) is proportional to the valve-top pneumatic pressure. A valve having this relation is called a *linear valve*.

Proportional Controller A rudimentary sketch of the mechanism of a pneumatic proportional controller is shown in Fig. 10.2. It should be emphasized at the start that this sketch and the ensuing discussion are considerably oversimplified. This is done in order to acquaint the reader with the fundamental ideas without considering too many details. The basic objection to the mechanism of Fig. 10.2 is that, while in principle it will produce a proportional action, it is much too sensitive to be used in practice. Methods for modifying the mechanism to correct this and other limitations are discussed in more detail in Chap. 22. In this later chapter, pneumatic mechanisms for accomplishing other modes of control, such as integral, will also be described.

With these restrictions in mind, we consider the mechanism of Fig. 10.2 to be a proportional controller and describe its physical behavior. This typical industrial controller provides for the recording and controlling of a process variable and is usually referred to as a *recorder-controller*. As shown, the value of the measured variable is traced by a pen on a circular chart (or possibly a strip chart) which is calibrated in terms of units of the controlled variable. This chart is driven at a known speed by a clock motor. The controller also contains a set-point knob which is used to move a pointer across the chart to the desired value (or set point) of the measured variable. In the case of a more realistic proportional controller, another knob would be present for adjusting the proportional constant K_c.

The controller of Fig. 10.2 consists of a *controller mechanism*, which includes a baffle-nozzle system, and an error-detecting mechanism, which produces a motion proportional to the error. In the baffle-nozzle system, a regulated supply of air at about 20 psig is allowed to flow through a small nozzle having a diameter of about 0.01 in. This air stream then impinges on a strip of metal called the baffle. As the baffle is moved toward the nozzle, the pressure p in the nozzle increases because the area for air discharge is reduced. The nozzle pressure becomes equal to the supply pressure when the nozzle is capped (or sealed) by the baffle, and the system is designed so that the nozzle pressure falls linearly as the baffle-to-nozzle distance is increased. This pressure variation in the nozzle is used to operate a pneumatic control valve. As an example in Fig. 10.2 we consider that cooling water is flowing through the valve. The cooling water is

---

[1] For a more detailed discussion of the flow characteristics of control valves, see D. P. Eckman, "Automatic Process Control," John Wiley & Sons, Inc., New York, 1958.

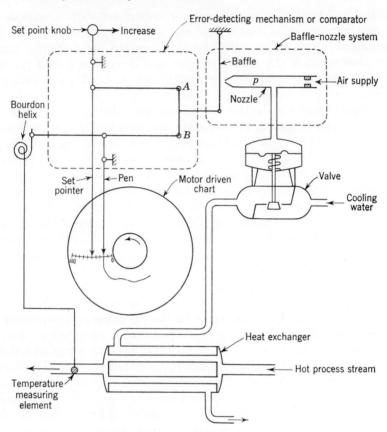

Fig. 10.2   Schematic diagram of high-sensitivity controller.

used to cool a hot stream for which the temperature is to be measured and controlled, and the nozzle pressure operates the cooling-water valve.

The error-detecting mechanism is constructed by pivoting the upper end of the baffle and moving the lower end, so that the overall baffle motion is proportional to the error.   Close study of Fig. 10.2 reveals the method by which the motion of the baffle is made dependent on the error.   As indicated in the figure, the lower end of link *AB* is attached to the measuring element.   For illustration, the measuring element is taken to be a Bourdon-type helix, which is attached to a mercury-filled bulb that senses temperature.   If the temperature of the controlled stream increases, the pressure of the mercury in the helix increases, causing it to unwind.   The motion of the helix moves the pen across the chart and moves the baffle toward the nozzle.   This baffle motion is made proportional to the pen motion by proper linkage and leverage.   The baffle motion results in a proportional increase in pressure in the nozzle and valve top, with a corresponding increase in cooling-water flow. Notice that the valve in Fig. 10.2 is of the pressure-to-open type.

To see how a set-point change affects the baffle motion, consider the temperature to remain fixed. An increase in set point (i.e., a request for a higher controlled temperature) causes point *A* to move to the left, and hence, the baffle moves away from the nozzle. This results in a closing of the cooling-water valve.

This concludes our brief introduction to pneumatic mechanisms. We now present transfer functions for such devices. These transfer functions, especially for controllers, are based on ideal devices which can be only approximated in practice. The degree of approximation is sufficiently good to warrant use of these transfer functions to describe the dynamic behavior of pneumatic mechanisms for ordinary design purposes. In addition, the transfer functions adequately describe the behavior of electronic controllers.

IDEAL TRANSFER FUNCTIONS

Control Valve A pneumatic valve always has some dynamic lag, which means that the stem motion does not respond instantaneously to a change in the applied pressure from the controller. From experiments conducted on pneumatic valves, it has been found that the relationship between flow and valve-top pressure for a linear valve can often be represented by a first-order transfer function; thus

$$\frac{Q(s)}{P(s)} = \frac{K_v}{\tau_v s + 1}$$

where K_r is the steady-state gain, i.e., the constant of proportionality between steady-state flow rate and valve-top pressure, and τ_r is the time constant of the valve.

In many practical systems, the time constant of the valve is very small when compared with the time constants of other components of the control system, and the transfer function of the valve can be approximated by a constant

$$\frac{Q(s)}{P(s)} = K_v$$

Under these conditions, the valve is said to contribute negligible dynamic lag.

To justify the approximation of a fast valve by a transfer function which is simply K_v, consider a first-order valve and a first-order process connected in series, as shown in Fig. 10.3.

Fig. 10.3 Block diagram for a first-order valve and a first-order process.

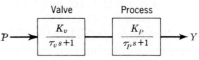

According to the discussion of Chap. 7, if we assume no interaction, the transfer function from $P(s)$ to $Y(s)$ is

$$\frac{Y(s)}{P(s)} = \frac{K_v K_P}{(\tau_v s + 1)(\tau_P s + 1)}$$

The assumption of no interaction is generally valid for this case.

For a unit-step change in P,

$$Y = \frac{1}{s} \frac{K_v K_P}{(\tau_v s + 1)(\tau_P s + 1)}$$

the inverse of which is

$$Y(t) = (K_v K_P) \left[1 - \frac{\tau_v \tau_P}{\tau_v - \tau_P} \left(\frac{1}{\tau_P} e^{-t/\tau_v} - \frac{1}{\tau_v} e^{-t/\tau_P} \right) \right]$$

If $\tau_v \ll \tau_P$, this equation is approximately

$$Y(t) = K_v K_P (1 - e^{-t/\tau_P})$$

The last expression is the unit-step response of the transfer function

$$\frac{Y(s)}{P(s)} = K_v \frac{K_P}{\tau_P s + 1}$$

so that the combination of process and valve is essentially first-order. This clearly demonstrates that, when the time constant of the valve is much smaller than that of the process, the valve transfer function can be taken as K_v.

A typical pneumatic valve has a time constant of the order of 10 sec. Many industrial processes behave as first-order systems or as a series of first-order systems having time constants which may range from a minute to an hour. For these systems we have shown that the lag of the valve is negligible, and we shall make frequent use of this approximation.

Controllers In this section, we shall present the transfer functions for the controllers frequently used in industrial processes. To simplify the discussion, only pneumatic controllers will be considered; however, the transfer functions which are obtained will be equally applicable to electronic controllers.

Proportional control. The proportional controller produces an output signal (pressure in the case of a pneumatic controller, current or voltage for an electronic controller) which is proportional to the error ϵ. This action may be expressed as

$$p = K_c \epsilon + p_s \tag{10.1}$$

where p = output pressure signal from controller, psi

K_c = gain, or sensitivity

ϵ = error = set point − measured variable

p_s = a constant

The error ϵ, which is the difference between the set point and the signal from the measuring element, may be in any suitable units. However, the units of set point and measured variable must be the same, since the error is the difference between these quantities.

In a controller having adjustable gain, the value of the gain K_c can be varied by moving a knob in the controller. The value of p_s is the value of the output pressure when ϵ is zero, and in most controllers p_s can be adjusted to obtain the required output signal when the control system is at steady state and $\epsilon = 0$. In terms of the system of Chap. 9, p_s would be set so that $T_s = T_{R_s}$.

To obtain the transfer function of Eq. (10.1), we first introduce the deviation variable

$$P = p - p_s$$

into Eq. (10.1). At time $t = 0$, we assume the error ϵ_s to be zero. Then ϵ is already a deviation variable. Equation (10.1) becomes

$$P(t) = K_c \epsilon(t) \tag{10.2}$$

Taking the transform of Eq. (10.2) gives the transfer function of an ideal proportional controller

$$\frac{P(s)}{\epsilon(s)} = K_c \tag{10.3}$$

The term *proportional band* is commonly used among process control engineers in place of the newer term *gain*. *Proportional band* is defined as the error (expressed as a percentage of the range of measured variable) required to move the valve from fully closed to fully open. A frequently used synonym is *bandwidth*. These terms will be most easily understood by considering the following example.

Example 10.1   A proportional controller is used to control temperature within the range of 60 to 100°F. The controller is adjusted so that the output pressure goes from 3 psi (valve fully open) to 15 psi (valve fully closed) as the measured temperature goes from 71 to 75°F with the set point held constant. Find the gain and the proportional band.

$$\text{Proportional band} = \frac{(75°F - 71°F)}{(100°F - 60°F)} \times 100$$

$$= 10\%$$

$$\text{Gain} = \frac{\Delta P}{\Delta \epsilon} = \frac{(15 \text{ psi} - 3 \text{ psi})}{(75°F - 71°F)} = 3 \text{ psi/°F}$$

Now assume that the proportional band of the controller is changed to 75 percent. Find the gain and the temperature change necessary to cause a valve to go from fully open to fully closed.

$$\Delta T = (\text{proportional band})(\text{range})$$

$$= 0.75 \, (40°F)$$

$$= 30°F$$

$$\text{Gain} = \frac{12 \text{ psi}}{30°F} = 0.4 \text{ psi/°F}$$

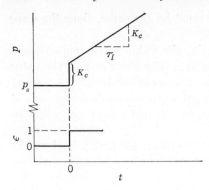

Fig. 10.4 **Response of a PI controller to a unit-step change in error.**

On-off control. A special case of proportional control is on-off control. If the gain K_c is made very high, the valve will move from one extreme position to the other if the pen deviates only slightly from the set point. This very sensitive action is called on-off action because the valve is either fully open (on) or fully closed (off); i.e., the valve acts like a switch. This is a very simple controller and is exemplified by the thermostat used in a home-heating system. The bandwidth of an on-off controller is approximately zero.

For various reasons, one of which was suggested in Chap. 1, it is often desirable to add other modes of control to the basic proportional action. These modes, integral and derivative action, are discussed below with the objective of obtaining the ideal transfer functions of the expanded controllers. The reasons for introducing these modes will be discussed briefly at the end of this chapter and in more detail in later chapters.

Proportional-integral (PI) control. This mode of control is described by the relationship

$$p = K_c\epsilon + \frac{K_c}{\tau_I} \int_0^t \epsilon \, dt + p_s \tag{10.4}$$

where K_c = gain
 τ_I = integral time, min
 p_s = constant
In this case, we have added to the proportional action term, $K_c\epsilon$, another term which is proportional to the integral of the error. The values of K_c and τ_I may be varied by two knobs in the controller.

To visualize the response of this controller, consider the response to a unit-step change in error, as shown in Fig. 10.4. This unit-step response is most directly obtained by inserting $\epsilon = 1$ into Eq. (10.4) which yields

$$p(t) = K_c + \frac{K_c}{\tau_I} t + p_s \tag{10.5}$$

Notice that p changes suddenly by an amount K_c, and then changes linearly with time at a rate K_c/τ_I.

To obtain the transfer function of Eq. (10.4), we again introduce the deviation variable $P = p - p_s$ into Eq. (10.4) and then take the transform to obtain

$$\frac{P(s)}{\epsilon(s)} = K_c \left(1 + \frac{1}{\tau_I s} \right) \tag{10.6}$$

Some manufacturers prefer to use the term *reset rate*, which is defined as the reciprocal of τ_I. The integral adjustment knob on a controller may be marked in terms of integral time or reset rate. The calibration of the proportional and integral knobs is often checked by observing the jump and slope of the step response shown in Fig. 10.4.

Proportional-derivative (PD) control. This mode of control may be represented by

$$p = K_c \epsilon + K_c \tau_D \frac{d\epsilon}{dt} + p_s \tag{10.7}$$

where K_c = gain

τ_D = derivative time, min

p_s = constant

In this case, we have added to the proportional term another term, $K_c \tau_D \, d\epsilon/dt$, which is proportional to the derivative of the error. The values of K_c and τ_D may be varied separately by knobs in the controller. Other terms which are used to describe the derivative action are *rate control* and *anticipatory control*.

The action of this controller can be visualized by considering the response to a linear change in error as shown in Fig. 10.5. This response is obtained by introducing the linear function $\epsilon(t) = At$ into Eq. (10.7) to obtain

$$p(t) = AK_c t + AK_c \tau_D + p_s$$

Notice that p changes suddenly by an amount $AK_c \tau_D$ as a result of the

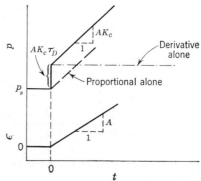

Fig. 10.5 Response of a PD controller to a linear input in error.

derivative action and then changes linearly at a rate AK_c. The effect of derivative action in this case is to anticipate the linear change in error by adding additional output $AK_c\tau_D$ to the proportional action.

To obtain the transfer function from Eq. (10.7), we introduce the deviation variable $P = p - p_s$ and then take the transform to obtain

$$\frac{P(s)}{\epsilon(s)} = K_c(1 + \tau_D s) \tag{10.8}$$

Proportional-derivative-integral (PID) control. This mode of control is a combination of the previous modes and is given by the expression

$$p = K_c\epsilon + K_c\tau_D \frac{d\epsilon}{dt} + \frac{K_c}{\tau_I} \int_0^t \epsilon \, dt + p_s \tag{10.9}$$

In this case, the controller contains three knobs for adjusting K_c, τ_D, and τ_I. The transfer function for this controller can be obtained from the Laplace transform of Eq. (10.9); thus

$$\frac{P(s)}{\epsilon(s)} = K_c\left(1 + \tau_D s + \frac{1}{\tau_I s}\right) \tag{10.10}$$

Motivation for Addition of Integral and Derivative Control Modes   Having introduced ideal transfer functions for integral and derivative modes of control, we now wish to indicate the practical motivation for use of these modes.   The curves of Fig. 10.6 show the behavior of a typical feedback control system using different kinds of control when it is subjected to a permanent disturbance.   This may be visualized in terms of the tank-temperature control system of Chap. 1 after a step change in T_i.   The value of the controlled variable is seen to rise at time zero owing to the disturbance. With no control, this variable continues to rise to a new steady-state value.

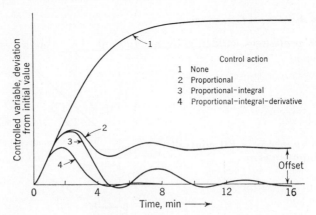

Fig. 10.6   **Response of a typical control system showing the effect of various modes of control.**

With control, after some time the control system begins to take action to try to maintain the controlled variable close to the value which existed before the disturbance occurred.

With proportional action only, the control system is able to arrest the rise of the controlled variable and ultimately bring it to rest at a new steady-state value. The difference between this new steady-state value and the original value is called *offset*. For the particular system shown, the offset is seen to be only 22 percent of the ultimate change which would have been realized for this disturbance in the absence of control.

As shown by the PI curve, the addition of integral action eliminates the offset; the controlled variable ultimately returns to the original value. This advantage of integral action is balanced by the disadvantage of a more oscillatory behavior.

The addition of derivative action to the PI action gives a definite improvement in the response. The rise of the controlled variable is arrested more quickly, and it is returned rapidly to the original value with little or no oscillation. Discussion of the PD mode is deferred to a later chapter.

The selection among the control systems whose responses are shown in Fig. 10.6 depends upon the particular application. If an offset of 22 percent is tolerable, proportional action would likely be selected, since a simple proportional controller is least expensive. If no offset were tolerable, integral action would be added. If excessive oscillations had to be eliminated, derivative action might be added. The addition of each mode means more initial expense and, as we shall see in later chapters, more difficult controller adjustment. The decision is always governed by economics. Our goal in forthcoming chapters will be to present the material which will enable the reader to develop curves such as those of Fig. 10.6 and thereby to design efficient, economic control systems.

SUMMARY

In this chapter we have presented a brief mechanistic discussion of pneumatic control valves and controllers. In addition, we have presented ideal transfer functions to represent their dynamic behavior and some typical results of using these controllers.

The ideal transfer functions actually describe the action of many types of nonpneumatic controllers, including hydraulic, mechanical, and electrical systems. Hence, the mathematical analyses of control systems to be presented in later chapters, which are based upon first- and second-order systems, transportation lags, and ideal controllers, generalize to many branches of the control field. After studying this text on process control, the reader should be able to apply his knowledge to, for example, problems in mechanical control systems. All that is required is a preliminary study of the physical nature of the systems involved.

PROBLEMS

10.1 A pneumatic PI controller has an output pressure of 10 psi when the set point and pen point are together. The set point and pen point are suddenly displaced by 0.5 in. (i.e., a step change in error is introduced) and the following data are obtained:

Time, sec	psi
0 −	10
0 +	8
20	7
60	5
90	3.5

Determine the actual gain (psi per inch displacement) and the integral time.

10.2 A unit-step change in error is introduced into a PID controller. If $K_c = 10$, $\tau_I = 1$, and $\tau_D = 0.5$, plot the response of the controller, $P(t)$.

10.3 An ideal PD controller has the transfer function

$$\frac{P}{\epsilon} = K_c(\tau_D s + 1)$$

As shown in Chap. 22, an actual PD controller has the transfer function

$$\frac{P}{\epsilon} = K_c \frac{\tau_D s + 1}{(\tau_D/\beta)s + 1}$$

where β is a large constant in an industrial controller.

If a unit-step change in error is introduced into a controller having the second transfer function, show that

$$P(t) = K_c(1 + A e^{-\beta t/\tau_D})$$

where A is a function of β which you are to determine. For $\beta = 5$ and $K_c = 0.5$, plot $P(t)$ versus t/τ_D. As $\beta \to \infty$, show that the unit-step response approaches that for the ideal controller.

10.4 A PID controller is at steady state with an output pressure of 9 psig. The set point and pen point are initially together. At time $t = 0$, the set point is moved away from the pen point at a rate of 0.5 in./min. The motion of the set point is in the direction of *lower* readings. If the knob settings are

$K_c = 2$ psi/in. of pen travel

$\tau_I = 1.25$ min

$\tau_D = 0.4$ min

plot the output pressure versus time.

11 *Block Diagram of a Chemical-reactor Control System*

To tie together the principles developed thus far and further to illustrate the procedure for reduction of a physical control system to a block diagram, we consider in this chapter the two-tank chemical-reactor control system of Fig. 11.1. This entire chapter is an example and may be omitted by the reader with no loss in continuity.

Description of System A liquid stream enters tank 1 at a volumetric flow rate F cfm and contains reactant A at a concentration of c_0 moles A/ft^3. Reactant A decomposes in the tanks according to the irreversible chemical reaction

$$A \rightarrow B$$

The reaction is first-order and proceeds at a rate

$$r = kc$$

where r = moles A decomposing$/(\text{ft}^3)(\text{time})$
c = concentration of A, moles A/ft^3
k = velocity constant, a function of temperature

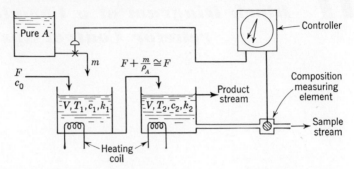

Fig. 11.1 **Control of a stirred-tank chemical reactor.**

The reaction is to be carried out in a series of two stirred tanks. The tanks are maintained at different temperatures. The temperature in tank 2 is to be greater than the temperature in tank 1, with the result that k_2, the velocity constant in tank 2, is greater than that in tank 1, k_1. We shall neglect any changes in physical properties due to chemical reaction.

The purpose of the control system is to maintain c_2, the concentration of A leaving tank 2, at some desired value in spite of variation in inlet concentration c_0. This will be accomplished by adding a stream of pure A to tank 1 through a control valve.

Reactor Transfer Functions We begin the analysis by making a material balance on A around tank 1; thus

$$V \frac{dc_1}{dt} = Fc_0 - \left(F + \frac{m}{\rho_A}\right) c_1 - k_1 V c_1 + m \tag{11.1}$$

where m = molar flow rate of pure A through the valve, lb moles/min
   ρ_A = density of pure A, lb moles/ft^3
   V = holdup volume of tank, a constant, ft^3

It is assumed that the volumetric flow of A through the valve m/ρ_A is much less than the inlet flow rate F with the result that Eq. (11.1) can be written

$$V \frac{dc_1}{dt} + (F + k_1 V)c_1 = Fc_0 + m \tag{11.2}$$

This last equation may be written in the form

$$\tau_1 \frac{dc_1}{dt} + c_1 = \frac{1}{1 + k_1 V/F} c_0 + \frac{1}{F(1 + k_1 V/F)} m \tag{11.3}$$

where

$$\tau_1 = \frac{V}{F + k_1 V}$$

At steady state, $dc_1/dt = 0$, and Eq. (11.3) becomes

$$c_{1_s} = \frac{1}{1 + k_1 V/F} c_{0_s} + \frac{1}{F(1 + k_1 V/F)} m_s \tag{11.4}$$

where s refers to steady state.

Subtracting Eq. (11.4) from (11.3) and introducing the deviation variables

$$C_1 = c_1 - c_{1_s}$$
$$C_0 = c_0 - c_{0_s}$$
$$M = m - m_s$$

give

$$\tau_1 \frac{dC_1}{dt} + C_1 = \frac{1}{1 + k_1 V/F} C_0 + \frac{1}{F(1 + k_1 V/F)} M \tag{11.5}$$

Taking the transform of Eq. (11.5) yields the transfer function of the first reactor:

$$C_1(s) = \frac{1/(1 + k_1 V/F)}{\tau_1 s + 1} C_0(s) + \frac{1/[F(1 + k_1 V/F)]}{\tau_1 s + 1} M(s) \tag{11.6}$$

A material balance on A around tank 2 gives

$$V \frac{dc_2}{dt} = F(c_1 - c_2) - k_2 V c_2 \tag{11.7}$$

As with tank 1, this last equation can be written in terms of deviation variables and arranged to give

$$\tau_2 \frac{dC_2}{dt} + C_2 = \frac{1}{1 + k_2 V/F} C_1 \tag{11.8}$$

where

$$\tau_2 = \frac{V}{F + k_2 V}$$
$$C_2 = c_2 - c_{2_s}$$

Taking the transform of Eq. (11.8) gives the transfer function for the second reactor:

$$C_2(s) = \frac{1/(1 + k_2 V/F)}{\tau_2 s + 1} C_1(s) \tag{11.9}$$

To obtain some numerical results, we shall assume the following data to apply to the system:

Molecular weight of A = 100 lb/lb mole
$$\rho_A = 0.8 \text{ lb mole/ft}^3$$
$$c_{0_s} = 0.1 \text{ lb mole } A/\text{ft}^3$$
$$F = 100 \text{ cfm}$$
$$m_s = 1.0 \text{ lb mole/min}$$
$$k_1 = \tfrac{1}{6} \text{ min}^{-1}$$
$$k_2 = \tfrac{2}{3} \text{ min}^{-1}$$
$$V = 300 \text{ ft}^3$$

Substituting these constants into the parameters of the problem yields the following values:

$$\tau_1 = 2 \text{ min}$$
$$\tau_2 = 1 \text{ min}$$
$$c_{1_s} = 0.0733 \text{ lb mole } A/\text{ft}^3$$
$$c_{2_s} = 0.0244 \text{ lb mole } A/\text{ft}^3$$
$$m_s/\rho_A = 1.25 \text{ cfm}$$

Control Valve Assume that the control valve selected for the process has the following characteristics: The flow of A through the valve varies linearly from zero to 2 cfm as the valve-top pressure varies from 3 to 15 psig. The time constant τ_v of the valve is so small compared with the other time constants in the system that its dynamics can be neglected.

From the data given, the valve sensitivity is computed as

$$K_v = \frac{2 - 0}{15 - 3} = \frac{1}{6} \text{ cfm/psi}$$

Since $m_s/\rho_A = 1.25$ cfm, the normal operating pressure on the valve is

$$p_s = 3 + \frac{1.25}{2} (15 - 3) = 10.5 \text{ psi} \tag{11.10}$$

The equation for the valve is therefore

$$m = [1.25 + K_v(p - 10.5)]\rho_A \tag{11.11}$$

In terms of deviation variables, this can be written

$$M = K_v\rho_A P \tag{11.12}$$

where

$$M = m - 1.25\rho_A$$
$$P = p - 10.5$$

Taking the transform of Eq. (11.12) gives

$$\frac{M(s)}{P(s)} = K_v\rho_A \tag{11.13}$$

as the valve transfer function.

Measuring Element For illustration, assume that the pen on the controller moves full scale (0 to 4.00 in.) as the concentration of A varies from 0.01 to 0.05 lb mole A/ft^3. We shall assume that the concentration measuring device is linear and has negligible lag. The sensitivity for the measuring device is therefore

$$K_m = \frac{4.00}{0.05 - 0.01} = 100 \text{ in. pen travel/(lb mole/ft}^3)$$

Since c_{2_s} is 0.0244 lb mole/ft^3, the normal pen reading is

$$\frac{0.0244 - 0.01}{0.05 - 0.01} (4.00) = 1.44 \text{ in.}$$

The equation for the measuring device is therefore

$$b = 1.44 + K_m(c_2 - 0.0244) \tag{11.14}$$

where b is the pen reading in inches of pen displacement. In terms of deviation variables, Eq. (11.14) becomes

$$B = K_m C_2 \tag{11.15}$$

where

$$B = b - 1.44$$

The transfer function for the measuring device is therefore

$$\frac{B(s)}{C_2(s)} = K_m \tag{11.16}$$

A measuring device which changes the units between input and output signals is called a *transducer*. In the present case, the concentration signal is transduced to a pen reading.

Controller For convenience, we shall assume the controller to have proportional action, in which case the relationship between controller output pressure and error is

$$p = p_s + K_c(c_R - b) = p_s + K_c \epsilon \tag{11.17}$$

where c_R = desired pen reading or set point, in.

K_c = controller sensitivity, psi/in.

ϵ = error = $c_R - b$

In terms of deviation variables, Eq. (11.17) becomes

$$P = K_c \epsilon \tag{11.18}$$

The transform of this equation gives the transfer function of the controller

$$\frac{P(s)}{\epsilon(s)} = K_c \tag{11.19}$$

Assuming the set point and the pen reading to be the same when the system is at steady state under normal conditions, we have for the reference value of the set point

$$c_{R_s} = b = 1.44 \text{ in.}$$

The corresponding deviation variable for the set point is

$$C_R = c_R - c_{R_s}$$

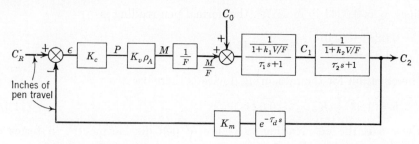

Fig. 11.2 Block diagram for a chemical-reactor control system.

Transportation Lag A portion of the liquid leaving tank 2 is continuously withdrawn through a sample line containing a concentration-measuring element, at a rate of 0.1 cfm. The measuring element must be remotely located from the process, because rigid ambient conditions must be maintained for accurate concentration measurements. The sample line has a length of 50 ft, and the cross-sectional area of the line is 0.001 ft².

The sample line can be represented by a transportation lag with parameter

$$\tau_d = \frac{\text{volume}}{\text{flow rate}} = \frac{(50)(0.001)}{0.1} = 0.5 \text{ min}$$

The transfer function for the sample line is, therefore,

$$e^{-\tau_d s} = e^{-0.5 s}$$

Block Diagram We have now completed the analysis of each component of the control system and have obtained a transfer function for each. These transfer functions can now be combined so that the overall system is represented by the block diagram in Fig. 11.2. An equivalent diagram is shown in Fig. 11.3 in which some of the blocks have been combined.

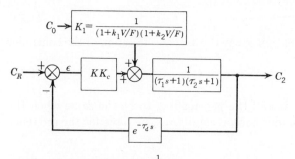

Fig. 11.3 Equivalent block diagram for a chemical-reactor control system. (C_R is now in concentration units.)

$$\tau_1 = 2, \quad \tau_2 = 1, \quad \tau_d = 0.5 \quad K_1 = \frac{1}{4.5}$$

$$\text{Open-loop gain} = KK_c = \frac{K_m K_v \rho_A}{F\left(1 + \dfrac{k_1 V}{F}\right)\left(1 + \dfrac{k_2 V}{F}\right)} \quad K_c = 0.03 K_c$$

Numerical quantities for the parameters in the transfer functions are given in Fig. 11.3. It should be emphasized that the block diagram is written for deviation variables. The true steady-state values, which are not given by the diagram, must be obtained from the analysis of the problem.

The example analyzed in this chapter will be used later in discussion of control system design. The design problem will be to select a value of K_c which gives satisfactory control of the composition C_2 despite the rather long transportation lag involved in getting information to the controller. In addition, we shall want to consider possible use of other modes of control for this system.

12 *Closed-loop Transfer Functions*

Standard Block-diagram Symbols In Chap. 9, a block diagram was developed for the control of a stirred-tank heater (Fig. 9.2). In Fig. 12.1, we have redrawn the block diagram and used some standard symbols for the variables and transfer functions, which are widely used in the control literature. These symbols are defined as follows:

R = set point or desired value
C = controlled variable
ϵ = error
B = variable produced by measuring element
M = manipulated variable
U = load variable or disturbance
G_c = transfer function of controller
G_1 = transfer function of final control element
G_2 = transfer function of process
H = transfer function of measuring element

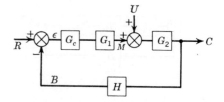

Fig. 12.1 Standard control system nomenclature.

In some cases, the blocks labeled G_c and G_1 will be lumped together into a single block as was done in Chap. 9. The series of blocks between the comparator and the controlled variable, which consist of G_c, G_1, and G_2, is referred to as the *forward path*. The block H between the controlled variable and the comparator is called the *feedback path*. The use of G for a transfer function in the forward path and H for one in the feedback path is a common convention.

In more complex control systems, the block diagram may contain several feedback paths and several loads. A typical example is shown in Fig. 12.2. Most of our examples will be represented by the simple diagram shown in Fig. 12.1.

Overall Transfer Function for Single-loop Systems Once a control system has been described by a block diagram, such as the one shown in Fig. 12.1, the next step is to determine the transfer function relating C to R or C to U. We shall refer to these transfer functions as *overall* transfer functions because they apply to the entire system. These overall transfer functions are used to obtain considerable information about the control system, as will be demonstrated in the succeeding chapters. For the present it is sufficient to note that they are useful in determining the response of C to any change in R and U. The response to a change in set point R, obtained by setting $U = 0$, represents the solution to the servo problem. The response to a change in load variable U, obtained by setting $R = 0$, is the solution to the regulator problem. A systematic approach for obtaining the overall transfer function for set-point change and load change will now be presented:

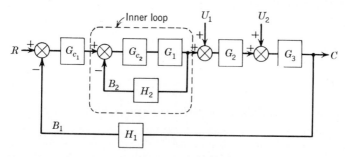

Fig. 12.2 Block diagram for a multiloop, multiload system.

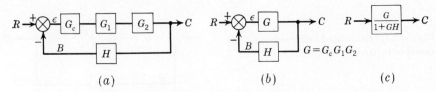

Fig. 12.3   **Block-diagram reduction to obtain overall transfer function.**

Overall transfer function for change in set point.   For this case, $U = 0$ and Fig. 12.1 may be simplified or reduced as shown in Fig. 12.3. In this reduction, we have made use of a simple rule of block-diagram reduction which states that a block diagram consisting of several transfer functions in series can be simplified to a single block containing a transfer function which is the product of the individual transfer functions.

This rule can be proved by considering two noninteracting blocks in series as shown in Fig. 12.4.   This block diagram is equivalent to the equations

$$\frac{Y}{X} = G_A \qquad \frac{Z}{Y} = G_B$$

Multiplying these equations gives

$$\frac{Y}{X}\frac{Z}{Y} = G_A G_B$$

which simplifies to

$$\frac{Z}{X} = G_A G_B$$

Thus, the intermediate variable Y has been eliminated, and we have shown the overall transfer function Z/X to be the product of the transfer functions $G_A G_B$.   This proof for two blocks can be easily extended to any number of blocks to give the rule for the general case.   This rule was developed in Chap. 7 for the specific case of several noninteracting, first-order systems in series.

With this simplification the following equations can be written directly from Fig. 12.3b.

$$C = G\epsilon \tag{12.1}$$
$$B = HC \tag{12.2}$$
$$\epsilon = R - B \tag{12.3}$$

Since there are four variables and three equations, we can solve the equations

$X \longrightarrow \boxed{G_A} \overset{Y}{\longrightarrow} \boxed{G_B} \longrightarrow Z$ *Fig. 12.4*

simultaneously for C in terms of R as follows:

$$C = G(R - B)$$
$$C = G(R - HC)$$
$$C = GR - GHC$$

or finally

$$\frac{C}{R} = \frac{G}{1 + GH} \tag{12.4}$$

This is the overall transfer function relating C to R and may be represented by an equivalent block diagram as shown in Fig. 12.3c.

Overall transfer function for change in load. In this case $R = 0$, and Fig. 12.1 is drawn as shown in Fig. 12.5. From the diagram we can write the following equations:

$$C = G_2(U + M) \tag{12.5}$$
$$M = G_cG_1\epsilon \tag{12.6}$$
$$B = HC \tag{12.7}$$
$$\epsilon = -B \tag{12.8}$$

Again the number of variables (C, U, M, B, ϵ) exceeds by one the number of equations, and we can solve for C in terms of U as follows:

$$C = G_2(U + G_cG_1\epsilon)$$
$$C = G_2[U + G_cG_1(-HC)]$$

or finally

$$\frac{C}{U} = \frac{G_2}{1 + GH} \tag{12.9}$$

where $G = G_cG_1G_2$. Notice that the transfer functions for load change or set-point change have denominators which are identical, $1 + GH$.

The following simple rule serves to generalize these results for the single-loop feedback system shown in Fig. 12.1: The transfer function relating

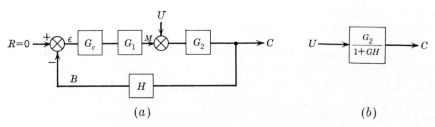

(a)   (b)

Fig. 12.5 Block diagram for change in load.

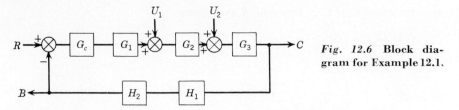

Fig. 12.6 Block diagram for Example 12.1.

any pair of variables X, Y is obtained by the relationship

$$\frac{Y}{X} = \frac{\pi_f}{1 + \pi_l} \qquad \text{negative feedback} \tag{12.10}$$

where π_f = product of transfer functions in the path between the locations of the signals X and Y

π_l = product of all transfer functions in the loop (i.e., in Fig. 12.1, $\pi_l = G_c G_1 G_2 H$)

If this rule is applied to finding C/R in Fig. 12.1, we obtain

$$\frac{C}{R} = \frac{G_c G_1 G_2}{1 + G_c G_1 G_2 H} = \frac{G}{1 + GH}$$

which is the same as before. For positive feedback, the reader should show that the following result is obtained:

$$\frac{Y}{X} = \frac{\pi_f}{1 - \pi_l} \qquad \text{positive feedback} \tag{12.10a}$$

Example 12.1 Determine the transfer functions C/R, C/U_1, and B/U_2 for the system shown in Fig. 12.6.

Using the rule given by Eq. (12.10), we obtain by inspection the results

$$\frac{C}{R} = \frac{G_c G_1 G_2 G_3}{1 + G}$$
$$\frac{C}{U_1} = \frac{G_2 G_3}{1 + G}$$
$$\frac{B}{U_2} = \frac{G_3 H_1 H_2}{1 + G}$$

where $G = G_c G_1 G_2 G_3 H_1 H_2$. The reader should check one or more of these results by the direct method of solution of simultaneous equations.

Overall Transfer Function for Multiloop Control System Obtaining the overall transfer function C/R for the system represented by Fig. 12.7a is straightforward if we first reduce the inner loop (or minor loop) involving G_{c_2}, G_1, and H_2 to a single block, as we have just done in the case of Fig. 12.1. For convenience, we may also combine G_2 and G_3 into a single block. These reductions are shown in Fig. 12.7b.

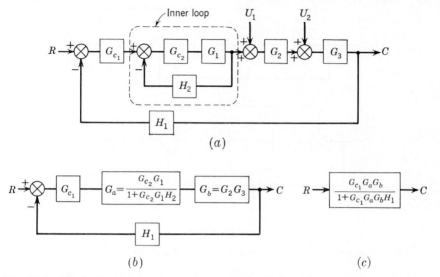

Fig. 12.7   Block-diagram reduction.   (*a*) **original diagram;** (*b*) **first reduction;**
(*c*) **final single-block diagram.**

Figure 12.7*b* is a single-loop block diagram which can be reduced to
one block as shown in Fig. 12.7*c*.

It should be clear, without a detailed example, that, to find any other
transfer function such as C/U_1 in Fig. 12.7*a*, we proceed in the same manner,
i.e., first reduce the inner loop to a single-block equivalent.

Summary   In this chapter, we have illustrated the procedure for
reducing the block diagram of a control system to a single block which
relates one input to one output variable.   This procedure consists of
writing, directly from the block diagram, a sufficient number of linear
algebraic equations and solving them simultaneously for the transfer function of the desired pair of variables.   For single-loop control systems, a
simple rule was developed for finding the transfer function between any
desired pair of input-output variables.   This rule is also useful in reducing
a multiloop system to a single-loop system.

It should be emphasized that, regardless of the pair of variables selected,
the denominator of the closed-loop transfer function will always contain the
same term, $1 + G$, where G is the open-loop transfer function of the single-
loop control system.   In the succeeding chapters, frequent use will be made
of the material in this chapter to determine the overall response of control
systems.

PROBLEMS

12.1   Determine the transfer function $Y(s)/X(s)$ for the block diagrams shown in
Fig. P12.1.   Express the results in terms of G_a, G_b, and G_c.

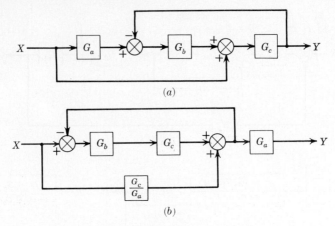

Fig. P12.1

12.2 Find the transfer function $Y(s)/X(s)$ of the system shown in Fig. P12.2.

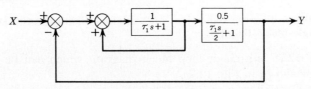

Fig. P12.2

13 Transient Response of Simple Control Systems

In this chapter the results of all the previous chapters will be applied to determining the transient response of a simple control system to changes in set point and load.[1]   Considerable use will be made of the results of Chaps. 5 to 8 because the overall transfer functions for the examples presented here reduce to first- and second-order systems.

Consider the control system for the heated stirred tank which has been discussed in Chaps. 1 and 9 and which is represented by Fig. 13.1.   The reader may want to refer to Chap. 9 for a description of this control system.

In Fig. 13.1a, the sketch of the apparatus is drawn in such a way that the source of heat (electricity or steam) is not specified.   To make this problem more realistic, we have shown in Fig. 13.1b that the source of heat is steam which is discharged directly into the water and in Fig. 13.1c the source of heat is electrical.   In the latter sketch, a device is shown sche-

---

[1] The reader who is interested in the simulation of control systems by an analog computer can begin the study of Chaps. 30 to 32 at this point.

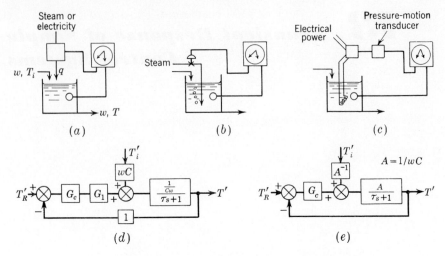

Fig. 13.1 **Block diagram of temperature-control system.**

matically in which a pneumatic valve motor adjusts the voltage from a variable transformer which in turn is connected to a resistance heater. The scheme using steam is usually more practical, but for the sake of generality, we admit the feasibility of the electrical system.

The block diagram is shown in Fig. 13.1d. The block representing the process is taken directly from Fig. 9.3. To reduce the number of symbols $1/wC$ has been replaced by A in Fig. 13.1e.

Throughout this chapter, we shall assume that the valve does not have any dynamic lag, for which case the transfer function of the valve (G_1 in Fig. 13.1) will be taken as a constant K_v. This assumption was shown to be reasonable in Chap. 10. To simplify the discussion further, K_v has been taken as 1. (If K_v were other than 1, we may simply replace G_c by G_cK_v in the ensuing discussion.)

In the first part of the chapter, we shall also assume that there is no dynamic lag in the measuring element ($\tau_m = 0$), so that it may be represented by a transfer function which is simply the constant 1. A bare thermocouple will have a response which is so fast that for all practical purposes it can be assumed to follow the slowly changing bath temperature without lag. When the feedback transfer function is unity, the system is called a *unity-feedback* system.

Introducing these assumptions leads to the simplified block diagram of Fig. 13.1e, for which we shall obtain overall transfer functions for changes in set point and load when proportional and proportional-integral control are used.

Proportional Control for Set-point Change (Servo Problem) For proportional control, $G_c = K_c$. Using the methods developed in the previous chapter, the overall transfer function in Fig. 13.1e is

$$\frac{T'}{T'_R} = \frac{K_c A/(\tau s + 1)}{1 + K_c A/(\tau s + 1)} = \frac{K_c A}{\tau s + 1 + K_c A} \tag{13.1}$$

This may be rearranged in the form of a first-order lag to give

$$\frac{T'}{T'_R} = \frac{A_1}{\tau_1 s + 1} \tag{13.2}$$

where

$$\tau_1 = \frac{\tau}{1 + K_c A}$$

$$A_1 = \frac{K_c A}{1 + K_c A} = \frac{1}{1 + 1/K_c A}$$

According to this result, the response of the tank temperature to change in set point is first-order. The time constant for the control system, τ_1, is less than that of the stirred tank itself, τ. This means that one of the effects of feedback control is to speed up the response. We may use the results of Chap. 5 to find the response to a variety of inputs.

The response of the system to a unit-step change in set point T'_R is shown in Fig. 13.2. (We have selected a unit change in set point for convenience; responses to steps of other magnitudes are obtained by superposition.) For this case of a unit-step change in set point, T' approaches $A_1 = K_c A/(1 + K_c A)$, a fraction of unity. The desired change is, of course, 1. Thus, the ultimate value of the temperature $T'(\infty)$ does not match the desired change. This discrepancy is called *offset* and is defined as

$$\text{Offset} = T'_R(\infty) - T'(\infty) \tag{13.3}$$

In terms of the particular control system parameters

$$\text{Offset} = 1 - \frac{K_c A}{1 + K_c A} = \frac{1}{1 + K_c A} \tag{13.4}$$

This discrepancy between set point and tank temperature at steady state is characteristic of proportional control. In some cases offset cannot be tolerated. However, notice from Eq. (13.4) that the offset decreases as K_c increases, and in theory the offset could be made as small as desired by increasing K_c to a sufficiently large value. To give a full answer to the problem of eliminating offset by high controller gain requires a discussion of stability and the response of the system when other lags, which have been neglected, are included in the system. Both these subjects are to be covered later. For the present we shall simply say that whether or not proportional control is satisfactory depends on the amount of offset which can be tolerated, the speed of response of the system, and the amount of

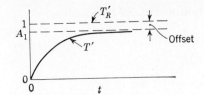

Fig. 13.2   **Unit-step response for set-point change (P control).**

gain which can be provided by the controller without causing the system to go unstable.

Proportional Control for Load Change (Regulator Problem)
The same control system shown in Fig. 13.1e is to be considered. This time the set point remains fixed; that is, $T'_R = 0$. We are interested in the response of the system to a change in the inlet stream temperature, i.e., to a load change.

Using the methods of Chap. 12, the overall transfer function becomes

$$\frac{T'}{T'_i} = \frac{A A^{-1}/(\tau s + 1)}{1 + K_c A/(\tau s + 1)} = \frac{1}{\tau s + 1 + K_c A} \tag{13.5}$$

This may be arranged in the form of the first-order lag; thus

$$\frac{T'}{T'_i} = \frac{A_2}{\tau_1 s + 1} \tag{13.6}$$

where

$$A_2 = \frac{1}{1 + K_c A}$$

$$\tau_1 = \frac{\tau}{1 + K_c A}$$

As for the case of set-point change, we have an overall response which is first-order. The overall time constant τ_1 is the same as for set-point changes. The response of the system to a unit-step change in inlet temperature T'_i is shown in Fig. 13.3. It may be seen that T' approaches $1/(1 + K_c A)$.

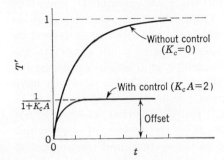

Fig. 13.3   **Unit-step response for load change (P control).**

To demonstrate the benefit of control, we have shown the response of the tank temperature (open-loop response) to a unit-step change in inlet temperature if no control were present; that is, $K_c = 0$. In this case, the major advantage of control is in reduction of offset. From Eq. (13.3), the offset becomes

$$\text{Offset} = T'_R(\infty) - T'(\infty) = 0 - \frac{1}{1 + K_c A}$$

$$= -\frac{1}{1 + K_c A} \tag{13.7}$$

As for the case of a step change in set point, the offset is reduced as controller gain K_c is increased.

Proportional-integral Control for Load Change   In this case, we replace G_c in Fig. 13.1e by $K_c(1 + 1/\tau_I s)$. The overall transfer function for load change is therefore

$$\frac{T'}{T'_i} = \frac{A A^{-1}/(\tau s + 1)}{1 + [K_c A/(\tau s + 1)](1 + 1/\tau_I s)} \tag{13.8}$$

Rearranging this gives

$$\frac{T'}{T'_i} = \frac{\tau_I s}{(\tau s + 1)(\tau_I s) + K_c A(\tau_I s + 1)}$$

or

$$\frac{T'}{T'_i} = \frac{\tau_I s}{\tau \tau_I s^2 + (K_c A \tau_I + \tau_I)s + K_c A}$$

Since the denominator contains a quadratic expression, the transfer function may be written in the standard form of the quadratic lag to give

$$\frac{T'}{T'_i} = \frac{(\tau_I/K_c A)s}{(\tau \tau_I/K_c A)s^2 + \tau_I(1 + 1/K_c A)s + 1}$$

or

$$\frac{T'}{T'_i} = \frac{A_1 s}{\tau_1^2 s^2 + 2\zeta \tau_1 s + 1} \tag{13.9}$$

where

$$A_1 = \frac{\tau_I}{K_c A}$$

$$\tau_1 = \sqrt{\frac{\tau \tau_I}{K_c A}}$$

$$\zeta = \frac{1}{2}\sqrt{\frac{\tau_I}{\tau}} \frac{1 + K_c A}{\sqrt{K_c A}}$$

For a unit-step change in load, $T'_i = 1/s$. Combining this with Eq. (13.9) gives

$$T' = \frac{A_1}{\tau_1^2 s^2 + 2\zeta\tau_1 s + 1} \tag{13.10}$$

Equation (13.10) shows that the response of the tank temperature is equivalent to the response of a second-order system to an impulse function of magnitude A_1. Since we have studied the impulse response of a second-order system in Chap. 8, the solution to the present problem is already known. This justifies in part our previous work on transients. Using Eq. (8.31), the impulse response for this system may be written for $\zeta < 1$ as

$$T' = A_1 \left(\frac{1}{\tau_1} \frac{1}{\sqrt{1 - \zeta^2}} e^{-\zeta t/\tau_1} \sin \sqrt{1 - \zeta^2} \frac{t}{\tau_1} \right) \tag{13.11}$$

Although the response of the system can be determined from Eq. (13.11) or Fig. 8.5, the effect of varying K_c and τ_I on the system response can be seen more clearly by plotting response curves, such as those shown in Fig. 13.4. From Fig. 13.4a, we see that an increase in K_c, for a fixed value of τ_I, improves the response by decreasing the maximum deviation and by making the response less oscillatory. The formula for ζ in Eq. (13.9) shows that ζ increases with K_c, which indicates that the response is less oscillatory. Figure 13.4b shows that, for a fixed value of K_c, a decrease in τ_I decreases the maximum deviation and period. However, a decrease in τ_I causes the response to become more oscillatory, which means that ζ decreases. This effect of τ_I on the oscillatory nature of the response is also given by the formula for ζ in Eq. (13.9).

For this case, the offset as defined by Eq. (13.3) is zero; thus

$$\text{Offset} = T'_R(\infty) - T'(\infty)$$
$$= 0 - 0 = 0$$

One of the most important advantages of PI control is the elimination of offset.

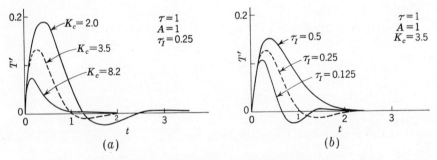

Fig. 13.4 **Unit-step response for load change (PI control).**

Proportional-integral Control for Set-point Change Again, the controller transfer function is $K_c(1 + 1/\tau_I s)$, and we obtain from Fig. 13.1e the transfer function

$$\frac{T'}{T'_R} = \frac{K_c A (1 + 1/\tau_I s)[1/(\tau s + 1)]}{1 + K_c A (1 + 1/\tau_I s)[1/(\tau s + 1)]} \tag{13.12}$$

This equation may be reduced to the standard quadratic form to give

$$\frac{T'}{T'_R} = \frac{\tau_I s + 1}{\tau_1^2 s^2 + 2\zeta \tau_1 s + 1} \tag{13.13}$$

where τ_1 and ζ are the same functions of the parameters as in Eq. (13.9). Introducing a unit-step change ($T'_R = 1/s$) into Eq. (13.13) gives

$$T' = \frac{1}{s} \frac{\tau_I s + 1}{\tau_1^2 s^2 + 2\zeta \tau_1 s + 1} \tag{13.14}$$

To obtain the response of T' in the time domain, Eq. (13.14) is expanded into two terms:

$$T' = \frac{\tau_I}{\tau_1^2 s^2 + 2\zeta \tau_1 s + 1} + \frac{1}{s} \frac{1}{\tau_1^2 s^2 + 2\zeta \tau_1 s + 1} \tag{13.15}$$

The first term on the right is equivalent to the response of a second-order system to an impulse function of magnitude τ_I. The second term is the unit-step response of a second-order system. It is convenient to use Figs. 8.2 and 8.5 to obtain the response for Eq. (13.15). For $\zeta < 1$, an analytic expression for T' is

$$T' = \frac{\tau_I}{\tau_1 \sqrt{1 - \zeta^2}} e^{-\zeta t/\tau_1} \sin \sqrt{1 - \zeta^2} \frac{t}{\tau_1}$$
$$+ 1 - \frac{1}{\sqrt{1 - \zeta^2}} e^{-\zeta t/\tau_1} \sin \left(\sqrt{1 - \zeta^2} \frac{t}{\tau_1} + \tan^{-1} \frac{\sqrt{1 - \zeta^2}}{\zeta} \right) \tag{13.16}$$

The last expression was obtained by combining Eqs. (8.17) and (8.31). A typical response for T' is shown in Fig. 13.5. The offset as defined by Eq. (13.3) is zero; thus

$$\text{Offset} = T'_R(\infty) - T'(\infty)$$
$$= 1 - 1 = 0$$

Again notice that the integral action in the controller has eliminated the offset.

Proportional Control of System with Measurement Lag In the previous examples the lag in the measuring element was assumed to be negligible, for which case the feedback transfer function was taken as 1. We now consider the same control system, the stirred-tank heater of Fig. 13.1, with a first-order measuring element having a transfer function

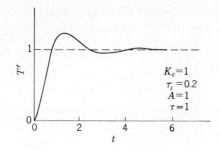

$K_c = 1$
$\tau_i = 0.2$
$A = 1$
$\tau = 1$

Fig. 13.5 Unit-step response for set point change (PI control).

$1/(\tau_m s + 1)$.   The block diagram for the modified system is now shown in Fig. 13.6.   By the usual procedure, the transfer function for set-point changes may be written

$$\frac{T'}{T'_R} = \frac{A_1(\tau_m s + 1)}{\tau_2^2 s^2 + 2\zeta_2\tau_2 s + 1} \tag{13.17}$$

where

$$A_1 = \frac{K_c A}{1 + K_c A}$$

$$\tau_2 = \sqrt{\frac{\tau\tau_m}{1 + K_c A}}$$

$$\zeta_2 = \frac{\tau + \tau_m}{2\sqrt{\tau\tau_m}} \frac{1}{\sqrt{1 + K_c A}}$$

We shall not obtain an expression for the transient response for this case, for it will be of the same form as Eq. (13.16).   Adding the first-order measuring lag to the control system of Fig. 13.1 produces a second-order system even for proportional control.   This means there will be an oscillatory response for an appropriate choice of the parameters τ, τ_m, K_c, and A. In order to understand the effect of gain K_c and measuring lag τ_m on the behavior of the system, response curves are shown in Fig. 13.7 for various combinations of K_c and τ_m for a fixed value of $\tau = 1$.   In general, the response becomes more oscillatory, or less stable, as K_c or τ_m increases.

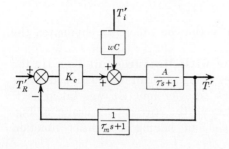

Fig. 13.6 Control system with measurement lag.

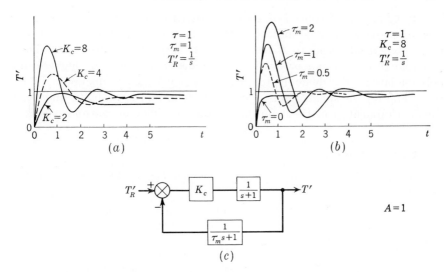

Fig. 13.7 **Effect of controller gain and measuring lag on system response for unit-step change in set point.**

For a fixed value of $\tau_m = 1$, Fig. 13.7a shows that the offset is reduced as K_c increases; however, this improvement in steady-state performance is obtained at the expense of a poorer transient response. As K_c increases, the overshoot becomes excessive and the response becomes more oscillatory. In general, we shall find that a control system having proportional control will require a value of K_c that is based on a compromise between low offset and satisfactory transient response.

For a fixed value of controller gain ($K_c = 8$), Fig. 13.7b shows that an increase in measurement lag produces a poorer transient response in that the overshoot becomes greater and the response more oscillatory as τ_m increases. This behavior illustrates a general rule that the measuring element in a control system should respond quickly if satisfactory response is to be achieved.

Summary   In this chapter, we have confined our attention to the response of simple control systems which were either first-order or second-order. This means that the transient response can be found by referring to Chaps. 5 and 8. However, if integral action were added to the controller in the system of Fig. 13.6, the overall transfer function would have a third-order polynomial in the denominator. Inversion would require factoring a cubic, which is generally a difficult task. Actually, systems with denominator polynomials of order greater than two are the rule rather than the exception. Hence, we shall develop in forthcoming chapters convenient techniques for studying the response of higher-order control systems. These techniques will be of direct use in control system design.

In Chap. 1, PI control of a heated stirred tank with measurement lag was discussed. It was indicated that incorrect selection of controller parameters could lead to a response with increasing amplitude. These unstable responses can occur in all systems with third- or higher-order polynomials in the denominator of the overall transfer function. In the next chapter, we shall present a concrete definition of stability and begin the development of methods for determining stability in control systems.

PROBLEMS

13.1 The set point of the control system shown in Fig. P13.1 is given a step change of 0.1 unit. Determine

    a. The maximum value of C and the time at which it occurs
    b. The offset
    c. The period of oscillation

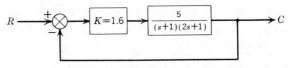

Fig. P13.1

Draw a sketch of $C(t)$ as a function of time.

13.2 The control system shown in Fig. P13.2 contains a three-mode controller.

    a. For the closed loop, develop formulas for the natural period of oscillation τ and the damping factor ζ in terms of the parameters K, τ_D, τ_I, and τ_1.

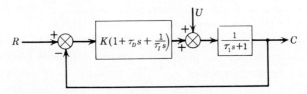

Fig. P13.2

For the following parts, $\tau_D = \tau_I = 1$ and $\tau_1 = 2$,

    b. Calculate ζ when K is 0.5 and when K is 2.
    c. Do ζ and τ approach limiting values as K increases, and if so, what are these values?
    d. Determine the offset for a unit-step change in load if K is 2.
    e. Sketch the response curve (C versus t) for a unit-step change in load when K is 0.5 and when K is 2.
    f. In both cases of part (*e*) determine the maximum value of C and the time at which it occurs.

13.3 The location of a load change in a control loop may affect the system response. In the block diagram shown in Fig. P13.3, a unit-step change in load enters at either location 1 or location 2.

a. What is the frequency of the transient response when the load enters at location 1 and when the load enters at location 2?

b. What is the offset when the load enters at location 1 and when it enters at location 2?

c. Sketch the transient response to a step change in U_1 and to a step change in U_2.

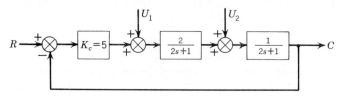

Fig. P13.3

13.4 Consider the liquid-level control system shown in Fig. P13.4. The tanks are noninteracting. The following information is known:

1. The resistances on the tanks are linear. These resistances were tested separately, and it was found that, if the steady-state flow rate q cfm is plotted against steady-state tank level h ft, the slope of the line dq/dh is 2 ft²/min.

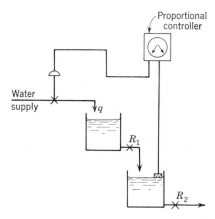

Fig. P13.4

2. The cross-sectional area of each tank is 2 ft².

3. The control valve was tested separately, and it was found that a change of 1 psi in pressure to the valve produced a change in flow of 0.1 cfm.

4. There is no dynamic lag in the valve or the measuring element.

a. Draw a block diagram of this control system, and in each block give the transfer function, with numerical values of the parameters.

b. Determine the controller gain K_c for a critically damped response.

c. If the tanks were connected so that they were interacting, what is the value of K_c needed for critical damping?

d. Using 1.5 times the value of K_c determined in part *c*, determine the response of the level in tank 2 to a step change in set point of 1 in. of level.

13.5   A PD controller is used in a control system having a first-order process and a measurement lag as shown in Fig. P13.5.

    a. Find expressions for ζ and τ for the closed loop response.

    b. If $\tau_1 = 1$ min, $\tau_m = 10$ sec, find K_c so that $\zeta = 0.7$ for the two cases: (1) $\tau_D = 0$, (2) $\tau_D = 3$ sec.

    c. Compare the offset and period realized for both cases, and comment on the advantage of adding the derivative mode.

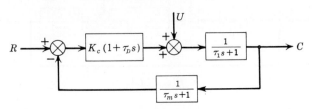

Fig. P13.5

13.6   The thermal system shown in Fig. P13.6 is controlled by a PD controller.
Data: $w = 250$ lb/min
    $\rho = 62.5$ lb/ft³
    $V_1 = 4$ ft³
    $V_2 = 5$ ft³
    $V_3 = 6$ ft³
    $C = 1$ Btu/(lb)(°F)
A change of 1 psi from the controller changes the flow rate of heat q by 500 Btu/min. The temperature of the inlet stream may vary.   There is no lag in the measuring element.

    a. Draw a block diagram of the control system with the appropriate transfer function in each block.   Each transfer function should contain numerical values of the parameters.

    b. From the block diagram, determine the overall transfer function relating the temperature in tank 3 to a change in set point.

    c. Find the offset for a unit-step change in inlet temperature if the controller gain K_c is 3 psi per °F of temperature error and the derivative time is 0.5 min.

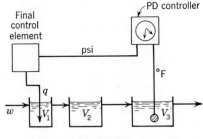

Fig. P13.6

14 *Stability*

CONCEPT OF STABILITY

In the previous chapter, the overall response of the control system was no higher than second-order. For these systems, the step response must resemble those of Fig. 5.6 or 8.2. Hence, the system is inherently *stable*. In this chapter we shall introduce the problem of stability by considering a control system (Fig. 14.1) only slightly more complicated than any studied previously. This system might represent proportional control of two stirred-tank heaters with measuring lag. In this discussion, only set-point

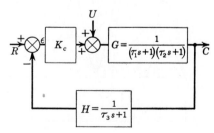

Fig. 14.1 **Third-order control system.**

changes are to be considered. From the methods developed in Chap. 12 for determining the overall transfer function, we have from Fig. 14.1

$$\frac{C}{R} = \frac{K_c G}{1 + K_c GH} \tag{14.1}$$

In terms of the particular transfer functions shown in Fig. 14.1, C/R becomes, after some rearrangement,

$$\frac{C}{R} = \frac{K_c(\tau_3 s + 1)}{(\tau_1 s + 1)(\tau_2 s + 1)(\tau_3 s + 1) + K_c} \tag{14.2}$$

The denominator of Eq. (14.2) is third-order. For a unit-step change in R, the transform of the response is

$$C = \frac{1}{s} \frac{K_c(\tau_3 s + 1)}{(\tau_1 s + 1)(\tau_2 s + 1)(\tau_3 s + 1) + K_c} \tag{14.3}$$

To obtain the transient response $C(t)$, it is necessary to find the inverse of Eq. (14.3). This requires obtaining the roots of the denominator of Eq. (14.2), which is third-order. We no longer can find these roots as easily as we did for the second-order systems by use of the quadratic formula. However, in principle they can always be obtained by algebraic methods.

It is seen that the roots of the denominator depend on the particular values of the time constants and K_c. These roots determine the nature of the transient response, according to the rules presented in Fig. 3.1 and Table 3.1. It is of interest to examine the nature of the response for the control system of Fig. 14.1 as K_c is varied, assuming the time constants τ_1, τ_2, τ_3 to be fixed. To be specific, consider the step response for $\tau_1 = 1$, $\tau_2 = \frac{1}{2}, \tau_3 = \frac{1}{3}$, for several values of K_c. Without going into the detailed calculations at this time, the results of inversion of Eq. (14.3) are shown as response curves in Fig. 14.2. From these response curves, it is seen that, as K_c increases, the system response becomes more oscillatory. In fact, beyond a certain value of K_c, the successive amplitudes of the response

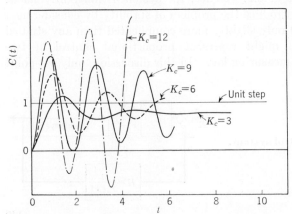

Fig. 14.2 **Response of control system of Fig. 14.1 for a unit-step change in set point.**

grow rather than decay; this type of response is called *unstable*. Evidently, for some values of K_c, there is a pair of roots corresponding to s_4 and s_4^* of Fig. 3.1. As control system designers, we are clearly interested in being able to determine quickly the values of K_c which give unstable responses such as that corresponding to $K_c = 12$ in Fig. 14.2.

If the order of Eq. (14.2) had been higher than three, the calculations necessary to obtain Fig. 14.2 would have been even more difficult. In the next chapter on root-locus methods, a powerful graphical tool for finding the necessary roots will be developed. In this chapter, we concentrate on developing a clearer understanding of the concept of stability. In addition, we shall develop a quick test for detecting roots having positive real parts, such as s_4 and s_4^* in Fig. 3.1.

Definition of Stability (Linear Systems) For our purposes, a stable system will be defined as one for which the output response is bounded for all bounded inputs. A system exhibiting an unbounded response to a bounded input is unstable. This definition, although somewhat loose, is adequate for most of the linear systems and simple inputs which we shall study.

A bounded input function is a function of time which always falls within certain bounds during the course of time. For example, the step function and sinusoidal function are bounded inputs. The function $f(t) = t$ is obviously unbounded.

Although the definition of an unstable system states that the output becomes unbounded, this is true only in the mathematical sense. An actual physical system always exhibits bounds or restraints. A linear mathematical model (set of linear differential equations describing the system) from which stability information is obtained is meaningful only over a certain range of variables. For example, a linear control valve gives a linear relation between flow and valve-top pressure only over the range of pressure (or flow) corresponding to values between which the valve is shut tight or wide open. When the valve is wide open, for example, further change in pressure to the diaphragm will not increase the flow. We often describe such a limitation by the term *saturation*. A physical system, when unstable, may not follow the response of its linear mathematical model beyond certain physical bounds but rather may saturate. However, the prediction of stability by the linear model is of utmost importance in a real control system, since operation with the valve shut tight or wide open is clearly unsatisfactory control.

STABILITY CRITERION

The purpose of this section is to translate the stability definition into a more simple criterion, which can be used to ascertain the stability of control systems of the form shown in Fig. 14.3.

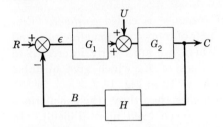

Fig. 14.3 Basic single-loop control system.

Characteristic Equation   From the block diagram of the control system (Fig. 14.3), we obtain by the methods of Chap. 12

$$C = \frac{G_1 G_2}{1 + G_1 G_2 H} R + \frac{G_2}{1 + G_1 G_2 H} U \tag{14.4}$$

In order to simplify the nomenclature, let $G = G_1 G_2 H$. We call G the *open-loop transfer function* because it relates the measured variable B to the set point R if the feedback loop of Fig. 14.3 is disconnected from the comparator (i.e., if the loop is opened). In terms of the open-loop transfer function G, Eq. (14.4) becomes

$$C = \frac{G_1 G_2}{1 + G} R + \frac{G_2}{1 + G} U \tag{14.5}$$

In principle, for given forcing functions $R(s)$ and $U(s)$, Eq. (14.5) may be inverted to give the control system response.

To determine under what conditions the system represented by Eq. (14.5) is stable, it is necessary to test the response to a bounded input. Suppose a unit-step change in set point is applied. Then

$$C(s) = \frac{G_1 G_2}{1 + G} \frac{1}{s} = \frac{G_1 G_2 F(s)}{s(s - r_1)(s - r_2) \cdots (s - r_n)} \tag{14.6}$$

where $r_1, r_2, \ldots, r_n$ are the n roots of the equation

$$1 + G(s) = 0 \tag{14.7}$$

and $F(s)$ is a function which arises in the rearrangement to the right-hand form of Eq. (14.6). Equation (14.7) is called the *characteristic equation* for the control system of Fig. 14.3. For example, for the control system of Fig. 14.1 the step response is

$$C(s) = \frac{G_1 G_2}{s(1 + G)}$$

$$= \frac{K_c}{(\tau_1 s + 1)(\tau_2 s + 1)} \bigg/ s \left[1 + \frac{K_c}{(\tau_1 s + 1)(\tau_2 s + 1)(\tau_3 s + 1)} \right]$$

which may be rearranged to

$$C(s) = \frac{K_c(\tau_3 s + 1)}{s[\tau_1\tau_2\tau_3 s^3 + (\tau_1\tau_2 + \tau_1\tau_3 + \tau_2\tau_3)s^2 + (\tau_1 + \tau_2 + \tau_3)s + (1 + K_c)]}$$

This last is equivalent to

$$C(s) = \frac{K_c(\tau_3 s + 1)/\tau_1\tau_2\tau_3}{s(s - r_1)(s - r_2)(s - r_3)}$$

where r_1, r_2, r_3 are the roots of the characteristic equation

$$\tau_1\tau_2\tau_3 s^3 + (\tau_1\tau_2 + \tau_1\tau_3 + \tau_2\tau_3)s^2 + (\tau_1 + \tau_2 + \tau_3)s + (1 + K_c) = 0$$

$$(14.8)$$

Evidently, for this case the function $F(s)$ in Eq. (14.6) is

$$F(s) = \frac{(\tau_1 s + 1)(\tau_2 s + 1)(\tau_3 s + 1)}{\tau_1\tau_2\tau_3}$$

In Chap. 3, the qualitative nature of the inverse transforms of equations such as Eq. (14.6) was discussed. It was shown that (see Fig. 3.1 and Table 3.1), if there are any of the roots $r_1, r_2, \ldots, r_n$ in the right half of the complex plane, then the response $C(t)$ will contain a term which grows exponentially in time and the system is unstable. If there are one or more roots of the characteristic equation at the origin, then there is an s^m in the denominator of Eq. (14.6) (where $m \geq 2$) and the response is again unbounded, growing as a polynomial in time. If there is a pair of conjugate roots on the imaginary axis, the contribution to the overall step response is a pure sinusoid which is bounded. However, if the bounded input is taken as $\sin \omega t$, where ω is the imaginary part of the conjugate roots, the contribution to the overall response is a sinusoid with an amplitude which increases as a polynomial in time.

It is evident from Eq. (14.5) that precisely the same considerations apply to a change in U. Therefore, the definition of *stability for linear systems* may be translated to the following criterion: A linear control system is unstable if any roots of its characteristic equation are on, or to the right of, the imaginary axis. Otherwise the system is stable.

It is important to note that the characteristic equation of a control system, which determines its stability, is the same for set-point or load changes. It depends only upon $G(s)$, the open-loop transfer function. Furthermore, although the rules derived above were based on a step input, they are applicable to any input. This is true, first, by the definition of stability and, second, because if there is a root of the characteristic equation in the right half plane, it contributes an unbounded term in the response to any input. This follows from Eq. (14.5) after it is rearranged to the form of Eq. (14.6) for the particular input.

Therefore, the stability of a control system of the type shown in Fig. 14.3 is determined solely by its open-loop transfer function through the roots of the characteristic equation.

Example 14.1 In terms of Fig. 14.3, a control system has the transfer functions

$$G_1 = 10\frac{0.5s + 1}{s} \quad \text{(PI controller)}$$

$$G_2 = \frac{1}{2s + 1} \quad \text{(stirred tank)}$$

$H = 1$ (measuring element without lag)

We have suggested a physical system by the components placed in parentheses. Find the characteristic equation and its roots, and determine if the system is stable.

The first step is to write the open-loop transfer function:

$$G = G_1G_2H = \frac{10(0.5s + 1)}{s(2s + 1)}$$

The characteristic equation is therefore

$$1 + \frac{10(0.5s + 1)}{s(2s + 1)} = 0$$

which is equivalent to

$$s^2 + 3s + 5 = 0$$

Solving by the quadratic formula gives

$$s = \frac{-3}{2} \pm \frac{\sqrt{9 - 20}}{2}$$

or

$$s_1 = \frac{-3}{2} + j\frac{\sqrt{11}}{2}$$

$$s_2 = \frac{-3}{2} - j\frac{\sqrt{11}}{2}$$

Since the real part of s_1 and s_2 is negative $(-\frac{3}{2})$, the system is stable.

ROUTH TEST FOR STABILITY

The Routh test is a purely algebraic method for determining how many roots of the characteristic equation have positive real parts; from this it can also be determined if the system is stable, for if there are no roots with positive real parts, the system is stable. The test is limited to systems which have polynomial characteristic equations. This means that it cannot be used to test the stability of a control system containing a transportation lag. We present the procedure for application of the Routh test without proof. The proof is available elsewhere[1] and is mathematically beyond the scope of this text.

The procedure for examining the roots is to write the characteristic equation in the form

$$a_0 s^n + a_1 s^{n-1} + a_2 s^{n-2} + \cdots + a_n = 0 \tag{14.9}$$

where a_0 is positive. (If a_0 is originally negative, both sides are multiplied by -1.) In this form, it is *necessary* that all the coefficients

$$a_0, a_1, a_2, \ldots, a_{n-1}, a_n$$

be positive if all the roots are to lie in the left half plane. If any coefficient is negative, the system is definitely unstable and the Routh test is not needed to answer the question of stability. (However, in this case, the Routh test will tell us the number of roots in the right half plane.) If all the coefficients are positive, the system may be stable or unstable. It is then necessary to apply the following procedure to determine stability.

Routh Array   Arrange the coefficients of Eq. (14.9) into the first two rows of the Routh array shown below:

Row				
1	a_0	a_2	a_4	a_6
2	a_1	a_3	a_5	a_7
3	b_1	b_2	b_3	
4	c_1	c_2	c_3	
5	d_1	d_2		
6	e_1	e_2		
7	f_1			
$n + 1$	g_1			

The array has been filled in for $n = 7$ in order to simplify the discussion For any other value of n, the array is prepared in the same manner. In general, there are $(n + 1)$ rows. For n even, the first row has one more element than the second row.

The elements in the remaining rows are found from the formulas

$$b_1 = \frac{a_1 a_2 - a_0 a_3}{a_1} \qquad b_2 = \frac{a_1 a_4 - a_0 a_5}{a_1} \cdots$$

$$c_1 = \frac{b_1 a_3 - a_1 b_2}{b_1} \qquad c_2 = \frac{b_1 a_5 - a_1 b_3}{b_1} \cdots$$

$$\cdots \cdots \cdots \cdots \cdots \cdots \cdots \cdots \cdots \cdots \cdots$$

[1] E. J. Routh, "Dynamics of a System of Rigid Bodies," Part II, Advanced, Macmillan & Co., Ltd., London, 1905.

The elements for the other rows are found from formulas which correspond to those given above.   The elements in any row are always derived from the elements of the two preceding rows.   During the computation of the Routh array, any row can be divided by a positive constant without changing the results of the test.   (The application of this rule often simplifies the arithmetic.)

Having obtained the Routh array, the following theorems are applied to determine stability:

Theorems of Routh Test

1. The necessary and sufficient condition for all the roots of the characteristic equation [Eq. (14.9)] to have negative real parts (stable system) is that all elements of the first column of the Routh array (a_0, a_1, b_1, c_1, etc.) be positive and nonzero.

2. If some of the elements in the first column are negative, the number of roots with positive real part (in the right half plane) is equal to the number of sign changes in the first column.

3. If *one* pair of roots is on the imaginary axis, equidistant from the origin, and all other roots are in the left half plane, all the elements of the nth row will vanish and none of the elements of the preceding row will vanish. The location of the pair of imaginary roots can be found by solving the equation

$$Cs^2 + D = 0$$

where the coefficients C and D are the elements of the array in the $(n - 1)$th row as read from left to right, respectively.   We shall find this last rule to be of value in the root-locus method presented in the next chapter.

The algebraic method for determining stability is limited in its usefulness in that all we can learn from it is whether or not a system is stable. It does not give us any idea of the degree of stability or the roots of the characteristic equation.

Example 14.2   Given the characteristic equation

$$s^4 + 3s^3 + 5s^2 + 4s + 2 = 0$$

Determine the stability by the Routh criterion.

Since all the coefficients are positive, the system may be stable.   To test this, form the Routh array shown below

Row			
1	1	5	2
2	3	4	
3	$1\frac{1}{3}$	$\frac{6}{3}$	
4	$2\frac{6}{11}$	0	
5	2		

The elements in the array are found by applying the formulas presented in the rules; for example, b_1, which is the element in the first column, third row, is obtained by

$$b_1 = \frac{a_1 a_2 - a_0 a_3}{a_1}$$

or in terms of numerical values

$$b_1 = \frac{(3)(5) - (1)(4)}{3} = \frac{15}{3} - \frac{4}{3} = \frac{11}{3}$$

Since there is no change in sign in the first column, there are no roots having positive real parts and the system is stable.

Example 14.3 Using $\tau_1 = 1$, $\tau_2 = \frac{1}{2}$, $\tau_3 = \frac{1}{3}$, determine values of K_c for which the control system of Fig. 14.1 is stable.

With these time constants, the characteristic equation is, according to Eq. (14.8),

$$\tfrac{1}{6}s^3 + s^2 + 1\tfrac{1}{6}s + (1 + K_c) = 0$$

The Routh array is therefore

Row		
1	$\frac{1}{6}$	$1\frac{1}{6}$
2	1	$(1 + K_c)$
3	$\dfrac{10 - K_c}{6}$	
4	$1 + K_c$	

Since K_c is positive, it is concluded that the control system of Fig. 14.1 will be stable if and only if $K_c < 10$, which is verified by Fig. 14.2.

SUMMARY AND GUIDE FOR FURTHER STUDY

A definition of stability for a control system has been presented and discussed. This definition was translated into a simple mathematical criterion relating stability to the location of roots of the characteristic equation. Briefly, it was found that a control system is stable if all the roots of its characteristic equation lie in the left half of the complex plane. The Routh criterion, a simple algebraic test for detecting roots of a polynomial lying in the right half of the complex plane, was presented and applied to control system stability analysis. This criterion suffers from two limitations: (1) It is applicable only to systems with polynomial characteristic equations, and (2) it gives no information about the actual location of the roots and, in particular, their proximity to the imaginary axis.

This latter point is quite important as can be seen from Fig. 14.2 and the results of Example 14.3. The Routh criterion tells us only that for $K_c < 10$ the system is stable. However, from Fig. 14.2 it is clear that the value $K_c = 9$ produces a response which is undesirable because it has a response time which is too long. In other words, the controlled variable oscillates too long before returning to steady state. It will be shown later that this happens because for $K_c = 9$ there is a pair of roots close to the imaginary axis.

In the next chapters we shall develop tools for obtaining more information about the actual location of the roots of the characteristic equation. This will enable us to predict the form of the curves of Fig. 14.2 for various values of K_c. The advantage of these tools is that they are graphical and are easy to apply compared with standard algebraic solution of the characteristic equation.

There are two distinct approaches to this problem: root-locus methods and frequency-response methods. The former are the subject of Chaps. 15 to 17 and the latter of Chaps. 18 to 21. These groups of chapters are written in parallel, and the reader may study one or both groups in either order. As a guide to making this decision, we offer the following general comments concerning the two approaches:

Root-locus methods allow rapid determination of the location of the roots of the characteristic equation as functions of parameters such as K_c of Fig. 14.1. In addition, they allow relatively easy determination of the transient response for higher-order systems by graphical means. However, they are difficult to apply to systems containing transportation lags. Also, they require a reasonably accurate knowledge of the theoretical process transfer function.

Frequency-response methods are an indirect solution to the location of the roots. They utilize the sinusoidal response of the open-loop transfer function to determine values of parameters such as K_c which keep these roots a "safe distance" from the right half plane. The actual transient response for a given value of K_c can be only crudely approximated. However, frequency-response methods are easily applied to systems containing transportation lags and may be used with only experimental knowledge of the unsteady-state process behavior.

A good mastery of control theory requires knowledge of both methods because they are complementary. However, the reader may choose to study only frequency response and still be adequately prepared for most of the material in the remainder of this book. The choice of studying only root locus will be more restrictive in terms of preparation for subsequent chapters. In addition, the bulk of literature on process dynamics relies heavily on frequency-response methods. However, as the newer root-locus methods are more widely understood, they will probably play an increasing role in the literature.

PROBLEMS

14.1 Write the characteristic equation and construct the Routh array for the control system shown in Fig. P14.1. Is the system stable for (*a*) $K_c = 9.5$, (*b*) $K_c = 11$, (*c*) $K_c = 12$?

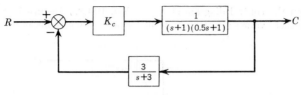

Fig. P14.1

14.2 By means of the Routh test, determine the stability of the system shown in Fig. P14.2 when $K_c = 2$.

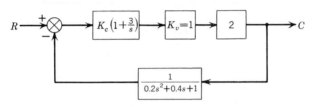

Fig. P14.2

14.3 In the control system of Prob. 13.6, determine the value of gain (psi/°F) which just causes the system to be unstable if (**a**) $\tau_D = 0.25$ min, (**b**) $\tau_D = 0.5$ min.

14.4 Prove that, if one or more of the coefficients $(a_0, a_1, \ldots, a_n)$ of the characteristic equation [Eq. (14.9)] is negative or zero, then there is necessarily an unstable root. *Hint:* First show that a_1/a_0 is minus the sum of all the roots, a_2/a_0 is plus the sum of all possible products of two roots, a_j/a_0 is $(-1)^j$ times the sum of all possible products of j roots, etc.

14.5 Prove that the converse statement of Prob. 14.4, i.e., that an unstable root implies that one or more of the coefficients will be negative or zero, is untrue for all $n > 2$. *Hint:* To prove that a statement is untrue, it is only necessary to demonstrate a single counterexample.

14.6 Deduce an extension of the Routh criterion which will detect the presence of roots with real parts greater than $-\sigma$ for any specified $\sigma > 0$.

14.7 Show that any complex number s satisfying $|s| < 1$ yields a value of

$$z = \frac{1 + s}{1 - s}$$

which satisfies

$$\text{Re}(z) > 0$$

(*Hint:* Let $s = x + jy$; $z = u + jv$. Rationalize the fraction, and equate real and imaginary parts of z and the rationalized fraction. Now consider what happens to the circle $x^2 + y^2 = 1$. To show that the *inside* of the circle goes over to the right half plane, consider a convenient point inside the circle.)

On the basis of this transformation, deduce an extension of the Routh criterion which will determine whether or not the system has roots inside the unit circle. Why might this information be of interest? How can the transformation be modified to consider circles of other radii?

14.8   Given the control diagram shown in Fig. P14.8. Deduce by means of the Routh criterion those values of τ_I for which the output C is stable for all inputs R and U.

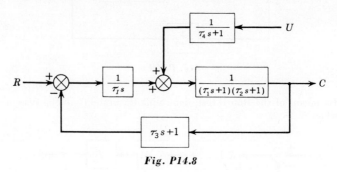

Fig. P14.8

part **IV** **Root-locus Methods**

15 *Root Locus*

In the previous chapter on stability, Routh's criterion was introduced to provide an algebraic method for determining the stability of a simple feedback control system (Fig. 14.3) from the characteristic equation of the system [Eq. (14.7)]. This criterion also yields the number of roots of the characteristic equation which are located in the right half of the complex plane. In this chapter, we shall develop a graphical method for finding the actual values of the roots of the characteristic equation, from which we can obtain the transient response of the system to an arbitrary forcing function.

CONCEPT OF ROOT LOCUS

In the previous chapter, the response of the simple feedback control system, shown again in Fig. 15.1, was given by the expression

$$C = \frac{G_1 G_2}{1 + G} R + \frac{G_2}{1 + G} U \tag{15.1}$$

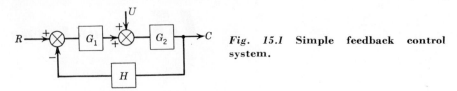

Fig. 15.1 Simple feedback control system.

where $G = G_1G_2H$. The factor in the denominator, $1 + G$, when set equal to zero, is called the characteristic equation of the closed-loop system. The roots of the characteristic equation determine the form (or character) of the response $C(t)$ to any particular forcing function $R(t)$ or $U(t)$.

The *root-locus* method is a graphical procedure for finding the roots of $1 + G = 0$, as one of the parameters of G varies continuously. In our work, the parameter which will be varied is the gain (or sensitivity) K_c of the controller. We can illustrate the concept of a root-locus diagram by considering the example presented in Fig. 14.1, which is represented by the block diagram of Fig. 15.1 with

$$G_1 = K_c$$

$$G_2 = \frac{1}{(\tau_1 s + 1)(\tau_2 s + 1)}$$

$$H = \frac{1}{\tau_3 s + 1}$$

For this case, the open-loop transfer function is

$$G = \frac{K_c}{(\tau_1 s + 1)(\tau_2 s + 1)(\tau_3 s + 1)}$$

which may be written in the alternate form

$$G(s) = \frac{K}{(s - p_1)(s - p_2)(s - p_3)} \tag{15.2}$$

where

$$K = \frac{K_c}{\tau_1 \tau_2 \tau_3}$$

$$p_1 = -\frac{1}{\tau_1} \qquad p_2 = -\frac{1}{\tau_2} \qquad p_3 = -\frac{1}{\tau_3}$$

The terms p_1, p_2, and p_3 are called the *poles* of the open-loop transfer function. A *pole* of $G(s)$ is any value of s for which $G(s)$ approaches infinity. For example, it is clear from Eq. (15.2) that, if $s = p_1$, the denominator of Eq. (15.2) is zero and therefore $G(s)$ approaches infinity. Hence $p_1 = -1/\tau_1$ is a pole of $G(s)$.

The characteristic equation for the *closed-loop* system is

$$1 + \frac{K}{(s - p_1)(s - p_2)(s - p_3)} = 0$$

This expression may be written

$$(s - p_1)(s - p_2)(s - p_3) + K = 0 \tag{15.3}$$

Using the same numerical values for the poles that were used at the beginning of Chap. 14 gives

$$(s + 1)(s + 2)(s + 3) + K = 0 \tag{15.4}$$

where

$$K = 6K_c$$

Expanding the product of this equation gives

$$s^3 + 6s^2 + 11s + (K + 6) = 0 \tag{15.5}$$

which is third-order. For any particular value of controller gain K_c, we can obtain the roots of the characteristic equation [Eq. (15.5)]. For example, if $K_c = 4.41$ ($K = 26.5$), Eq. (15.5) becomes

$$s^3 + 6s^2 + 11s + 32.5 = 0$$

Solving[1] this equation for the three roots gives

$$r_1 = -5.10$$
$$r_2 = -0.45 - j2.5$$
$$r_3 = -0.45 + j2.5$$

By selecting other values of K, other sets of roots are obtained as shown in Table 15.1.

[1] The procedure for obtaining the roots of a higher-order equation, such as Eq. (15.5), is covered in any text on advanced algebra. In a later section of this chapter we shall find the roots by a graphical technique called the root-locus method.

Table 15.1 **Roots of the characteristic equation**
$(s + 1)(s + 2)(s + 3) + K = 0$

$K = 6K_c$	r_1	r_2	r_3
0	−3	−2	−1
0.23	−3.10	−1.75	−1.15
0.39	−3.16	−1.42	−1.42
1.58	−3.45	−1.28 − j0.75	−1.28 + j0.75
6.6	−4.11	−0.95 − j1.5	−0.95 + j1.5
26.5	−5.10	−0.45 − j2.5	−0.45 + j2.5
60.0	−6.00	0.0 − j3.32	0.0 + j3.32
100.0	−6.72	0.35 − j4	0.35 + j4

For convenience, we may plot the roots r_1, r_2, and r_3 on the complex plane as K changes continuously. Such a plot is called a *root-locus* diagram and is shown in Fig. 15.2. Notice that there are three loci or *branches* corresponding to the three roots and that they "emerge" or begin (for $K = 0$) at the poles of the open-loop transfer function $(-1, -2, -3)$. The direction of increasing K is indicated on the diagram by an arrow. Also the values of K are marked on each locus. The root-locus diagram for this system and others to follow is symmetrical with respect to the real axis, and only the portion of the diagram in the upper half plane need be drawn. This follows from the fact that the characteristic equation for a physical system contains coefficients which are real, and therefore complex roots of such an equation must appear in conjugate pairs.

The root-locus diagram has the distinct advantage of giving at a glance the character of the response as the gain of the controller is continuously changed. The diagram of Fig. 15.2 reveals two critical values of K; one is at K_2 where two of the roots become equal, and the other is at K_3 where two of the roots are pure imaginary. It should be clear from the discussion in Chap. 14 that the nature of the response $C(t)$ will depend only on the roots r_1, r_2, r_3. Thus, if the roots are all real, which occurs for $K < K_2$ in Fig. 15.2, the response will be nonoscillatory. If two of the roots are complex and have negative real parts $(K_2 < K < K_3)$, the response will include damped sinusoidal terms which will produce an oscillatory response. If $K > K_3$, two of the roots are complex and have positive real parts, and the response is a growing sinusoid. Some of these types of response were shown in Fig. 14.2. After discussing in detail the methods for rapidly plotting root-locus diagrams such as Fig. 15.2, we shall consider the method of obtaining the transient response of a system from the root-locus plot.

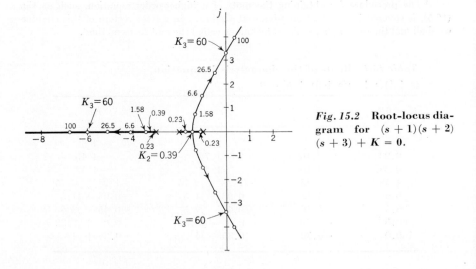

Fig. 15.2 **Root-locus diagram for** $(s + 1)(s + 2)$ $(s + 3) + K = 0$.

PLOTTING THE ROOT-LOCUS DIAGRAM

Having introduced the concept of root locus by a simple example for which the characteristic equation was only third-order, we shall describe some rules which were first introduced by Evans[1] for plotting root-locus diagrams of characteristic equations of any order. Without these rules, the time and effort needed to plot root-locus diagrams would be too great to render them useful in engineering computations.

The first step in applying the root-locus technique to determine the roots of the characteristic equation of the closed-loop control system is to write the open-loop transfer function $(G = G_1G_2H)$ in the standard form

$$G = K \frac{N}{D} \tag{15.6}$$

where

$$K = \text{constant}$$
$$N = (s - z_1)(s - z_2) \cdots (s - z_m)$$
$$D = (s - p_1)(s - p_2) \cdots (s - p_n)$$

The term z_i is called a *zero* of the open-loop transfer function. The term p_i is called a *pole* of the open-loop transfer function and was defined earlier in this chapter. A *zero* of $G(s)$ is any value of s for which $G(s)$ equals zero. The factored terms $(s - z_i)$ and $(s - p_i)$ in N/D arise naturally in the open-loop transfer function. For example, in the control system considered at the beginning of this chapter, Eq. (15.2) was written in the standard form with

$$K = \frac{K_c}{\tau_1\tau_2\tau_3}$$
$$D = (s - p_1)(s - p_2)(s - p_3)$$
$$N = 1$$

To obtain a situation for which N is other than 1, we may consider the same basic control system of Fig. 15.1 with $G_1 = K_c(1 + 1/\tau_I s)$. In this case, the open-loop transfer function takes the form

$$G = \frac{K(s - z_1)}{(s - p_1)(s - p_2)(s - p_3)(s - p_4)}$$

where

$$K = \frac{K_c}{\tau_1\tau_2\tau_3}$$
$$z_1 = -\frac{1}{\tau_I}$$
$$p_1 = -\frac{1}{\tau_1} \qquad p_2 = -\frac{1}{\tau_2} \qquad p_3 = -\frac{1}{\tau_3} \qquad p_4 = 0$$

[1] W. R. Evans, Graphical Analysis of Control Systems, *Trans. AIEE*, **67**:547–551 (1948), and "Control-system Dynamics," McGraw-Hill Book Company, New York, 1954.

The reader should verify this result. For this case, G has one zero at z_1 and four poles at p_1, p_2, p_3, and p_4.

Using the form of G given by Eq. (15.6), the characteristic equation $1 + G = 0$ may be written in the alternate form

$$1 + K \frac{N}{D} = 0$$

or

$$D + KN = 0 \tag{15.7}$$

It is assumed in the remainder of this chapter that $n \geqq m$, which is true for all physical systems. This being the case, the characteristic equation will be of nth order and have n roots, r_1, r_2, . . . , r_n.

To develop the graphical method for determining the root locus, the characteristic equation is rewritten as

$$K \frac{N}{D} = -1 \tag{15.8}$$

In terms of the poles and zeros of the open-loop transfer function, Eq. (15.8) becomes

$$K \frac{(s - z_1)(s - z_2) \cdot \cdot \cdot (s - z_m)}{(s - p_1)(s - p_2) \cdot \cdot \cdot (s - p_n)} = -1 \tag{15.9}$$

Since the left-hand member is in general complex, we may write Eq. (15.9) in the equivalent form involving magnitude and phase angle; thus

$$K \frac{|s - z_1| \, |s - z_2| \cdot \cdot \cdot |s - z_m|}{|s - p_1| \, |s - p_2| \cdot \cdot \cdot |s - p_n|} = 1 \tag{15.10}$$

$$\angle (s - z_1) + \angle (s - z_2) + \cdot \cdot \cdot + \angle (s - z_m)$$
$$- [\angle (s - p_1) + \cdot \cdot \cdot + \angle (s - p_n)] = (2i + 1)\pi \tag{15.11}$$

where i is any integer (positive or negative) or zero. Equations (15.10) and (15.11) may be used to find the root locus by trial and error as follows: The trace of the locus is found entirely from the *angle criterion* of Eq. (15.11), which is independent of K. After the locus is established, the gain K for any point on it may be obtained from Eq. (15.10), which we shall refer to as the *magnitude criterion*.

To understand the procedure for determining the root locus from the angle criterion [Eq. (15.11)], consider the simple example

$$K \frac{N}{D} = \frac{K(s - z_1)}{(s - p_1)(s - p_2)}$$

for which the poles and zeros are located as shown in Fig. 15.3. (It is convenient to indicate open-loop poles by $\times$ and open-loop zeros by $\bigcirc$

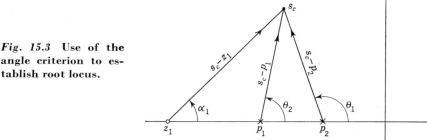

Fig. 15.3 Use of the angle criterion to establish root locus.

in root-locus diagrams.) To plot the root locus, a trial point (labeled s_c in Fig. 15.3) is selected and the vectors representing $(s_c - z_1)$, $(s_c - p_1)$, and $(s_c - p_2)$ are drawn. If the trial point is correct, all the angles associated with these vectors (labeled θ_1, θ_2, and α_1 in Fig. 15.3), when substituted into Eq. (15.11), will yield an odd multiple of π. For this example, the trial point s_c is correct if

$$\alpha_1 - \theta_1 - \theta_2 = (2i + 1)\pi$$

for some value of i. The trial point is moved until the angle criterion [Eq. (15.11)] is satisfied. After a sufficient number of trial points have been established as correct, the root locus is drawn by connecting them with a smooth curve. The gains K associated with various points on the locus are determined by use of the magnitude criterion [Eq. (15.10)]. Again with reference to the example shown in Fig. 15.3, if we find the point s_c to be on the root locus by using the angle criterion, then the gain is obtained from Eq. (15.10); thus

$$\frac{K|s_c - z_1|}{|s_c - p_1|\,|s_c - p_2|} = 1$$

Solving for K gives

$$K = \frac{|s_c - p_1|\,|s_c - p_2|}{|s_c - z_1|}$$

It should be emphasized that the root-locus plot is symmetrical with respect to the real axis (i.e., complex roots occur as conjugate pairs). For this reason, the trial-and-error procedure for finding points on the loci need be done for only the upper half plane. The loci in the lower half plane can be drawn from symmetry.

In principle, the trial-and-error method will produce the root-locus plot; however, to save time it should be used only after applying the following rules which give a rapid guide to the general location of the loci. These rules are proved in the Appendix unless they are self-evident. We state them below and then illustrate their use with examples. It will probably be expedient first to glance over the list of rules and then study them more carefully in conjunction with Examples 15.1 to 15.3.

Rules for Plotting Root-locus Diagrams (Negative Feedback)   In the following rules $n \geq m$.

Rule 1   The number of loci or branches is equal to the number of open-loop poles, n.

Rule 2   The root loci begin at open-loop poles and terminate at open-loop zeros. The termination of $(n - m)$ of the loci will occur at the zeros at infinity along asymptotes to be described later. In the case of a qth-order pole,[1] q loci emerge from it. For a qth-order zero, q loci terminate there.

Rule 3   Locus on real axis.   The real axis is part of the root locus when the sum of the number of poles and zeros to the right of a point on the real axis is odd. It is necessary to consider only the real poles and zeros in applying this rule, for the complex poles and zeros always occur in conjugate pairs and their effects cancel in checking the angle criterion for points on the real axis. Furthermore, a qth-order pole (or zero) must be counted q times in applying the rule.

Rule 4   Asymptotes.   There are $(n - m)$ loci which approach (as $K \to \infty$) asymptotically $(n - m)$ straight lines, radiating from the *center of gravity* of the poles and zeros of the open-loop transfer function. The center of gravity is given by

$$\gamma = \frac{\sum\limits_{j=1}^{n} p_j - \sum\limits_{i=1}^{m} z_i}{n - m} \tag{15.12}$$

These asymptotic lines make angles of $\pi[(2k + 1)/(n - m)]$ with the real axis and are, therefore, equally spaced at angles $2\pi/(n - m)$ to each other ($k = 0, 1, 2, \ldots, n - m - 1$).

Rule 5   Breakaway point.   The point at which two root loci, emerging from adjacent poles (or moving toward adjacent zeros) on the real axis, intersect and then leave (or enter) the real axis is determined by the solution of the equation

$$\sum_{i=1}^{m} \frac{1}{s - z_i} = \sum_{j=1}^{n} \frac{1}{s - p_j} \tag{15.13}$$

These loci leave (or enter) the real axis at angles of $\pm \pi/2$. Equation (15.13) is solved by trial by checking it for various test points, $s = s_c$, on the real

---

[1] A pole p_a of order q is present in the open-loop transfer function if the denominator of G contains $(s - p_a)^q$. A zero z_a of order q is present if the numerator of G contains $(s - z_a)^q$.

axis between the poles (or zeros) of interest. For real poles or zeros, the terms in the denominator of Eq. (15.13) are obtained by simply measuring distances along the real axis between the test point and the poles and zeros. If a pair of complex poles, $p_i = a_i \pm jb_i$, are present, add to the right side of Eq. (15.13) the term

$$\frac{2(s - a_i)}{(s - a_i)^2 + b_i^2}$$

(This term accounts for both poles of the complex pair.) This term is merely the result of simplifying the sum

$$\frac{1}{s - a_i - jb_i} + \frac{1}{s - a_i + jb_i}$$

For a pair of complex zeros, add a similar term to the left-hand side of Eq. (15.13).

Rule 6 *Angle of departure or approach.* There are q loci emerging from each qth-order open-loop pole at angles determined by

$$\theta = \frac{1}{q} \left[(2k + 1)\pi + \sum_{i=1}^{m} \angle (p_a - z_i) - \sum_{\substack{j=1 \\ j \neq a}}^{n} \angle (p_a - p_j) \right] \qquad (15.14)$$

$$k = 0, 1, 2, \ldots, q - 1$$

where p_a is a particular pole of order q. Each of the m loci which do not approach the asymptotes will terminate at one of the m zeros. They will approach their particular zeros at angles

$$\theta = \frac{1}{v} \left[(2k + 1)\pi + \sum_{j=1}^{n} \angle (z_b - p_j) - \sum_{\substack{i=1 \\ i \neq b}}^{m} \angle (z_b - z_i) \right] \qquad (15.15)$$

$$k = 0, 1, 2, \ldots, v - 1$$

where z_b is a particular zero of order v. For simple poles (or zeros) on the real axis, the angle of departure (or approach) will be 0 or π.

An analog from potential theory is useful in plotting a root-locus diagram. It may be shown that the loci correspond to the paths taken by a positively charged particle in an electrostatic field which is established by poles (positive charges) and zeros (negative charges). In general, we may expect a locus to be repelled by a pole and attracted toward a zero.

Another general aid to plotting the loci is to be aware of the fact that for $n - m \geq 2$, the sum of the roots $(r_1 + r_2 + \cdots + r_n)$ is constant, real, and independent of K. This requires that motion of branches to the right be counterbalanced by the motion of other branches to the left.

Most of the open-loop transfer functions encountered in single-loop chemical process control systems will have all their poles on the real axis. In exceptional cases where the feedback path includes second-order measuring elements, such as a pressure transmitter, the open-loop transfer function will contain complex poles, but very often they will be located so far from the remaining dominant poles that they can be ignored. A discussion of dominant poles and neglecting of certain terms in the transient response will be given in the next chapter.

These rules and guides will now be explained by applying them to specific examples.

Example 15.1 Plot the root-locus diagram for the open-loop transfer function:[1]

$$G = \frac{K}{(s + 1)(s + 2)(s + 3)}$$

In general, our step-wise procedure will follow the same order in which the rules were presented.

1. Plot the open-loop poles as shown in Fig. 15.4a. The poles are indicated by $\times$. There are no open-loop zeros for this example.
2. (Rule 1) Since we have three poles, there are three branches.
3. (Rule 3) A portion of the locus is on the real axis between -1 and -2 and another portion is to the left of -3.
4. (Rule 4) Since $n - m = 3$, we have three asymptotes and the center of gravity is $\gamma = (-3 - 2 - 1)/3 = -2$. Angles which the asymptotes make with the real axis are $\pi/3$, $3\pi/3$, and $5\pi/3$. These asymptotes are shown in Fig. 15.4a.

With these few steps completed, a rough sketch of the root-locus diagram can be made as follows: Since the real axis to the left of -3 is an asymptote and one branch emerges from the pole at -3, it should be clear that one entire branch is the real axis to the left of -3. Furthermore, from the fact that two loci must emerge from the poles -1 and -2 and that the real axis between these poles is part of the locus, we see that two loci move toward each other along the real axis between -1 and -2 and eventually meet at some common point. Since the location of the asymptotes is known, it is therefore necessary that the two loci which meet on the real axis must break away and eventually follow the asymptotes. From these observations, we could sketch a root-locus diagram which closely resembles that of Fig. 15.4c. If the breakaway point and the crossings of the imaginary axis were known, the sketch could be made with considerable accuracy. We now continue the example by applying Rule 5 to find the breakaway point and the Routh test to find the crossings of the imaginary axis.

5. *Breakaway point.* (Rule 5) The roots emerging from -1 and -2 move toward each other until they meet, at which point the loci leave the real axis at angles of $\pm\pi/2$. The breakaway point is found from Eq. (15.13) as follows

$$0 = \frac{1}{s - p_1} + \frac{1}{s - p_2} + \frac{1}{s - p_3}$$

or

$$0 = \frac{1}{s + 1} + \frac{1}{s + 2} + \frac{1}{s + 3}$$

---

[1] To grasp more easily the graphical procedure for plotting the root locus, the reader should actually plot these examples according to the steps given in the solution. Also note that this is the same example which was treated by algebraic methods at the beginning of this chapter.

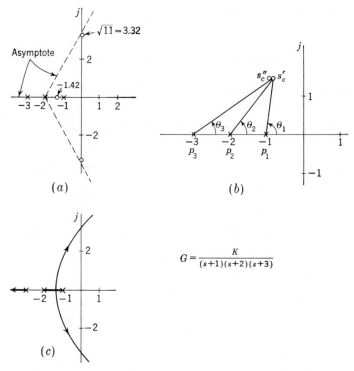

$$G = \frac{K}{(s+1)(s+2)(s+3)}$$

Fig. 15.4 **Root-locus construction for Example 15.1.**

Solving this gives

$$s = -1.42$$

6. To find the points at which the loci cross the imaginary axis, the Routh test (theorem 3) of Chap. 14 may be used. Writing the characteristic equation $D + KN = 0$ in polynomial form gives

$$D + KN = (s + 1)(s + 2)(s + 3) + K = 0$$

or

$$s^3 + 6s^2 + 11s + K + 6 = 0$$

from which we can write the Routh array:

Row		
1	1	11
2	6	$K + 6$
3	b_1	

The theorem states that, if one pair of roots are on the imaginary axis and all others in the left half plane, all the elements of the nth row must be zero. From this we obtain for the element b_1

$$b_1 = \frac{(6)(11) - (K + 6)}{6} = 0$$

Solving for K,

$$K = 60$$

A root on the imaginary axis is expressed as simply ja. Substituting $s = ja$ and $K = 60$ into the polynomial gives

$$-ja^3 - 6a^2 + 11aj + 66 = 0$$
$$(66 - 6a^2) + (11a - a^3)j = 0$$

Equating the real part or the imaginary part to zero gives

$$a = \pm \sqrt{11} = \pm 3.32$$

Therefore the loci intersect the imaginary axis at $+ j \sqrt{11}$ and $- j \sqrt{11}$.

7. Having found these general features of the root-locus plot, we can sketch the root locus. If it is desirable to have a more accurate plot of the loci, the construction is continued by the trial-and-error method described earlier in this chapter. A spirule[1] can be of considerable help in sketching the locus. To illustrate the method, suppose the trial point, $s_c' = -0.75 + 1.5j$ of Fig. 15.4b, is selected. This point is checked by the angle criterion [Eq. (15.11)] which for this example may be written

$$\angle (s + 1) + \angle (s + 2) + \angle (s + 3) = (2i + 1)\pi$$

or

$$\theta_1 + \theta_2 + \theta_3 = (2i + 1)\pi$$

From Fig. 15.4b, these angles are found to be

$$\theta_1 = 81° \qquad \theta_2 = 51° \qquad \theta_3 = 34°$$

and we have

$$81° + 51° + 34° = 166° \neq (2i + 1)\pi$$

Shifting the trial point horizontally to the left will increase the sum of the angles. As a second trial point, $s_c'' = -0.95 + 1.5j$ gives for the sum of the angles

$$88° + 56° + 37° = 181° \cong \pi$$

This result is sufficiently close to π, which is $(2i + 1)\pi$ with $i = 0$, and we accept the point as one on the locus. In this manner, more points on the locus can be found and a curve drawn through them.

8. *Gain.* To determine the gain at various points along the loci, the magnitude criterion [Eq. (15.10)] is used. For example, if the gain at $s = -0.95 + j1.5$ (labeled s_c'' in Fig. 15.4b) is wanted, we measure the distances directly with a ruler; thus

$$|s - p_1| = 1.50$$
$$|s - p_2| = 1.82$$
$$|s - p_3| = 2.52$$

---

[1] The spirule is essentially a drawing instrument which can be used to add angles by rotating an arm with respect to a disk. It can also be used to multiply together the lengths of line segments in a manner similar to that of a slide rule. The spirule is of great assistance in constructing root-locus diagrams. Its operation will not be described here, for complete instructions are given with each one. It is manufactured by the Spirule Company, 9728 El Venado, Whittier, Calif., and may be obtained directly from the manufacturer or through bookstores.

(It is important to measure the vector lengths in units which are consistent with those used on the axes of the graph.)

Substituting these values into Eq. (15.10) gives

$$\frac{K}{(1.50)(1.82)(2.52)} = 1$$

or $K = (1.50)(1.82)(2.52) = 6.8$. To find the point corresponding to $K = 6.8$ on the branch along the real axis to the left of p_3 requires a trial-and-error solution if the graphical approach is used. For example, if $s = -4.5$ is tried, we obtain

$$|s - p_1| = 3.5$$
$$|s - p_2| = 2.5$$
$$|s - p_3| = 1.5$$

from which we get

$$K = (1.5)(2.5)(3.5) = 13.1$$

We see that $s = -4.5$ does not correspond to a gain of 6.8. It is therefore necessary to try other values of s greater than -4.5 until the desired value of $K = 6.8$ is obtained. Although this procedure may seem very tedious, the actual calculations go quite quickly as the reader will discover for himself as he works out this example.

We also may find the root on the real axis more directly by applying the following theorem from algebra:

The sum of the roots $(r_1 + r_2 + \cdots + r_n)$ of the nth-order polynomial equation

$$a_0 x^n + a_1 x^{n-1} + \cdots + a_n = 0$$

is given by

$$(r_1 + r_2 + \cdots + r_n) = -\frac{a_1}{a_0}$$

In this case, we have just found the complex roots for $K = 6.8$ to be

$$r_2, r_3 = -0.95 \pm j1.5$$

The polynomial equation is

$$(s + 1)(s + 2)(s + 3) + K = 0$$

which can be expanded into

$$s^3 + 6s + 11s + (K + 6) = 0$$

According to the theorem

$$r_1 + (-0.95 + j1.5) + (-0.95 - j1.5) = -\frac{6}{1}$$

or

$$6 = -[r_1 - 2(0.95)]$$

or

$$r_1 = -4.10$$

All the detailed steps needed to plot the root locus for this problem have been discussed. The complete locus is shown in Fig. 15.4c. This same plot is also shown in more detail in Fig. 15.2. This root-locus plot will be used in the next chapter when we discuss the transient response of a control system by means of a root-locus diagram.

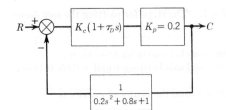

Fig. 15.7 **Block diagram for Example 15.3.**

Example 15.3 Consider the block diagram for the control system shown in Fig. 15.7. We may consider this system[1] to consist of a process having negligible lag, an underdamped second-order measuring element, and a PD controller. The open-loop transfer function is

$$G = \frac{0.2K_c(1 + \tau_D s)}{0.2s^2 + 0.8s + 1}$$

Rearranging this into the standard form of KN/D gives

$$G = \frac{K(s - z_1)}{(s - p_1)(s - p_2)} \tag{15.16}$$

where

$$K = K_c \tau_D$$
$$z_1 = \frac{-1}{\tau_D}$$
$$p_1 = -2 + j \qquad p_2 = -2 - j$$

Note that the quadratic term in the denominator of Eq. (15.16) contributes two poles which are complex conjugates. It is desired to plot the root-locus diagram for variation in K, for $\tau_D = \frac{1}{3}$. The procedure for obtaining the root-locus diagram is as follows:

1. Plot the open-loop poles $p_1 = -2 + j$ and $p_2 = -2 - j$ and the open-loop zero $z_1 = -3$ in Fig. 15.8.

[1] The system shown in Fig. 15.7 may approximate the control of flow rate. In this case the block labeled K_p would represent a valve having no dynamic lag. The feedback element would represent a flow-measuring device, such as a mercury manometer placed across an orifice plate. Mercury manometers are known to have underdamped second-order dynamics.

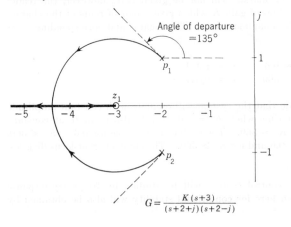

Fig. 15.8 **Root-locus diagram for Example 15.3.**

2. Since there are two poles, there are two branches. Each branch emerges from a complex pole.

3. A portion of the locus is on the real axis to the left of z_1.

4. Since $n - m = 1$, there is one asymptote, which is the real axis. At this stage, we know that two loci must emerge from the complex poles p_1 and p_2 and in some way enter the real axis, one branch moving along the real axis toward the zero z_1 and the other branch moving along the real axis toward $-\infty$. Before a reasonably accurate sketch can be made, we must know the point at which the branches enter the real axis and the angle at which the branches emerge from the poles. Rules 5 and 6 will give this information.

5. *Breakaway point.* The point at which the loci enter the real axis is obtained by Eq. (15.13), which may be written for this problem

$$\frac{1}{s + 3} = \frac{2(s + 2)}{(s + 2)^2 + 1^2}$$

Cross-multiplying this equation and rearranging give

$$s^2 + 6s + 7 = 0$$

Solving for s gives

$$s = -4.41, -1.59$$

From step 3, which shows that the root locus is on the real axis to the left of $z_1 = -3$, we see that the correct point of entry is $s = -4.41$.

6. *Angle of departure (Rule 6).* From Eq. (15.14), we have for the pole p_1

$$\theta = 1\left(\pi + \frac{\pi}{4} - \frac{\pi}{2}\right) = \frac{3\pi}{4} = 135°$$

Because of symmetry, only the locus emerging from p_1 need be considered.

7. *Sketching the root locus.* Points on the root locus are found by trial and error using the angle criterion [Eq. (15.11)]. This locus is more difficult to plot than the one of the preceding example because the asymptote, which is the real axis, does not guide the plotting of the locus between a pole and the point where the locus enters the real axis. The gains (K) along the locus are found by the same procedure used in the previous example; the details are not given here.

From the root-locus diagram we see that the control system never becomes unstable. In fact, the response becomes more damped as K increases.

SUMMARY

In this chapter, the rules for plotting root-locus diagrams have been presented and applied to several control systems. It should be emphasized that the basic advantage of this method is the speed and ease with which a rough sketch of the loci can be obtained. This sketch frequently gives much of the desired information on stability. A few further calculations of points on the locus are usually all that are necessary to obtain accurate, quantitative behavior of the roots.

The root locus for variation of parameters other than K_c, such as τ_D, has not been discussed here. The method of constructing this type of diagram is similar to that presented here and is discussed in detail in other texts.[1]

In the next chapter, we shall make further use of the root-locus diagram by using it to determine the transient response for typical control systems.

PROBLEMS

15.1 Draw the root-locus diagram for the system shown in Fig. P15.1 where $G_c = K_c(1 + 0.5s + 1/s)$.

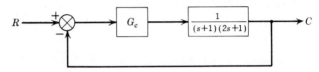

$$R \xrightarrow{\quad} \underset{-}{\overset{+}{\bigotimes}} \xrightarrow{\quad} \boxed{G_c} \xrightarrow{\quad} \boxed{\dfrac{1}{(s+1)(2s+1)}} \xrightarrow{\quad} C$$

Fig. P15.1

15.2 Draw the root-locus diagram for the system shown in Fig. P13.4 for (**a**) $\tau_I = 0.4$ min and (**b**) $\tau_I = 0.2$ min. (The proportional controller is replaced by a PI controller.) Determine the controller gain which makes the system just unstable. The values of parameters of the system are:

K_v = valve constant 0.070 cfm/psi
K_m = transducer constant 6.74 (in. pen travel)/(ft of tank level)
R_2 = 0.55 ft level/cfm
τ_1 = time constant of tank 1 = 2.0 min
τ_2 = time constant of tank 2 = 0.5 min

The controller gain K_c has the units of pounds per square inch per inch of pen travel.

15.3 Sketch the root-locus diagram for the system shown in Fig. P14.2. If the system is unstable at higher values of K_c, find the roots on the imaginary axis and the corresponding value of K_c.

15.4 Sketch the root loci for the following equations:

$$\textbf{\textit{a.}} \quad 1 + \frac{K}{(s + 1)(2s + 1)} = 0$$

$$\textbf{\textit{b.}} \quad 1 + \frac{K}{s(s + 1)(2s + 1)} = 0$$

$$\textbf{\textit{c.}} \quad 1 + \frac{K(4s + 1)}{s(s + 1)(2s + 1)} = 0$$

$$\textbf{\textit{d.}} \quad 1 + \frac{K(1.5s + 1)}{s(s + 1)(2s + 1)} = 0$$

$$\textbf{\textit{e.}} \quad 1 + \frac{K(0.5s + 1)}{s(s + 1)(2s + 1)} = 0$$

On your sketch you should locate quantitatively all poles, zeros, and asymptotes. In addition show the parameter which is being varied along the locus and the direction in which the loci travel as this parameter is increased.

[1] C. H. Wilts, "Principles of Feedback Control," Addison-Wesley Publishing Company, Inc., Reading, Mass., 1960.

15.5   For the control system shown in Fig. P15.5

Case 1: $\tau_D = \frac{2}{3}$
Case 2: $\tau_D = \frac{1}{9}$

      a. Sketch the root-locus diagram in each case.
      b. If the system can go unstable, find the value of K_c which just causes instability.
      c. Using Theorem 3 (Chap. 14) of the Routh test, find the locations (if any) at which the loci cross into the unstable region.

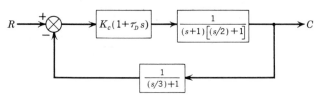

Fig. P15.5

16 *Transient Response from Root Locus*

The transient response of a control system to a simple input (step, ramp, sinusoid, etc.) can be found from the root-locus diagram entirely by graphical procedures. The method can be most easily described by means of an example.

Step Response

Example 16.1 Determine the unit-step response of the control system shown in Fig. 16.1 for $K_c = 6$. This system is the same as that described in Chap. 14. In this example, we shall show how the curves of Fig. 14.2 can be derived from the root-locus diagram.

For a change in set point, we can write

$$\frac{C}{R} = \frac{K_c/[(\tau_1 s + 1)(\tau_2 s + 1)]}{1 + K_c/[(\tau_1 s + 1)(\tau_2 s + 1)(\tau_3 s + 1)]} \tag{16.1}$$

This can be rearranged to give

$$\frac{C}{R} = \frac{(K_c/\tau_1\tau_2\tau_3)\tau_3(s + 1/\tau_3)}{(s + 1/\tau_1)(s + 1/\tau_2)(s + 1/\tau_3) + (K_c/\tau_1\tau_2\tau_3)} \tag{16.2}$$

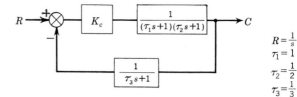

Fig. 16.1 Control system for Example 16.1.

$$R = \frac{1}{s}$$
$$\tau_1 = 1$$
$$\tau_2 = \frac{1}{2}$$
$$\tau_3 = \frac{1}{3}$$

Equation (16.2) is in the form of Eq. (14.6); the denominator is the characteristic function, which is written $D + KN$ in Eq. (15.7). Writing the denominator of Eq. (16.2) as $(s - r_1)(s - r_2) \cdots (s - r_n)$, where $r_1, r_2, \ldots, r_n$ are the roots of the characteristic equation, gives

$$\frac{C}{R} = \frac{(K_c/\tau_1\tau_2\tau_3)\tau_3(s + 1/\tau_3)}{(s - r_1)(s - r_2)(s - r_3)} \tag{16.3}$$

For a unit-step change in R and for $K_c = 6$, $\tau_1 = 1$, $\tau_2 = \frac{1}{2}$, and $\tau_3 = \frac{1}{3}$, Eq. (16.3) becomes

$$C(s) = \frac{12(s + 3)}{s(s - r_1)(s - r_2)(s - r_3)} \tag{16.4}$$

To obtain the inverse of $C(s)$, the right side of Eq. (16.4) may be expanded into partial fractions to give

$$C(s) = \frac{12(s + 3)}{s(s - r_1)(s - r_2)(s - r_3)} = \frac{A_0}{s} + \frac{A_1}{s - r_1} + \frac{A_2}{s - r_2} + \frac{A_3}{s - r_3} \tag{16.5}$$

The inverse of $C(s)$ can be written

$$C(t) = A_0 + A_1 e^{r_1 t} + A_2 e^{r_2 t} + A_3 e^{r_3 t} \tag{16.6}$$

The roots are found from the root-locus plot which was constructed in the previous chapter (Fig. 15.2) and which is presented again in Fig. 16.2. For $K_c = 6(K = 36)$, we obtain from Fig. 16.2

$$r_1 = -5.4$$
$$r_2 = -0.30 - j2.8$$
$$r_3 = -0.30 + j2.8$$

If the coefficients ($A_0, A_1, \ldots$) are evaluated, the response $C(t)$ can be obtained. Since r_2 and r_3 are complex conjugates, we know that A_2 and A_3 are complex conjugates (see Chap. 3) and it is necessary to find only A_2. At this stage, A_0, A_1, A_2 could be evaluated by the algebraic method which was described in Chap. 3; however, we shall use the graphical approach here in order to illustrate another method.

To evaluate A_1, we write from Eq. (16.5)

$$A_1 = [(s - r_1)C(s)] \Big|_{s = r_1} \tag{16.7}$$

or

$$A_1 = \frac{12(r_1 + 3)}{(r_1 - 0)(r_1 - r_2)(r_1 - r_3)} \tag{16.8}$$

The members in the right-hand side of Eq. (16.8) ($r_1 + 3$, $r_1 - 0$, etc.) can be interpreted as vectors in the complex plane. For example, $r_1 - r_2$ is the vector joining the points

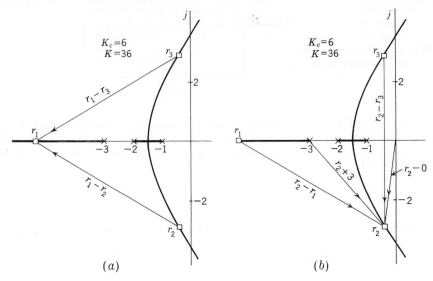

Fig. 16.2   Determining coefficients from root-locus diagram (Example 16.1).

r_1 and r_2.   For the present, A_1 is assumed to be complex and is written in the form of magnitude and angle; thus

$$A_1 = |A_1| \, \measuredangle A_1$$

The magnitude $|A_1|$ can be determined from Eq. (16.8) as

$$|A_1| = \frac{12|r_1 + 3|}{|r_1 - 0| \, |r_1 - r_2| \, |r_1 - r_3|} \tag{16.9}$$

and the angle $\measuredangle A_1$ can be written

$$\measuredangle A_1 = \measuredangle (r_1 + 3) - [\measuredangle (r_1 - 0) + \measuredangle (r_1 - r_2) + \measuredangle (r_1 - r_3)] \tag{16.10}$$

Where convenient, these vectors are shown in Fig. 16.2a.   By direct measurement of lengths and angles we obtain

$$
\begin{aligned}
|r_1 + 3| &= 2.4 & \measuredangle (r_1 + 3) &= 180° \\
|r_1 - 0| &= 5.4 & \measuredangle (r_1 - 0) &= 180° \\
|r_1 - r_2| &= 5.83 & \measuredangle (r_1 - r_2) &= +151° \\
|r_1 - r_3| &= 5.83 & \measuredangle (r_1 - r_3) &= -151°
\end{aligned}
$$

Substituting these magnitudes and angles into Eqs. (16.9) and (16.10) gives

$$
\begin{aligned}
|A_1| &= \frac{(12)(2.4)}{(5.4)(5.83)(5.83)} = 0.16 \\
\measuredangle A_1 &= 180 - (180 - 151 + 151) = 0
\end{aligned}
$$

Therefore

$$A_1 = 0.16$$

Actually, we should know that A_1 is real and that only the magnitude expression of Eq.

⟨16.9) is needed to determine its value.   We have determined the angle only to illustrate the graphical procedure that is used when the coefficient may be complex.

To evaluate A_2, we write from Eq. (16.5)

$$A_2 = [(s - r_2)C(s)] \Big|_{s=r_2} \tag{16.11}$$

or

$$A_2 = \frac{12(r_2 + 3)}{(r_2 - 0)(r_2 - r_1)(r_2 - r_3)} \tag{16.12}$$

Since A_2 is complex, it is written as

$$A_2 = |A_2| \, \measuredangle A_2 \tag{16.13}$$

We obtain $|A_2|$ and $\measuredangle A_2$ from Eq. (16.12) by writing

$$|A_2| = \frac{12|r_2 + 3|}{|r_2 - 0| \, |r_2 - r_1| \, |r_2 - r_3|} \tag{16.14}$$

$$\measuredangle A_2 = \measuredangle (r_2 + 3) - [\measuredangle (r_2 - 0) + \measuredangle (r_2 - r_1) + \measuredangle (r_2 - r_3)] \tag{16.15}$$

The construction of the vectors is shown in Fig. 16.2*b*.   The magnitudes and angles associated with each vector are found to be

$$
\begin{aligned}
|r_2 + 3| &= 3.85 & \measuredangle (r_2 + 3) &= -46° \\
|r_2 - 0| &= 2.82 & \measuredangle (r_2 - 0) &= -96° \\
|r_2 - r_1| &= 5.80 & \measuredangle (r_2 - r_1) &= -28° \\
|r_2 - r_3| &= 5.60 & \measuredangle (r_2 - r_3) &= -90°
\end{aligned}
$$

Substituting into Eqs. (16.14) and (16.15) gives

$$|A_2| = \frac{12(3.85)}{(2.82)(5.80)(5.60)} = 0.51$$

$$\measuredangle A_2 = -46 - (-96 - 28 - 90) = +168$$

or

$$A_2 = 0.51 e^{j168°} = -0.5 + j0.105$$

Since A_3 is the conjugate of A_2, we have

$$A_3 = 0.51 e^{-j168°} = -0.5 - j0.105$$

To find A_0, we write

$$A_0 = [sC(s)] \Big|_{s=0} \tag{16.16}$$

By either the algebraic or graphical approach, there is obtained

$$A_0 = 0.84$$

The complete response $C(t)$ can now be written as

$$C(t) = 0.84 + 0.16 e^{-5.4t} + e^{-0.3t}(-\cos 2.8t + 0.21 \sin 2.8t) \tag{16.17}$$

To obtain the last term of Eq. (16.17), we have used Eq. (3.12), which states that the inverse of the sum

$$\frac{A_2}{s + k_1 + jk_2} + \frac{A_3}{s + k_1 - jk_2} \tag{16.18}$$

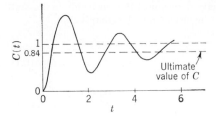

Fig. 16.3   Transient response to step change (Example 16.1).

where $A_2 = a_1 + jb_1$, $A_3 = a_1 - jb_1$, is

$$e^{-k_1 t}(2a_1 \cos k_2 t + 2b_1 \sin k_2 t) \qquad (16.19)$$

An alternate form of Eq. (16.19) follows from the identity

$$p \cos A + q \sin A = r \cos (A - \phi)$$
$$r = \sqrt{p^2 + q^2}$$
$$\tan \phi = \frac{q}{p}$$

Using this identity, Eq. (16.19) can be written

$$e^{-k_1 t}\left[2 \sqrt{a_1{}^2 + b_1{}^2} \cos (k_2 t - \phi)\right] \qquad (16.20)$$

where

$$\phi = \tan^{-1} \frac{b_1}{a_1} = \measuredangle A_2$$

Equation (16.20) can be rewritten in terms of the sine

$$e^{-k_1 t}\left[2 \sqrt{a_1{}^2 + b_1{}^2} \sin (k_2 t - \phi + 90°)\right] \qquad (16.21)$$

Since $|A_2| = |A_3| = \sqrt{a_1{}^2 + b_1{}^2}$, Eq. (16.21) can be written

$$e^{-k_1 t}\left[2|A_2| \sin (k_2 t - \phi + 90°)\right] \qquad (16.22)$$

The last form is more convenient when A_2 and A_3 are found graphically, since $|A_2|$ and ϕ, which are found directly from the graphical construction, can be used in Eq. (16.22).   If the response $C(t)$ is written so that the last term is of the form of Eq. (16.22), we have

$$C(t) = 0.84 + 0.16e^{-5.4t} + e^{-0.3t}[1.02 \sin (2.8t - 168° + 90°)]$$

or

$$C(t) = 0.84 + 0.16e^{-5.4t} + e^{-0.3t}[1.02 \sin (2.8t - 78°)] \qquad (16.23)$$

The response $C(t)$ is plotted in Fig. 16.3.   The response for other values of K_c can be obtained by this procedure, which results in the curves of Fig. 14.2.

Dominant Roots   The responses of many higher-order systems resemble that of a second-order system.   When this is true, one can often obtain the *nature* of the response by inspection of the root-locus diagram without having to obtain the complete expression for the transient response as we did in the previous example.   Before illustrating this approach, we shall first discuss the root location for the second-order system.

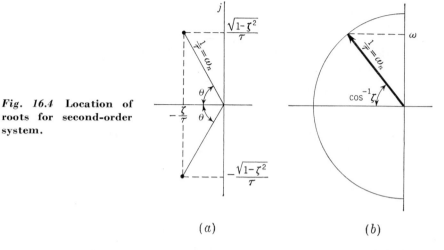

Fig. 16.4 Location of roots for second-order system.

(*a*)   (*b*)

Root location for a second-order system. The transfer function of the second-order system

$$\frac{Y(s)}{X(s)} = \frac{1}{\tau^2 s^2 + 2\zeta\tau s + 1}$$

can be written

$$\frac{Y(s)}{X(s)} = \frac{1/\tau^2}{\left[s - \left(-\frac{\zeta}{\tau} - j\frac{\sqrt{1-\zeta^2}}{\tau}\right)\right]\left[s - \left(-\frac{\zeta}{\tau} + j\frac{\sqrt{1-\zeta^2}}{\tau}\right)\right]}$$

$$(16.24)$$

The denominator of Eq. (16.24) contains the roots

$$\left(-\frac{\zeta}{\tau} - j\frac{\sqrt{1-\zeta^2}}{\tau}\right) \qquad \left(-\frac{\zeta}{\tau} + j\frac{\sqrt{1-\zeta^2}}{\tau}\right)$$

of the characteristic equation for this system, which can be plotted on the complex plane as shown in Fig. 16.4a. As ζ varies, τ remaining constant, the roots fall on a circle of radius $\omega_n = 1/\tau$, as shown in Fig. 16.4b. This follows from the geometry of Fig. 16.4a and the fact that the radial distance is

$$\left[\left(\frac{\sqrt{1-\zeta^2}}{\tau}\right)^2 + \left(-\frac{\zeta}{\tau}\right)^2\right]^{1/2} = \left(\frac{1-\zeta^2}{\tau^2} + \frac{\zeta^2}{\tau^2}\right)^{1/2} = \frac{1}{\tau}$$

Furthermore, the angle θ between the negative real axis and the vector joining the origin and either root is

$$\theta = \cos^{-1}\left(\frac{\zeta/\tau}{1/\tau}\right)$$

or

$$\theta = \cos^{-1} \zeta \tag{16.25}$$

Also notice from Fig. 16.4 that the imaginary part of the root is the actual radian frequency,

$$\omega = \frac{\sqrt{1 - \zeta^2}}{\tau}$$

From these observations, we can readily locate the roots of a second-order system from the parameters ζ and τ. They are on radial lines which form an angle with the negative real axis of $\theta = \cos^{-1} \zeta$ and at a radial distance $1/\tau$ from the origin.

   Application.   Returning to Example 16.1, we see that, for $K_c = 6$, there are one real root ($r_1 = -5.4$) and two complex roots (see Fig. 16.2). From the previous discussion of the root location of a second-order system, we can describe the complex roots in terms of the natural frequency ($\omega_n = 1/\tau$) and the damping coefficient by measuring the radial distance to obtain ω_n and by measuring the angle θ to obtain ζ. From Fig. 16.2, for the roots r_2 and r_3

$$\omega_n = \frac{1}{\tau} = 2.82 \text{ rad/sec}$$
$$\theta = 84°$$

or

$$\zeta = \cos 84° = 0.104$$

If the real root r_1 is far removed from the complex roots, it will contribute to the transient response $C(t)$ a term ($A_1 e^{r_1 t}$) which is highly damped. If the coefficient A_1 is also small relative to A_2 and A_3, the term may be neglected, for which case the transient response is determined solely by the pair of complex roots (r_2 and r_3) and is approximately second-order. If the response is approximately second-order and the parameters ζ and τ are known, then we know such features of the transient response as period of oscillation, percent overshoot, etc. To make this clear, consider again the system of Example 16.1.

   Example 16.2   It is desired to estimate the unit-step response of the control system shown in Fig. 16.1. An inspection of the root-locus diagram (Fig. 16.2) shows that the term $A_1 e^{r_1 t}$ in the transient response will be highly damped because r_1 is greater in magnitude than the other roots r_2 and r_3. It is also possible to show directly from Fig. 16.2a that $|A_1|$ is less than $|A_2|$. To do this, we first write these coefficients in terms of vector magnitude as given in Eqs. (16.9) and (16.14); thus

$$|A_1| = \frac{12|r_1 + 3|}{|r_1 - 0| \, |r_1 - r_2| \, |r_1 - r_3|}$$
$$|A_2| = \frac{12|r_2 + 3|}{|r_2 - 0| \, |r_2 - r_1| \, |r_2 - r_3|}$$

or

$$\frac{|A_1|}{|A_2|} = \frac{|r_1 + 3|\ |r_2 - 0|\ |r_2 - r_3|}{|r_2 + 3|\ |r_1 - 0|\ |r_1 - r_3|}$$

Inspection of Fig. 16.2 shows that

$$|r_1 + 3| < |r_2 + 3|$$
$$|r_1 - 0| > |r_2 - 0|$$
$$|r_1 - r_3| \cong |r_2 - r_3|$$

Using the above approximation and inequalities in the expression for $|A_1|/|A_2|$ gives

$$\frac{|A_1|}{|A_2|} \cong \frac{|r_1 + 3|\ |r_2 - 0|}{|r_2 + 3|\ |r_1 - 0|} < 1$$

or

$$|A_1| < |A_2|$$

From rough estimates of the vector lengths, we see that

$$\frac{|A_1|}{|A_2|} \cong \frac{1}{4}$$

Therefore, the term $A_1 e^{r_1 t}$ will be small compared with the others, and the response may be approximated by that of a second-order system having only the roots r_2 and r_3. From Fig. 16.2, we find that for these roots $\tau = 1/2.82$ and $\zeta = 0.104$. To find the ultimate value of C, the final-value theorem is applied to Eq. (16.4):

$$C(\infty) = \lim_{s \to 0} [sC(s)]$$
$$= \frac{(12)(3)}{(-r_1)(-r_2)(-r_3)}$$

Introducing the values of r_1, r_2, and r_3 gives

$$C(\infty) = \frac{(12)(3)}{(5.4)(+0.3 - j2.8)(+0.3 + j2.8)} = 0.84$$

Knowing ζ, τ, and $C(\infty)$ completely describes the transient response, and the step-response curves of Fig. 8.2 may be used to obtain a response curve for this system. The curve for the proper value of ζ is multiplied by $C(\infty)$, and the time scale is interpreted in terms of the value of τ. If the second-order approximate response obtained in Example 16.2 is compared with the true response of Eq. (16.23) or with Fig. 16.3, it can be seen that the two responses show much agreement. The ultimate value $C(\infty) = 0.84$ is the same for each response; however, this is to be expected, because $C(\infty)$ was calculated for either response by considering all the roots. From Eq. (16.23), the true frequency of oscillation ω is 2.8, whereas the frequency for the approximate response is

$$\omega = \omega_n \sqrt{1 - \zeta^2} = (2.82) \sqrt{1 - (0.104)^2} = 2.81$$

From Fig. 16.3, the overshoot for the time response can be found to be 93 percent. The overshoot for the approximate response having $\zeta = 0.104$ is found from Fig. 8.4 to be 73 percent. Although there are some discrepancies between the true response and the approximate response, they are small and may be disregarded in many practical problems.

In more complex problems for which there are several pairs of complex roots and several roots on the real axis, one can often neglect the terms in the transient response contributed by complex roots and real roots which are located far from the origin.   The roots which are not neglected are called the *dominant* roots because the transient terms arising from these roots are the most important.   By neglecting some terms in the transient response, one can save the work required to evaluate the coefficients of these terms and also the work involved in calculating the response $C(t)$ from an expression involving all terms.   However, care must be taken to ensure that the terms dropped have only a small effect on the transient.

Summary   In Chap. 15, the graphical techniques for rapidly sketching the root locus of the characteristic equation were established.   Once the roots are available, the response of the system to any forcing function can be obtained by the usual procedures of partial fractions and inversion given in Chap. 3.   In this chapter, we have described the method for obtaining the transient response by *graphical construction on the root-locus diagram*. Although only the response to a step input was considered, the general procedure for obtaining the response to any other forcing function is the same.   The possibility of obtaining an approximation to the time response by considering only the dominant roots was also illustrated.   In the next chapter, we shall consider the transient behavior of some simple control systems in terms of their root-locus diagrams.

PROBLEMS

16.1   Draw the root-locus diagram for the control system shown in Fig. P16.1.

    a. Determine the value of K_c needed to obtain a root of the characteristic equation of the closed-loop response which has an imaginary part 0.75.
    b. Using the value of K_c found in part *a*, determine all the other roots of the characteristic equation from the root-locus diagram.
    c. If a unit impulse is introduced into the set point, determine the response of the system, $C(t)$.

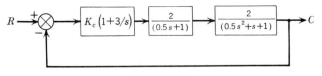

Fig. P16.1

16.2   Given the control system shown in Fig. P16.2.

$\tau = 1$ min
$\tau_m = 10$ sec

a. If $\tau_D = 0$, find K_c so that $\zeta = 0.7$ for the closed loop.

b. If $\tau_D = 10$ sec, find K_c so that $\zeta = 0.7$ for the closed loop.

c. Compare the offset for a unit-step change in load for each of the above cases and comment.

d. Find the natural period in both cases.

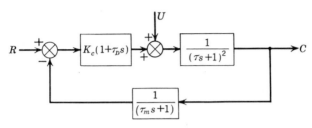

Fig. P16.2

Note: ζ refers to the damping coefficient associated with the closed-loop system.

17 Application of Root Locus to Control Systems

Having developed the root-locus technique, we shall now use it to explain the effect of various modes of control on the response of a typical process. For this discussion, we shall use the system shown in Fig. 17.1 which controls the composition of a two-tank chemical reactor. In Chap. 11, this system was analyzed and the corresponding block diagram developed

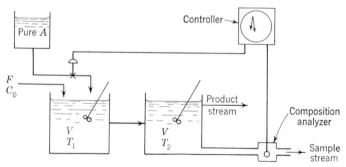

Fig. 17.1 **Two-tank chemical reactor with transportation lag.**

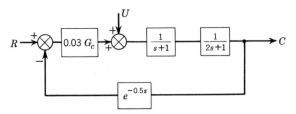

Fig. 17.2   Block diagram of chemical reactor of Fig. 17.1.

(Fig. 11.3).   The reader may wish to review the details of Chap. 11 which led to the block diagram of Fig. 17.2, or this block diagram may be accepted as simply that of a typical system,[1] the response of which is to be investigated by root locus.

In the first part of this chapter, we shall study by means of root locus the effect of the mode of control on the response of the system when the transportation lag is absent.   In a later section, the effect of the presence of transportation lag on the response of the system will be considered.   The root-locus diagram for a system containing a transportation lag requires a special approach which has not yet been discussed.

EFFECT OF CONTROL ACTION

In this section, the root-locus diagrams will be presented for the control system of Fig. 17.2 when various controller modes are used and when the transportation lag is absent.   The block diagram for this simplified system is shown in Fig. 17.3.   The open-loop transfer function is

$$G = \frac{0.03G_c}{(2s + 1)(s + 1)} \tag{17.1}$$

where G_c will depend on the mode of control being used.   Only a unit-step change in set point will be considered.

Proportional Control   When proportional control is used, Eq. (17.1) is

$$G = \frac{0.03K_c}{(2s + 1)(s + 1)}$$

[1] One may consider the block diagram of Fig. 17.2 to represent some system other than a stirred-tank chemical reactor.   For example, it may represent two stirred-tank heaters in series, with a long pipe separating the second tank from the measuring element.

Fig. 17.3 Block dia-gram for system shown in Fig. 17.2 without transportation lag.

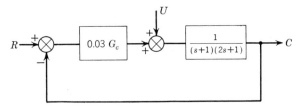

or

$$G = \frac{K}{(s + 0.5)(s + 1)}$$

where

$$K = \frac{0.03K_c}{2}$$

The root-locus diagram for this system is shown in Fig. 17.4a.   Note that there are only two roots.   For higher values of K, these roots become complex but they always remain in the left half plane.   From these facts, we know that the response of C is second-order and that the period of oscillation and the damping decrease with gain K.   Step-response curves are shown in Fig. 17.4b for several values of K.

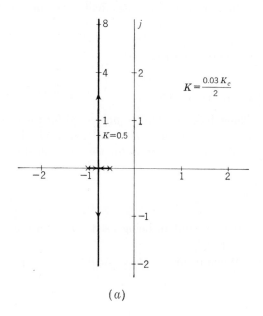

$$K = \frac{0.03 K_c}{2}$$

(a)

Fig. 17.4 Root-locus diagram and step response for proportional control.

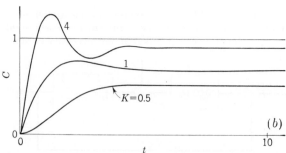

(b)

Proportional-integral Control   When PI control is used, the open-loop transfer function of Eq. (17.1) becomes

$$G = \frac{0.03K_c}{(2s + 1)(s + 1)}\left(1 + \frac{1}{\tau_I s}\right)$$

This can be written for root-locus construction in the form

$$G = \frac{K(s + 1/\tau_I)}{s(s + 0.5)(s + 1)}$$

where

$$K = \frac{0.03K_c}{2}$$

Note that PI action contributes a pole at the origin and a zero at $-1/\tau_I$. In Fig. 17.5a, the root-locus diagram is shown for $\tau_I = 0.5$.   For this value, the integral action has produced a system which is unstable for high values of K.   The step response is shown in Fig. 17.5b for two values of K which

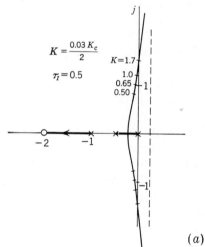

Fig. 17.5   Root-locus diagram and step response for PI control.

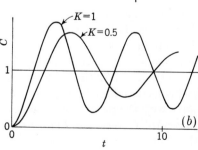

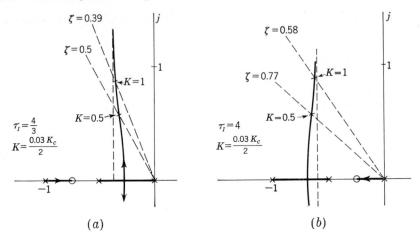

Fig. 17.6 **Effect of integral action on root-locus diagram.**

yield a stable system. The root-locus diagram correctly predicts that the dominant roots are lightly damped for these values of K, even though the system is stable.

It will be instructive to study the effect of integral time τ_I on the response of this control system by means of root locus. As τ_I changes, the zero contributed by integral action moves along the real axis.

In Fig. 17.6a the root-locus diagram is shown for the case where the zero falls between the poles at -1 and -0.5. For this case, we see that the system does not become unstable as controller gain increases. Furthermore, by comparing Fig. 17.5a with Fig. 17.6a, it can be seen that, for the same controller gain, the system represented by the latter diagram is not so lightly damped. This means that the transient response to a step change should be less oscillatory than those shown in Fig. 17.5b.

The effect of still larger integral time is shown by the root-locus diagram of Fig. 17.6b in which the zero is located between the origin and the pole at -0.5. For this case, the system does not become unstable for any value of gain. Comparison of Figs. 17.5a and 17.6b shows that, for the same gain, the system represented by the latter diagram is more damped. For the same gain, the system of Fig. 17.6b is somewhat more damped than that of Fig. 17.6a. Also, it may be seen that the negative root which is on the branch close to the origin in Fig. 17.6b will contribute to the transient response an exponential term $e^{-t/\tau}$, which decays relatively slowly.

In Fig. 17.7, the step responses are shown for the three values of τ_I just considered for a fixed gain, $K = 1$. Notice that these transient curves agree with the predictions made from the root-locus diagrams in the previous paragraphs. For the highly damped response for $\tau_I = 4$, it can be seen that there is no overshoot even though $\zeta = 0.58$. The cause of this is

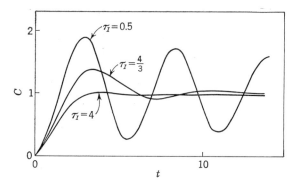

Fig. 17.7 Effect of integral action on transient response to a unit-step change in set point ($K = 0.03K_c/2 = 1$).

the slowly decaying exponential term which corresponds to the root near the origin. This is an example where the complex roots cannot be taken as the dominant roots. All three roots are important in the transient response.

This serves as a good example of the use of root-locus diagrams for control system design. Comparison of Figs. 17.5a and 17.6a and b clearly shows the effect of changing τ_I and indicates a suitable range of values for this parameter.

Proportional-derivative Control When PD control is used, the open-loop transfer function of Eq. (17.1) becomes

$$G = \frac{(0.03)K_c(1 + \tau_D s)}{(2s + 1)(s + 1)}$$

or

$$G = \frac{K(s + 1/\tau_D)}{(s + 0.5)(s + 1)}$$

where

$$K = \frac{0.03K_c\tau_D}{2}$$

With PD control, a zero at $-1/\tau_D$ is included in the open-loop transfer function. The root-locus diagram for $\tau_D = 0.5$ is shown in Fig. 17.8a. Notice that the system is stable for all values of gain. Furthermore, the damping coefficient ζ always exceeds 0.79, which means that the system is well damped.

The effect of τ_D on the system response can also be readily seen by means of root locus. As the derivative action increases, the zero will move along the real axis toward the origin. In Fig. 17.9 are shown root-locus diagrams for the case where the zero falls between the poles and for the case where the zero is between the pole at -0.5 and the origin. These diagrams indicate a nonoscillatory response which will resemble the second-order responses of Fig. 8.2 for $\zeta > 1$. In general, we would probably prefer

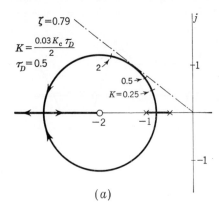

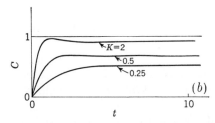

Fig. 17.8 Root-locus diagram and step response for PD control.

a response which is less sluggish than these.   The type of response indicated by Fig. 17.8 is preferable, and the root-locus diagram therefore suggests a τ_D of less than 1.

Proportional-integral-derivative Control   When three-mode control is used, the open-loop transfer function becomes

$$G = 0.03 K_c \left(1 + \tau_D s + \frac{1}{\tau_I s}\right) \frac{1}{(s + 1)(2s + 1)}$$

which can be written

$$G = K \frac{s^2 + (1/\tau_D)s + (1/\tau_I \tau_D)}{s(s + 0.5)(s + 1)}$$

where

$$K = \frac{0.03 K_c \tau_D}{2}$$

The quadratic term in the numerator will yield a pair of zeros; the location of these zeros will depend on the values of τ_I and τ_D.   For example, if $\tau_I = 1$ and $\tau_D = 0.5$, we obtain for the quadratic expression

$$s^2 + \frac{1}{\tau_D} s + \frac{1}{\tau_I \tau_D} = s^2 + 2s + 2 = 0$$

Solving for s by the quadratic formula gives

$$s = -1 \pm j$$

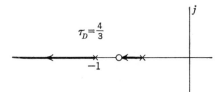

$$\tau_D = \frac{4}{3}$$

(a)

Fig. 17.9 Effect of derivative action on root-locus diagram.

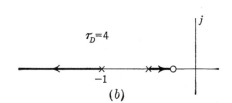

$$\tau_D = 4$$

(b)

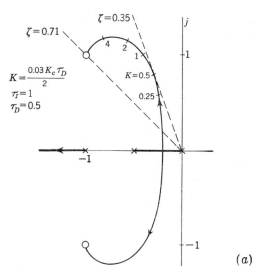

$$K = \frac{0.03\,K_c\,\tau_D}{2}$$
$$\tau_I = 1$$
$$\tau_D = 0.5$$

(a)

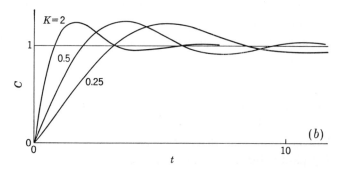

(b)

Fig. 17.10 Root-locus diagram and step response for PID control.

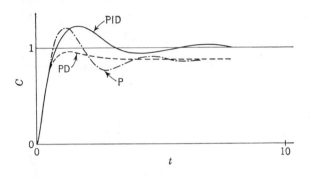

Fig. 17.11 **Comparison of step response for various controllers. (For each curve, 0.03K$_c$ = 8.)**

The root-locus diagram for these values of τ_I and τ_D is shown in Fig. 17.10a. The step-response curves are shown in Fig. 17.10b. Notice from Fig. 17.10a that the damping coefficient ζ always exceeds 0.35.

For this three-mode control, the root-locus diagram can be changed considerably by changing both τ_I and τ_D. Since three parameters (K_c, τ_I, and τ_D) can be changed, the problem of studying the effect of controller settings on the root-locus diagram requires more effort. In Chaps. 19 and 23, some empirical rules will be presented for selecting controller settings for one-, two-, and three-mode controllers which will give satisfactory response. Once controller settings are obtained from such rules, the response of the system can be checked by the root-locus method illustrated here.

Comparison of Response   In order to compare the quality of control for different modes of control, the step-response curves of Figs. 17.4, 17.8, and 17.10 for $0.03K_c = 8$ have been placed on one graph as shown in Fig. 17.11.

Comparing the response for proportional control with that for PD control, we see that there is less overshoot for the PD case. Of course, the offset remains the same for either mode of control.

The response for three-mode control has about the same maximum as that for proportional control, but the period of oscillation is longer. The advantage of three-mode control over proportional or PD control is clearly the elimination of offset. In Chaps. 19 and 23 we shall discuss further the effect of the mode of control and the controller settings on the transient response.

EFFECT OF DISTANCE VELOCITY LAG ON RESPONSE

We shall now consider the system of Fig. 17.2 for which the transportation lag parameter is $\tau = 0.5$ min. For this case, the block diagram

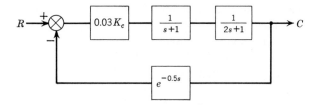

Fig. 17.12 Control system with transportation lag.

is shown in Fig. 17.12. In this discussion, assume that proportional control is used and that only a step change in set point is introduced.

The open-loop transfer function is

$$G = \frac{0.03 K_c e^{-0.5s}}{(s + 1)(2s + 1)}$$

For convenience in root-locus construction, $G(s)$ may be written in the alternate form

$$G = K \frac{e^{-0.5s}}{(s + 1)(s + 0.5)}$$

where

$$K = \frac{0.03 K_c}{2}$$

Up to this point, the transfer function $e^{-\tau s}$ has not been encountered in root-locus construction. We cannot express $e^{-\tau s}$ in rational form, and as a result the plotting of the root-locus diagram requires a different approach. One approach is to expand $e^{-\tau s}$ into an infinite series and to retain only a finite number of terms. For example, first write $e^{-\tau s}$ as

$$e^{-\tau s} = \frac{e^{-\tau s/2}}{e^{\tau s/2}}$$

Writing numerator and denominator in terms of a Taylor series gives

$$e^{-\tau s} = \frac{1 - (\tau s/2) + \frac{1}{2}(\tau s/2)^2 - \cdots}{1 + (\tau s/2) + \frac{1}{2}(\tau s/2)^2 + \cdots}$$

If all terms in s which are greater than first power are discarded, the approximation becomes

$$e^{-\tau s} = \frac{1 - \tau s/2}{1 + \tau s/2} = -\frac{s - 2/\tau}{s + 2/\tau} \tag{17.2}$$

In Chap. 32 on analog computer simulation, other possible expansions which can be used to approximate $e^{-\tau s}$ will be investigated.

The true root-locus diagram of a transfer function involving $e^{-\tau s}$ contains an infinite number of branches because $e^{-\tau s}$ is a periodic function.

To obtain this true root-locus diagram requires considerable effort. A graphical procedure is explained by Chu,[1] and a detailed example is worked out. In many practical situations, the true root-locus plot will contain one branch relatively close to the origin, and the remaining branches will be so far removed from the origin that the roots on these branches can be ignored.

We shall now continue the example by approximating $e^{-0.5s}$ by the expression in Eq. (17.2). The block diagram using this approximation of $e^{-\tau s}$ is shown in Fig. 17.13. The open-loop transfer function now becomes

$$G = -0.03K_c \frac{s-4}{s+4} \frac{1}{(s+1)(2s+1)}$$

Writing $G(s)$ in the alternate form which is used in plotting the root locus gives

$$G = -K \frac{s-4}{(s+0.5)(s+1)(s+4)} \tag{17.3}$$

where

$$K = \frac{0.03K_c}{2}$$

The minus sign on K in Eq. (17.3) means that we have a positive feedback system which has a zero in the right half plane at $+4$. This does not mean that the system is unstable for all values of K_c. To discover the nature of the system response, we shall now plot the root-locus diagram for the open-loop transfer function

$$G = K \frac{s-4}{(s+0.5)(s+1)(s+4)} \tag{17.4}$$

when we have positive feedback.

To indicate the reason for consideration of positive feedback, Fig. 17.13 has been redrawn in Fig. 17.14 in an equivalent form which more clearly shows the presence of positive feedback.

Before continuing the example, it will be necessary to discuss the root-locus approach for a positive feedback system. To make this discussion more general, consider the block diagram in Fig. 17.14 to contain the

[1] Yaohan Chu, *Trans. AIEE* part II, **71**:291–296 (1952).

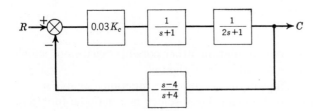

Fig. 17.13 **Block diagram of control system with approximation of $e^{-\tau s}$.**

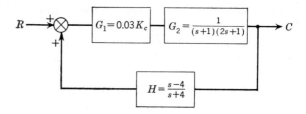

transfer functions G_1, G_2, and H which are rational functions of s. Using Eq. (12.10a) for determining the overall transfer function for positive feedback gives the following result:

$$\frac{C}{R} = \frac{G_1 G_2}{1 - G_1 G_2 H}$$

or

$$\frac{C}{R} = \frac{G_1 G_2}{1 - G}$$

where $G = G_1 G_2 H$ is the open-loop transfer function. The characteristic equation for positive feedback is therefore

$$1 - G = 0$$

and if G is written in the standard form used in Chap. 15, we have

$$G = K \frac{N}{D}$$

where

$$\frac{N}{D} = \frac{(s - z_1)(s - z_2) \cdots (s - z_m)}{(s - p_1)(s - p_2) \cdots (s - p_n)}$$

The characteristic equation can now be written

$$1 - K \frac{N}{D} = 0$$

or

$$K \frac{N}{D} = +1$$

Using arguments parallel to those used in Chap. 15, the following angle criterion is obtained:

$$\angle (s - z_1) + \angle (s - z_2) + \cdots + \angle (s - z_m)$$
$$- [\angle (s - p_1) + \angle (s - p_2) + \cdots + \angle (s - p_n)] = 2i\pi$$

where $i = 0, \pm 1, \pm 2, \ldots$. The magnitude criterion is

$$K \frac{|s - z_1| \, |s - z_2| \cdots |s - z_m|}{|s - p_1| \, |s - p_2| \cdots |s - p_n|} = 1$$

Thus, we see that the angle criterion differs for negative feedback and positive feedback but the magnitude criterion is the same. One can develop a set of rules for positive feedback, similar to those of Chap. 15, which are useful in sketching the root-locus plot. For convenience, these rules are presented below.

Rules for Plotting Root-locus Diagrams (Positive Feedback)

(In the following rules, $n \geq m$)

Rule 1   Branches   Same as Rule 1, Chap. 15.

Rule 2   Same as Rule 2, Chap. 15.

Rule 3   Locus on real axis   The real axis is on the root locus when the sum of the number of real poles and real zeros to the right of the branch is *even*. (In this statement "even" is to be interpreted as 0, 2, 4, . . . ; this means that the real axis to the right of *all* real poles and real zeros is part of the locus.)

Rule 4   Asymptotes   Same as Rule 4, Chap. 15, with the exception that the asymptotes make angles of $2\pi i/(n - m)$ with the real axis ($i = 0$, 1, 2, . . . , $n - m - 1$).

Rule 5   Breakaway point   Same as Rule 5, Chap. 15.

Rule 6   Angle of departure   Replace $(2k + 1)\pi$ in Eqs. (15.14) and (15.15) by $2k\pi$.

Returning now to the example for which the open-loop transfer function is given by Eq. (17.4), we can apply these rules for positive feedback to obtain the root-locus plot of Fig. 17.15. This plot shows clearly that the system becomes unstable for $K > 3.4$.

The roots of the characteristic equation having been obtained from the root-locus diagram, the step response for a change in set point can be

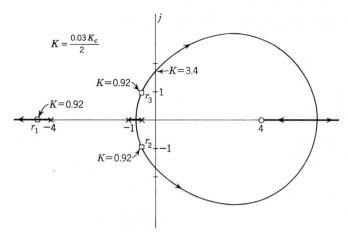

Fig. 17.15   Root-locus diagram for system of Fig. 17.14.

determined. For positive feedback, we have from Fig. 17.14

$$\frac{C}{R} = \frac{0.03K_c/[(s+1)(2s+1)]}{1 - 0.03K_c(s-4)/[(s+1)(2s+1)(s+4)]}$$

This expression can be written in the form KN/D to give

$$\frac{C}{R} = \frac{K(s+4)}{(s+0.5)(s+1)(s+4) - K(s-4)}$$

where

$$K = \frac{0.03K_c}{2}$$

The denominator on the right-hand side of the last expression is the characteristic function, which can be written in terms of the roots of the characteristic equation as follows:

$$\frac{C}{R} = \frac{K(s+4)}{(s-r_1)(s-r_2)(s-r_3)}$$

For a unit-step change in R, C becomes

$$C = \frac{K(s+4)}{s(s-r_1)(s-r_2)(s-r_3)} \tag{17.5}$$

Expanding by partial fractions gives

$$C = \frac{A_0}{s} + \frac{A_1}{s-r_1} + \frac{A_2}{s-r_2} + \frac{A_3}{s-r_3} \tag{17.6}$$

To illustrate the step response for a particular controller gain, let $K = 0.92$, which corresponds to $K_c = 61.5$. The roots for $K = 0.92$ obtained from Fig. 17.15 are

$$r_1 = -4.55$$
$$r_2 = -0.5 - j$$
$$r_3 = -0.5 + j$$

Evaluating the coefficients A_0, A_1, A_2, A_3 by the method described in the previous chapter gives

$$A_0 = 0.65$$
$$A_1 = 0.0064$$
$$A_2 = 0.36e^{j203°}$$
$$A_3 = 0.36e^{-j203°}$$

Substituting these coefficients into Eq. (17.6) and expressing the last two terms in the form of Eq. (16.22) give

$$C(t) = 0.65 + 0.0064e^{-4.55t} + e^{-0.5t}0.72 \sin (t - 203 + 90)$$

or

$$C(t) = 0.65 + 0.0064e^{-4.55t} + 0.72e^{-0.5t} \sin (t - 113°) \qquad (17.7)$$

This response is plotted in Fig. 17.16.

For the first $\frac{1}{2}$ min following the introduction of the unit-step change in R, the response can be determined exactly because during this time, the measured variable will not change and $C(t)$ will be simply the unit-step response for the transfer function

$$G = \frac{0.03K_c}{(s + 1)(2s + 1)}$$

which is

$$C(t) = 0.03K_c[1 - 2(e^{-0.5t} - 0.5e^{-t})] \qquad (17.8)$$

This response is plotted in Fig. 17.16 in order to compare the true response with the approximate response during the first $\frac{1}{2}$ min. Notice that there is hardly any difference between the true response and the approximate response of Eq. (17.7) over the time interval ($\frac{1}{2}$ min) for which the true response is known.

It should be mentioned here that the accuracy of the calculated response, when $e^{-\tau s}$ is approximated by an expression such as the one used in this example, will depend on the magnitude of τ relative to the other lags in the system. For this example, τ was one-fourth the largest time constant. As τ increases relative to the other lags in the system, we can expect the difference between the true response and the approximate response to increase. In Chap. 32 other approximations to $e^{-\tau s}$ will be

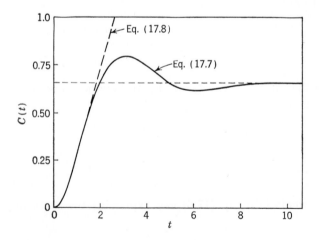

Fig. 17.16 Step response of control system of Fig. 17.12.

considered. At that time the transient response for the same control system of Fig. 17.12 will be obtained by using a more accurate approximation to $e^{-\tau s}$. (See Fig. 32.19, curve 1, for comparison with Fig. 17.16.)

SUMMARY

In the last three chapters, the root-locus method has been developed and applied to the analysis of several control systems. The root-locus approach was used only for variation in controller gain and for discrete changes in τ_I and τ_D. This approach can be extended to cover the case of continuous variation in other parameters, such as τ_D and τ_I, and to solve multiloop systems; however the procedure becomes more tedious.[1]

The root-locus method is one of the basic tools, along with frequency-response methods, that a control engineer can use to solve control problems. The root-locus diagram has a distinct advantage in giving at a glance the stability characteristics of the system. Furthermore, the root-locus diagram provides the roots of the characteristic equation, from which the transient response can be obtained. However, the root-locus approach is difficult to apply to system transfer functions containing $e^{-\tau s}$. As will be shown in the next chapter, such a transfer function is more easily handled by the frequency-response approach.

PROBLEMS

17.1 The liquid-level system shown in Fig. P17.1 is to be controlled by several types of controllers. The following data apply:

1. The resistance of the first tank is linear and has a value of $R = 0.3$ ft/cfm. The cross-sectional area of the tank is 1.67 ft².

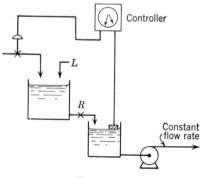

Fig. P17.1

[1] The application of the root-locus method for variation in one of the time constants is presented in several textbooks on servomechanisms. For example, see C. H. Wilts, "Principles of Feedback Control," Addison-Wesley Publishing Company, Inc., Reading, Mass., 1960.

2. The flow rate from the second tank is maintained constant by means of a pump. The cross-sectional area of this tank is 0.5 ft².

3. The constant of the valve is 0.25 cfm/psi.

    a. Draw a block diagram of the control system, and show in each block the numerical values of the transfer function.

    b. By means of root-locus sketches, discuss the stability of the control system for each type of controller that is listed in the following table. For each case, determine, when possible, the value of controller gain K_c for which the system just becomes unstable and the value of controller gain for which $\zeta = 0.5$.

Case	Controller	τ_D	τ_I	Value of K_c for which system just becomes unstable	K_c for which $\zeta = 0.5$
I	Proportional				
II	PD	0.25			
IIIa	PI		0.25		
IIIb	PI		1.0		

17.2 For the control system shown in Fig. P17.2

    a. Use the root-locus plot to select settings for

        1. Proportional
        2. Proportional-integral
        3. Proportional-derivative
        4. Proportional-integral-derivative control modes

    b. Calculate step load responses for your design settings.
    c. Compare the results of *b* with those obtained from an analog simulation.
    d. Vary the controller settings in a typical case on the analog computer. Compare the effect realized on the system response with that predicted by the root-locus diagram.

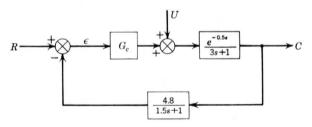

Fig. P17.2

17.3 On the basis of the root-locus diagrams of Figs. 17.5a and 17.6a and b, recommend suitable controller settings for PI control of the reactor system.

part **V** **Frequency-response Methods**

18 Introduction to Frequency Response

In Chaps. 5 and 8, we discussed briefly the response of first- and second-order systems to sinusoidal forcing functions. These frequency responses were derived by using the standard Laplace transform technique. In this chapter, we shall establish a convenient graphical technique for obtaining the frequency response of linear systems. The motivation for doing so will become apparent in the following chapters, where it will be found that frequency response is a valuable tool in the analysis and design of control systems.

SUBSTITUTION RULE

A Fortunate Circumstance Consider a simple first-order system with transfer function

$$G(s) = \frac{1}{\tau s + 1} \tag{18.1}$$

Substituting the quantity $j\omega$ for s in Eq. (18.1) gives

$$G(j\omega) = \frac{1}{j\omega\tau + 1} \tag{18.2}$$

Converting this complex number to polar form results in

$$G(j\omega) = \frac{1}{\sqrt{\omega^2\tau^2 + 1}} \; \measuredangle \; \tan^{-1}(-\omega\tau) \tag{18.3}$$

The details of this conversion are left as an exercise for the reader. The quantities on the right side of Eq. (18.3) are familiar. In Chap. 5 we showed that, after sufficient time had elapsed, the response of a first-order system to a sinusoidal input of frequency ω is also a sinusoid of frequency ω. Furthermore, we saw that the ratio of the amplitude of the response to that of the input is $1/\sqrt{\omega^2\tau^2 + 1}$ and the phase difference between output and input is $\tan^{-1}(-\omega\tau)$. Hence, we have shown here that for the frequency response of a first-order system,

$$\text{AR} = |G(j\omega)|$$
$$\text{Phase angle} = \measuredangle G(j\omega)$$

That is, to obtain the amplitude ratio (AR) and phase angle, one merely substitutes $j\omega$ for s in the transfer function and then takes the magnitude and argument of the resulting complex number, respectively.

Example 18.1   Rework Example 5.2. The pertinent transfer function is

$$G(s) = \frac{1}{0.1s + 1}$$

The frequency of the bath-temperature variation is given as $10/\pi$ cycles/min which is equivalent to 20 rad/min.
Hence, let

$$s = 20j$$

to obtain

$$G(20j) = \frac{1}{2j + 1}$$

In polar form, this is

$$G(20j) = \frac{1}{\sqrt{5}} \; \measuredangle -63.5°$$

which agrees with the previous result.

Generalization   At this point, it is necessary to ascertain whether or not we may generalize the result of the last section to other systems. This can be done by checking the result for second-order systems, third-order systems, etc. However, it is more satisfying to prove the general validity of the result as follows: (The reader may, if desired, accept the result as general and skip to Example 18.2. We remark here that an important restriction on this rule is that it applies only to systems whose transfer functions yield stable responses.)

An nth-order linear system is characterized by an nth-order differential equation:

$$a_n \frac{d^n Y}{dt^n} + a_{n-1} \frac{d^{n-1} Y}{dt^{n-1}} + \cdots + a_1 \frac{dY}{dt} + a_0 Y = X(t) \qquad (18.4)$$

where Y is the output variable and $X(t)$ is the forcing function or input variable. For specific cases of Eq. (18.4), refer to Eq. (5.5) for a first-order system and Eq. (8.4) for a second-order system. If $X(t)$ is sinusoidal

$$X(t) = A \sin \omega t$$

the solution of Eq. (18.4) will consist of a complementary solution, and a particular solution of the form

$$Y_p(t) = C_1 \sin \omega t + C_2 \cos \omega t \qquad (18.5)$$

If the system is stable, the roots of the characteristic equation of (18.4) all lie to the left of the imaginary axis and the complementary solution will vanish exponentially in time. Then Y_p is the quantity previously defined as the sinusoidal or *frequency response*. If the system is not stable, the complementary solution grows exponentially and the term frequency response has no physical significance because $Y_p(t)$ is inconsequential.

The problem is now evaluation of C_1 and C_2 in Eq. (18.5). Since we are interested in the amplitude and phase of $Y_p(t)$, Eq. (18.5) is rewritten as

$$Y_p = D_1 \sin (\omega t + D_2) \qquad (18.6)$$

as was done previously [cf. Eq. (5.23) *et seq.*].

It will be convenient to change $X(t)$ and $Y_p(t)$ from trigonometric to exponential form, using the identity

$$\sin \theta = \frac{e^{j\theta} - e^{-j\theta}}{2j}$$

Thus,

$$X(t) = \frac{A}{2j} (e^{j\omega t} - e^{-j\omega t}) \qquad (18.7)$$

and from Eq. (18.6)

$$Y_p(t) = \frac{D_1}{2j} [e^{j(\omega t + D_2)} - e^{-j(\omega t + D_2)}] \qquad (18.8)$$

Substitution of Eqs. (18.7) and (18.8) into Eq. (18.4) yields:

$$\frac{D_1 e^{j(\omega t + D_2)}}{2j} [a_n(j\omega)^n + a_{n-1}(j\omega)^{n-1} + \cdots + a_1(j\omega) + a_0]$$

$$- \frac{D_1 e^{-j(\omega t + D_2)}}{2j} [a_n(-j\omega)^n + a_{n-1}(-j\omega)^{n-1} + \cdots + a_1(-j\omega) + a_0]$$

$$= \frac{A}{2j}(e^{j\omega t} - e^{-j\omega t}) \quad (18.9)$$

The coefficients of $e^{j\omega t}$ on both sides of Eq. (18.9) must be equal. Hence,

$$D_1 e^{jD_2}[a_n(j\omega)^n + a_{n-1}(j\omega)^{n-1} + \cdots + a_1(j\omega) + a_0] = A \quad (18.10)$$

Equation (18.10) will be satisfied if and only if

$$\left| \frac{1}{a_n(j\omega)^n + a_{n-1}(j\omega)^{n-1} + \cdots + a_1(j\omega) + a_0} \right| = \frac{D_1}{A} \quad (18.11)$$

$$\measuredangle \frac{1}{a_n(j\omega)^n + a_{n-1}(j\omega)^{n-1} + \cdots + a_1(j\omega) + a_0} = D_2$$

But D_1/A and D_2 are the AR and phase angle of the response, respectively, as may be seen from Eq. (18.6) and the forcing function. Furthermore, from Eq. (18.4) the transfer function relating X and Y is

$$\frac{Y(s)}{X(s)} = \frac{1}{a_n s^n + a_{n-1}s^{n-1} + \cdots + a_1 s + a_0} \quad (18.12)[1]$$

Equations (18.11) and (18.12) establish the general result.

Example 18.2 Find the frequency response of the system with the general second-order transfer function and compare the results with those of Chap. 8. The transfer function is

$$\frac{1}{\tau^2 s^2 + 2\zeta\tau s + 1}$$

Putting $s = j\omega$ yields

$$\frac{1}{1 - \tau^2\omega^2 + j2\zeta\omega\tau}$$

which may be converted to the polar form

$$\frac{1}{\sqrt{(1 - \omega^2\tau^2)^2 + (2\zeta\omega\tau)^2}} \measuredangle \tan^{-1}\left(\frac{-2\zeta\omega\tau}{1 - \omega^2\tau^2}\right)$$

Hence,

$$\text{AR} = \frac{1}{\sqrt{(1 - \omega^2\tau^2)^2 + (2\zeta\omega\tau)^2}} \quad (18.13)$$

$$\text{Phase angle} = \tan^{-1}\frac{-2\zeta\omega\tau}{1 - \omega^2\tau^2}$$

which agree with Eq. (8.40).

[1] In writing this equation, it is assumed that X and Y have been written as deviation variables, so that initial conditions are zero.

Transportation Lag The response of a transportation lag is not described by Eq. (18.4). Rather, a transportation lag is described by the relation

$$Y(t) = X(t - \tau) \tag{18.14}$$

which states that the output Y lags the input X by an interval of time τ. If X is sinusoidal,

$$X = A \sin \omega t$$

then from Eq. (18.14)

$$Y = A \sin \omega(t - \tau) = A \sin (\omega t - \omega\tau)$$

It is apparent that the AR is unity and the phase angle is $(-\omega\tau)$.

To check the substitution rule of the previous section, recall that the transfer function is given by

$$G(s) = \frac{Y(s)}{X(s)} = e^{-\tau s}$$

Putting $s = j\omega$,

$$G(j\omega) = e^{-j\omega\tau}$$

Then,

$$\text{AR} = |e^{-j\omega\tau}| = 1$$
$$\text{Phase angle} = \measuredangle e^{-j\omega\tau} = -\omega\tau \tag{18.15}$$

and the validity of the rule is verified.

Example 18.3 The stirred-tank heater of Chap. 1 has a capacity of 15 gal. Water is entering and leaving the tank at the constant rate of 600 lb/min. The heated water which leaves the tank enters a well-insulated section of 6-in.-ID pipe. Two feet from the tank, a thermocouple is placed in this line for recording the tank temperature, as shown in Fig. 18.1. The electrical heat input is held constant at 1,000 kw.

If the inlet temperature is varied according to the relation

$$T_i = 75 + 5 \sin 46t$$

where T_i is in degrees Fahrenheit and t is in minutes, find the eventual behavior of the thermocouple reading T_m. Compare this with the behavior of the tank temperature T.

Fig. 18.1 Tank-temperature system for Example 18.3.

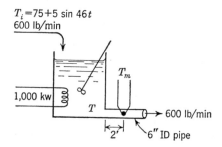

$T_i = 75 + 5 \sin 46t$
600 lb/min

1,000 kw

T_m

T

600 lb/min

2'

6" ID pipe

It may be assumed that the thermocouple has a very small time constant and effectively measures the true fluid temperature at all times.

The problem is to find the frequency response of T_m to T_i. Deviation variables must be used. Define the deviation variable T'_i as

$$T'_i = T_i - 75 = 5 \sin 46t$$

To define a deviation variable for T_m, note that, if T_i were held at 75°F, T_m would come to the steady state satisfying

$$q_s = wC(T_{m_s} - T_{i_s})$$

This may be solved for T_{m_s}:

$$T_{m_s} = \frac{q_s}{wC} + T_{i_s} = \frac{(1,000)(1,000)(0.0569)}{(600)(1.0)} + 75 = 170°F$$

Hence, define

$$T'_m = T_m - 170$$

Now the overall system between T'_i and T'_m is made up of two components in series: the tank and the 2-ft section of pipe. The transfer function for the tank is

$$G_1(s) = \frac{1}{\tau_1 s + 1}$$

where, as we have seen before, τ_1 is given by

$$\tau_1 = \frac{\rho V}{w} = \frac{(60.3)(15)}{(600)(7.48)} = 0.202 \text{ min}$$

The transfer function of the 2-ft section of pipe, which corresponds to a transportation lag, is

$$G_2(s) = e^{-\tau_2 s}$$

where τ_2 is the length of time required for the fluid to transverse the length of pipe. This is

$$\tau_2 = \frac{L}{v} = \frac{(2)(60.3)(0.197)}{600} = 0.0396 \text{ min}$$

The factor 0.197 is the cross-sectional area of the pipe in square feet.

Since the two systems are in series, the overall transfer function between T'_i and T'_m is

$$\frac{T'_m}{T'_i} = \frac{e^{-\tau_2 s}}{\tau_1 s + 1} = \frac{e^{-0.0396 s}}{0.202 s + 1}$$

To find the AR and phase lag, we merely substitute $s = 46j$ and take the magnitude and argument of the resulting complex number. However, note that we have previously derived the individual frequency responses for the first-order system and transportation lag. The overall transfer function is the product of the individual transfer functions; hence, its magnitude will be the product of the magnitudes and its argument the sum of the arguments of the individual transfer functions. In general, if

$$G(s) = G_1(s)G_2(s) \cdots G_n(s)$$

then

$$|G(j\omega)| = |G_1(j\omega)| \, |G_2(j\omega)| \ldots |G_n(j\omega)|$$
$$\angle G(j\omega) = \angle G_1(j\omega) + \angle G_2(j\omega) + \cdots + \angle G_n(j\omega)$$

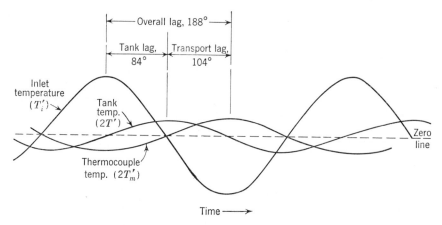

Fig. 18.2   **Temperature variation in Example 18.3.**

This rule makes it very convenient to find the frequency response of a number of systems in series.

Using Eq. (18.3) for the tank,

$$\text{AR} = \frac{1}{\sqrt{(46 \times 0.202)^2 + 1}} = \frac{1}{9.35} = 0.107$$

Phase angle $= \tan^{-1}[(-46)(0.202)] = -84°$

For the section of pipe, the AR is unity, so that the overall AR is just 0.107. The phase lag due to the pipe may be obtained from Eq. (18.15) as

Phase angle $= -\omega\tau_2 = -(46)(0.0396) = -1.82$ rad $= -104°$

The overall phase lag from T'_i to T'_m is the sum of the individual lags,

$$\angle \frac{T'_m}{T'_i} = -84 - 104 = -188°$$

Hence

$$T_m = 170 + 0.535 \sin(46t - 188°)$$

For comparison, a plot of T'_i, T'_m, and T' is given in Fig. 18.2, where

$$T' = \text{tank temperature} - 170°F$$

It should be emphasized that this plot applies only after sufficient time has elapsed for the complementary solution to become negligible. This restriction applies to all the forthcoming work on frequency response. Also, note that, for convenience of scale, the tank and thermocouple temperatures have been plotted as $2T'$ and $2T'_m$, respectively.

A Control Problem   An interesting conclusion may be reached from a study of Fig. 18.2. Suppose that we are trying to control the tank temperature, using the deviation between the thermocouple reading and the set point as the error. A block diagram for proportional control might appear as in Fig. 18.3, where T'_i is replaced by U, T' by C, and T'_m by B to conform with our standard block-diagram nomenclature. The variable R denotes

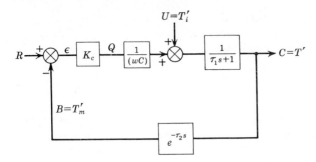

Fig. 18.3 Proportional control of heated stirred tank.

the deviation of the set point from 170°F and is the desired value of the deviation C. The value of R is assumed to be zero in the following analysis (control at 170°F). The following arguments, while not rigorous, serve to give some insight regarding application of frequency response to control system analysis.

The heat being added to the tank is given in deviation variables as $-K_cB$. With reference to Fig. 18.2, which shows the response of the *uncontrolled* tank to a sinusoidal variation in U, it can be seen that the peaks of U and B are almost exactly opposite because the phase difference is 188°. This means that, if the loop were closed, the control system would have a tendency to add *more* heat when the inlet temperature T_i is at its high peak, because B is then negative and $-K_cB$ becomes positive. (Recall that the set point R is held constant at zero.) Conversely, when the inlet temperature is at a low point, the *tendency* will be for the control system to add less heat because B is positive. This is precisely opposite to the way the heat input should be controlled.

Therefore, the possibility of an unstable control system exists for this particular sinusoidal variation in frequency. Indeed, we shall demonstrate in Chap. 19 that, *if K_c is taken too large*, the tank temperature will oscillate with increasing amplitude for *all* variations in U and hence we have an unstable control system. The fact that such information may be obtained by study of the *frequency response* (i.e., the particular solution for a sinusoidal forcing function) justifies further study of this subject.

BODE DIAGRAMS

Thus far, it has been necessary to calculate AR and phase lag by direct substitution of $s = j\omega$ into the transfer function for the particular frequency of interest. It can be seen from Eqs. (18.3), (18.13), and (18.15) that the AR and phase lag are functions of frequency. There is a convenient graphical representation of their dependence on the frequency, which largely eliminates direct calculation. *This is called a Bode diagram and consists of two graphs: logarithm of AR versus logarithm of frequency, and phase angle versus logarithm of frequency.* The Bode diagram will be shown in Chap. 19 to be a convenient tool for analyzing control problems such as the one discussed in the preceding section. The remainder of the present chapter is devoted to developing this tool and presenting Bode diagrams for the basic components of control loops.

First-order System The AR and phase angle for the sinusoidal response of a first-order system are

$$\text{AR} = \frac{1}{\sqrt{\tau^2\omega^2 + 1}} \tag{18.16}$$

$$\text{Phase angle} = \tan^{-1}(-\omega\tau) \tag{18.17}$$

It is convenient to regard these as functions of $\omega\tau$ for the purpose of generality. From Eq. (18.16)

$$\log \text{AR} = -\tfrac{1}{2} \log [(\omega\tau)^2 + 1] \tag{18.18}$$

The first part of the Bode diagram is a plot of Eq. (18.18). The true curve is shown as the solid line on the upper part of Fig. 18.4. Some asymptotic considerations can simplify this plot. As $(\omega\tau) \rightarrow 0$, Eq. (18.16) shows that $\text{AR} \rightarrow 1$. This is indicated by the low-frequency asymptote on Fig. 18.4. As $(\omega\tau) \rightarrow \infty$, Eq. (18.18) becomes asymptotic to

$$\log \text{AR} = -\log (\omega\tau)$$

which is a line of slope -1, passing through the point

$$\omega\tau = 1 \qquad \text{AR} = 1$$

This line is indicated as the high-frequency asymptote in Fig. 18.4. The frequency $\omega_c = 1/\tau$, where the two asymptotes intersect, is known as the *corner frequency;* it may be shown that the deviation of the true AR curve from the asymptotes is a maximum at the corner frequency. Using $\omega_c = 1/\tau$ in Eq. (18.16) gives

$$\text{AR} = \frac{1}{\sqrt{2}} = 0.707$$

as the true value, whereas the intersection of the asymptotes occurs at $\text{AR} = 1$. Since this is the maximum deviation and is an error of less than 30 percent, for engineering purposes it is often sufficient to represent the curve entirely by the asymptotes. Alternately, the asymptotes and the value of 0.707 may be used to sketch the curve if more accuracy is required.

In the lower half of Fig. 18.4, we have shown the phase curve as given by Eq. (18.17). Since

$$\phi = \tan^{-1}(-\omega\tau) = -\tan^{-1}(\omega\tau)$$

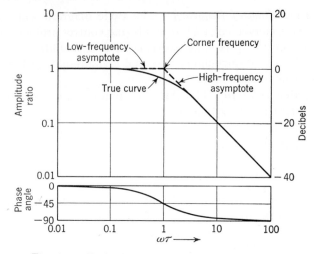

Fig. 18.4 **Bode diagram for first-order system.**

it is evident that ϕ approaches $0°$ at low frequencies and $-90°$ at high frequencies. This verifies the low- and high-frequency portions of the phase curve. At the corner frequency, $\omega_c = 1/\tau$,

$$\phi_c = -\tan^{-1}(\omega_c \tau) = -\tan^{-1}(1) = -45°$$

There are asymptotic approximations available for the phase curve, but they are not so accurate or so widely used as those for the AR. Instead, it is convenient to note that the curve is symmetric about $-45°$ and to use the following table for sketching the curve.

Table 18.1 **Values of ϕ at Intermediate Values of $\omega\tau$; First-order System**

$\omega\tau$	ϕ, *Eq.* *(18.17)*
0.1	$-5.7°$
0.5	$-26.6°$
1.0	$-45°$
2.0	$-63.5°$
10.0	$-84.3°$

It should be stated that, in a great deal of the literature on control theory, amplitude ratios (or gains) are reported in decibels. The decibel is defined by

$$\text{Decibels} = 20 \log_{10}(\text{AR})$$

Thus, an AR of unity corresponds to zero decibels and an amplitude ratio of 0.1 corresponds to -20 decibels.   The abbreviation for the decibel is db. The value of the AR in decibels is given on the right-hand ordinate of Fig. 18.4.

First-order Systems in Series   The advantages of the Bode plot become evident when we wish to plot the frequency response of systems in series.   As shown in Example 18.3, the rules for multiplication of complex numbers indicate that the AR for two first-order systems in series is the product of the individual AR's:

$$\text{AR} = \frac{1}{\sqrt{\omega^2\tau_1{}^2 + 1}\,\sqrt{\omega^2\tau_2{}^2 + 1}} \tag{18.19}$$

Similarly, the phase angle is the sum of the individual phase angles

$$\phi = \tan^{-1}(-\omega\tau_1) + \tan^{-1}(-\omega\tau_2) \tag{18.20}$$

Since the AR is plotted on a logarithmic basis, multiplication of the AR's is accomplished by addition of logarithms on the Bode diagram.   The phase angles are added directly.   The procedure is best illustrated by an example.

Example 18.4   Plot the Bode diagram for the system whose overall transfer function is

$$\frac{1}{(s + 1)(s + 5)}$$

To put this in the form of two first-order systems in series, it is rewritten as

$$\frac{\frac{1}{5}}{(s + 1)(\frac{1}{5}s + 1)} \tag{18.21}$$

The time constants are $\tau_1 = 1$ and $\tau_2 = \frac{1}{5}$.   The factor $\frac{1}{5}$ in the numerator corresponds to the steady-state gain.

From Eqs. (18.21) and (18.19)

$$\text{AR} = \frac{\frac{1}{5}}{\sqrt{\omega^2 + 1}\,\sqrt{(\omega/5)^2 + 1}}$$

Hence,

$$\log \text{AR} = \log \frac{1}{5} - \frac{1}{2}\log(\omega^2 + 1) - \frac{1}{2}\log\left[\left(\frac{\omega}{5}\right)^2 + 1\right]$$

or

$$\log \text{AR} = \log \frac{1}{5} + \log(\text{AR})_1 + \log(\text{AR})_2 \tag{18.22}$$

where $(\text{AR})_1$ and $(\text{AR})_2$ are the AR's of the individual first-order systems, each with unity gain.   Equation (18.22) shows that the overall AR is obtained, on logarithmic coordinates, by adding the individual AR's and a constant corresponding to the steady-state gain.

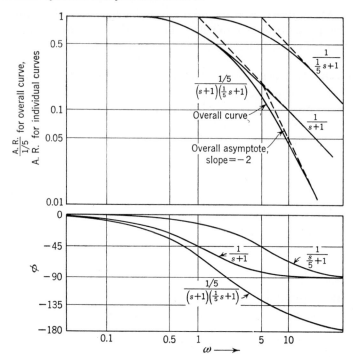

Fig. 18.5 Bode diagram for $\frac{1}{5}/[(s+1)(\frac{1}{5}s+1)]$.

The individual AR's must be plotted as functions of $\log \omega$ rather than $\log (\omega\tau)$ because of the different time constants. This is easily done by shifting the curves of Fig. 18.4 to the right or left so that the corner frequency falls at $\omega = 1/\tau$. Thus, the individual curves of Fig. 18.5 are placed so that the corner frequencies fall at $\omega_{c_1} = 1$ and $\omega_{c_2} = 5$. These curves are added to obtain the overall curve shown. Note that in this case the logarithms are negative and the addition is downward. To complete the AR curve, the factor $\log \frac{1}{5}$ should be added to the overall curve. This would have the effect of shifting the entire curve down by a constant amount. Instead of doing this, the factor $\frac{1}{5}$ is incorporated by plotting the overall curve as $AR/\frac{1}{5}$ instead of AR. This procedure is usually more convenient.

Asymptotes have also been indicated on Fig. 18.5. The sum of the individual asymptotes gives the overall asymptote, which is seen to be a good approximation to the overall curve. The overall asymptote has a slope of zero below $\omega = 1$, -1 for ω between 1 and 5, and -2 above $\omega = 5$. Its slope is obtained by simply adding the slopes of the individual asymptotes.

To obtain the phase angle, the individual phase angles are plotted and added according to Eq. (18.20). The factor $\frac{1}{5}$ has no effect on the phase angle, which approaches $-180°$ at high frequency.

Graphical Rules for Bode Diagrams Before proceeding to a development of the Bode diagram for other systems, it is desirable to summarize the graphical rules which were utilized in Example 18.4.

Consider a number of systems in series. As shown in Example 18.3, the overall AR is the product of the individual AR's, and the overall phase angle is the sum of the individual phase angles. Therefore,

$$\log (AR) = \log (AR)_1 + \log (AR)_2 + \cdots + \log (AR)_n \qquad (18.23)$$

and

$$\phi = \phi_1 + \phi_2 + \cdots + \phi_n$$

where n is the total number of systems. Therefore, the following rules apply to the true curves or to the asymptotes on the Bode diagram:

1. The overall AR is obtained by adding the individual AR's. For this graphical addition, curves above unity AR are taken as positive and curves below are negative. (In later sections curves will be given which lie above unity.)
2. The overall phase angle is obtained by addition of the individual phase angles.
3. The presence of a constant in the overall transfer function shifts the entire AR curve vertically by a constant amount and has no effect on the phase angle. It is usually more convenient to include a constant factor in the definition of the ordinate.

These rules will be of considerable value in later examples. We now proceed to develop Bode diagrams for other control system components.

The Second-order System As shown in Example 18.2, the frequency response of a system with a second-order transfer function

$$G(s) = \frac{1}{\tau^2 s^2 + 2\zeta\tau s + 1}$$

is given by Eq. (18.13), repeated here for convenience,

$$AR = \frac{1}{\sqrt{(1 - \omega^2\tau^2)^2 + (2\zeta\omega\tau)^2}}$$

$$\text{Phase angle} = \tan^{-1} \frac{-2\zeta\omega\tau}{1 - (\omega\tau)^2} \qquad (18.13)$$

If $\omega\tau$ is used as the abscissa for the general Bode diagram, it is clear that ζ will be a parameter. That is, there is a different curve for each value of ζ. These curves appear as in Fig. 18.6.

The qualitative aspects of the curves are easily verified. Thus, considering first the phase angle for $\omega\tau \ll 1$, Eq. (18.13) indicates that

$$\tan \phi \cong -2\zeta\omega\tau$$

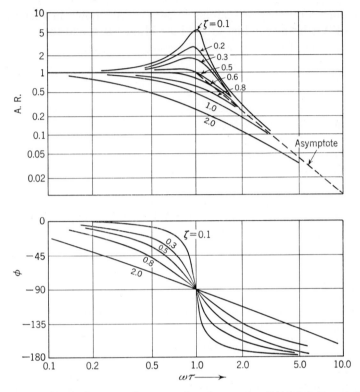

Fig. 18.6   **Bode diagram for second-order system** $1/(\tau^2 s^2 + 2\zeta\tau s + 1)$.

so that ϕ approaches zero as $\omega\tau$ becomes small. For $\omega\tau = 1$,

$$\tan \phi = -\infty$$

which means ϕ is $-90°$ independently of ζ. This verifies that all phase curves intersect at $-90°$ as shown. As $\omega\tau$ increases above 1, $\tan \phi$ starts to approach zero but always assumes positive values. This means that ϕ varies from -90 to -180 as $\omega\tau$ goes from 1 to infinity, as shown.

The AR curves may be similarly checked. For $\omega\tau \ll 1$, the AR or gain approaches unity. For $\omega\tau \gg 1$, the AR becomes asymptotic to the line

$$AR = \frac{1}{(\omega\tau)^2}$$

This asymptote has slope -2 and intersects the line AR $= 1$ at $\omega\tau = 1$. The asymptotic lines are indicated on Fig. 18.6. For $\zeta \geq 1$, we have shown that the second-order system is equivalent to two first-order systems in series. The fact that the gain for $\zeta \geq 1$ (as well as for $\zeta < 1$) attains a slope of -2 and phase of -180 is, therefore, consistent.

Figure 18.6 also shows that, for $\zeta < 0.707$, the AR curves attain maxima in the vicinity of $\omega\tau = 1$. This can be checked by differentiating the expression for the AR with respect to $\omega\tau$ and setting the derivative to zero. The result is

$$(\omega\tau)_{max} = \sqrt{1 - 2\zeta^2} \qquad \zeta < 0.707 \tag{18.24}$$

for the value of $\omega\tau$ at which the maximum AR occurs. The value of the maximum AR, obtained by substituting $(\omega\tau)_{max}$ into Eq. (18.13), is

$$(AR)_{max} = \frac{1}{2\zeta \sqrt{1 - \zeta^2}} \qquad \zeta < 0.707$$

A plot of the maximum AR against ζ is given in Fig. 18.7. The frequency at which the maximum AR is attained is called the resonant frequency and is obtained from Eq. (18.24),

$$\omega_r = \frac{1}{\tau} \sqrt{1 - 2\zeta^2} \tag{18.25}$$

The phenomenon of resonance is frequently observed in our everyday experience. A vase may vibrate severely when the hi-fi set plays a particular note. As a car decelerates, perceptible vibrations may occur at a particular speed.

It may be seen that AR values exceeding unity are attained by systems for which $\zeta < 0.707$. This is in sharp contrast to the first-order system, for which the AR is always less than unity.

The curves of Fig. 18.6 for $\zeta < 1$ are not simple to construct, particularly in the vicinity of the resonant frequency. Fortunately, almost all second-order control system components for which we shall want to construct Bode diagrams have $\zeta > 1$. That is, they are composed of two first-order systems in series. Actually, the curves of Fig. 18.6 are presented primarily because they are useful in analyzing the *closed-loop frequency*

Fig. 18.7 **Maximum AR versus damping for second-order system.**

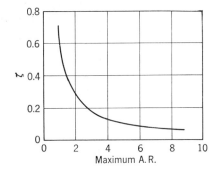

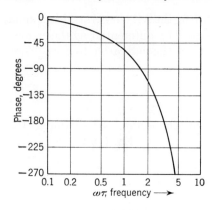

Fig. 18.8 **Phase characteristic of transportation lag.**

response of many control systems. An example of this will be discussed in Chap. 20.

Transportation Lag As shown by Eq. (18.15), the frequency response for $G(s) = e^{-\tau s}$ is

$$AR = 1$$
$$\phi = -\omega\tau \qquad \text{rad}$$

There is no need to plot the AR. On logarithmic coordinates, the phase angle appears as in Fig. 18.8, where $\omega\tau$ is used as the abscissa to make the figure general. The transportation lag contributes a phase lag which increases without bound as ω increases. Note that it is necessary to convert $\omega\tau$ from radians to degrees to prepare Fig. 18.8. Representative points are listed in Table 18.2 for convenience in sketching the phase-angle curve for a transportation lag.

Proportional Controller A proportional controller with transfer function K_c has amplitude ratio K_c and phase angle zero at all frequencies. No Bode diagram is necessary for this component.

Table 18.2 **Phase Angle of a Transportation Lag**

$\omega\tau$	$-\phi$, *Eq. (18.15)*, *deg*
0.1	5.7
0.2	11.5
0.5	28.7
1	57.3
2	114.6
5	287
10	573

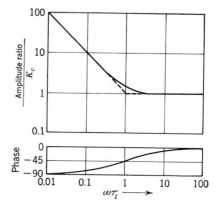

Fig. 18.9 Bode diagram for PI controller.

Proportional-integral Controller This component has the ideal transfer function

$$G(s) = K_c\left(1 + \frac{1}{\tau_I s}\right)$$

Accordingly, the frequency response is given by

$$\text{AR} = |G(j\omega)| = K_c\left|1 + \frac{1}{\tau_I j\omega}\right| = K_c\sqrt{1 + \frac{1}{(\omega\tau_I)^2}}$$

$$\text{Phase} = \angle G(j\omega) = \angle\left(1 + \frac{1}{\tau_I j\omega}\right) = \tan^{-1}\left(-\frac{1}{\omega\tau_I}\right)$$

The Bode plot of Fig. 18.9 uses $(\omega\tau_I)$ as the abscissa. The constant factor K_c is included in the ordinate for convenience. Asymptotes with a corner frequency of $\omega_c = 1/\tau_I$ are indicated. The verification of Fig. 18.9 is recommended as an exercise for the reader.

Proportional-derivative Controller The transfer function is

$$G(s) = K_c(1 + \tau_D s)$$

The reader should show that this has amplitude and phase behavior which is just the inverse of the first-order system

$$\frac{1}{\tau s + 1}$$

Hence, the Bode plot is as shown in Fig. 18.10. The corner frequency is $\omega_c = 1/\tau_D$.

This system is important because it introduces phase lead. Thus, it can be seen that using PD control for the tank temperature-control system of Example 18.3 would decrease the phase lag at all frequencies.

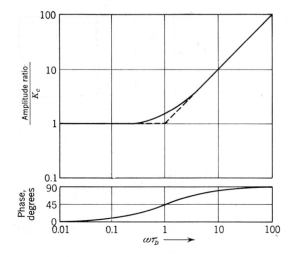

Fig. 18.10 **Bode diagram for PD controller.**

In particular, 180° of phase lag would not occur until a higher frequency. This may exert a stabilizing influence on the control system. In the next chapter, we shall look in detail at designing stabilizing controllers using Bode diagram analysis. It is appropriate to conclude this chapter with a summarizing example.

Example 18.5 Plot the Bode diagram for the open-loop transfer function of the control system of Fig. 18.11. This system might represent PD control of three tanks in series, with a transportation lag in the measuring element.

The open-loop transfer function is

$$G(s) = \frac{10(0.5s + 1)e^{-s/10}}{(s + 1)^2(0.1s + 1)}$$

The individual components are plotted as dashed lines in Fig. 18.12. Only the asymptotes are used on the AR portion of the graph. Here it is easiest to plot the factor $(s + 1)^{-2}$ as a line of slope -2 through the corner frequency of 1. For the phase-angle graph, the factor $(s + 1)^{-1}$ is plotted and added in twice to form the overall curve. The overall curves are obtained by the graphical rules previously presented. For comparison, the overall curves obtained without derivative action [i.e., by not adding in the curves corresponding to $(0.5s + 1)$] are also shown. It should be noted, that, on the asymptotic AR diagram, the slopes of the individual curves are added to obtain the slope of the overall curve.

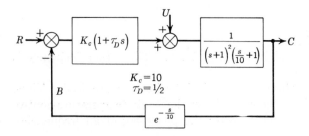

Fig. 18.11 **Block diagram of control system for Example 18.5.**

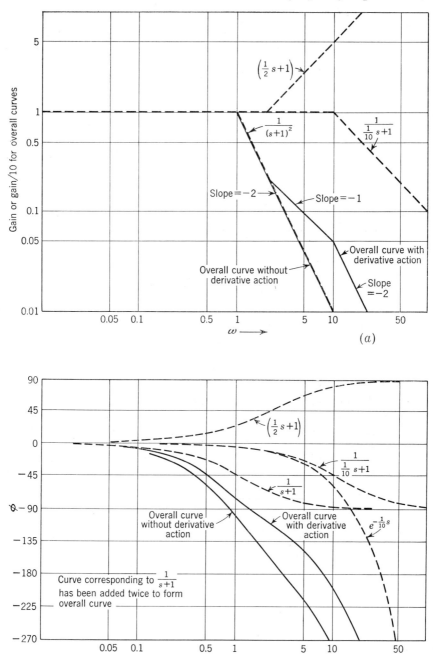

Fig. 18.12 **Bode diagram for Example 18.5.** (*a*) **Amplitude ratio;** (*b*) **phase angle.**

PROBLEMS

18.1 For each of the following transfer functions, sketch the gain versus frequency asymptotic Bode diagram.   For each case, find the actual gain and phase angle at $\omega = 10$. *Note:* It is not necessary to use log-log paper; simply rule off decades on rectangular paper.

a. $\dfrac{100}{(10s + 1)(s + 1)}$

b. $\dfrac{10s}{(s + 1)(0.1s + 1)^2}$

c. $\dfrac{s + 1}{(0.1s + 1)(10s + 1)}$

d. $\dfrac{s - 1}{(0.1s + 1)(10s + 1)}$

e. $(10s + 1)^2$

f. $(10 + s)^2$

18.2 A temperature bath in which the temperature varies sinusoidally at various frequencies is used to measure the frequency response of a temperature-measuring element B.   The apparatus is shown in Fig. P18.2.   A standard thermocouple A, for which

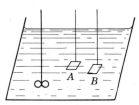

Fig. P18.2

the time constant is 0.1 min for the arrangement shown in the sketch, is placed near the element to be measured.   The response of each temperature-measuring element is recorded simultaneously on a two-channel recorder.   The phase lag between the two chart records at different frequencies is shown in the table.   From these data, show that it is reasonable to consider element B as a first-order process and calculate the time constant.   Describe your method clearly.

Frequency, cycles/min	Phase lag of B behind A, deg
0.1	7.1
0.2	12.9
0.4	21.8
0.8	28.2
1.0	29.8
1.5	26.0
2.0	23.6
3.0	18.0
4.0	14.2

18.3 Plot the asymptotic Bode diagram for the PID controller:

$$G(s) = K_c \left(1 + \tau_D s + \frac{1}{\tau_I s} \right)$$

where $K_c = 10$, $\tau_I = 1$, $\tau_D = 100$. Label corner frequencies and give slopes of asymptotes.

18.4 One way of experimentally measuring frequency response is to plot the output sine wave versus the input sine wave. The results of such a plot look like the figure shown in Fig. P18.4. This is the sinusoidal *deviation* in output versus sinusoidal *deviation* in input and appears as an ellipse centered at the origin. Show how to obtain the AR and phase lag from this plot.

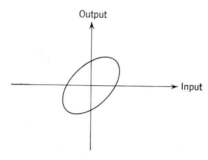

Output

Input

Fig. P18.4

19 *Control System Design by Frequency Response*

The purpose of this chapter is twofold. First, it will be indicated that the stability of a control system can usually be determined from the Bode diagram of its open-loop transfer function. Then methods will be presented for rational selection of controller parameters based on this Bode diagram. The material to be presented here is one of the more useful design aspects of the subject of frequency response.

Tank-temperature Control System It was indicated in the discussion following Example 18.3 that the control system of Fig. 19.1 might

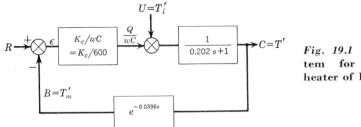

Fig. 19.1 Control system for stirred-tank heater of Example 18.3.

offer stability problems because of excessive phase lag. To review, this system represents proportional control of tank temperature with a delay in the feedback loop. The factor $1/600$ is the process sensitivity $1/wC$, which gives the ultimate change in tank temperature per unit change in heat input Q. The proportional sensitivity K_c, in Btu per hour per degree of temperature error, is to be specified by the designer.

The open-loop transfer function for this system is

$$G(s) = \frac{(K_c/600)e^{-0.0396s}}{0.202s + 1} \qquad (19.1)$$

The Bode diagram for $G(s)$ is plotted in Fig. 19.2. As usual, the constant factor $K_c/600$ is included in the definition of the ordinate for AR. At the

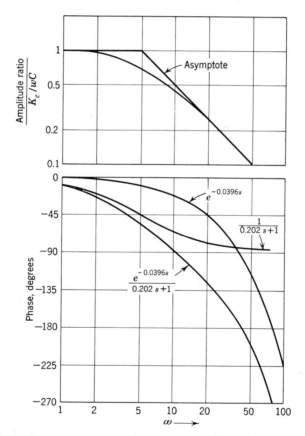

Fig. 19.2 Bode diagram for open-loop transfer function of control system for stirred-tank heater: $(K_c/wC)e^{-\tau_2 s}[1/(\tau_1 s + 1)]$. (Block diagram shown in Fig. 19.1.)

frequency of 43 rad/min, the phase lag is exactly 180° and

$$\frac{\text{AR}}{K_c/600} = 0.12$$

Therefore, if a proportional gain of 5000 Btu/(hr)(°F) is used,

$$\text{AR} = 0.12 \frac{5000}{600} = 1$$

This is the AR between the signals ϵ and B. Note that it is dimensionless as it must be, since ϵ and B both have the units of temperature.

The control system is redrawn for $K_c = 5000$ in Fig. 19.3a, with the loop opened. That is, the feedback signal B is disconnected from the comparator. It is imagined that a set point disturbance

$$R = \sin 43t$$

is applied to the opened loop. Then, since the open-loop AR and phase lag are unity and 180°,

$$B = \sin (43t - 180°) = -\sin 43t$$

Now imagine that, at some instant of time, R is set to zero and simultaneously the loop is closed. Figure 19.3b indicates that the closed loop continues to oscillate indefinitely. *This oscillation is theoretically sustained even though both R and U are zero.*

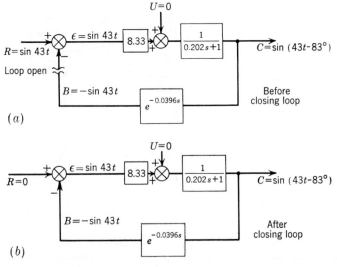

Fig. 19.3 **Sustained closed-loop oscillation.**

Now suppose K_c is set to a slightly higher value and the same experiment repeated. This time, the signal ϵ is amplified slightly each time it passes around the loop. Thus, if K_c is set to 5001, after the first time around the loop the signal ϵ becomes $(5001/5000) \sin 43t$. After the second time, it is $(5001/5000)^2 \sin 43t$, etc. The phase-angle relations are not affected by changing K_c. We thus conclude that, for $K_c > 5000$, the response is unbounded, since it oscillates with *increasing amplitude*.

Using the definition of stability presented in Chap. 14, it is concluded that the control system is unstable for $K_c > 5000$ because it exhibits an unbounded response to the bounded input described above. (The bounded input is zero in this case, for $U = R = 0$.) The condition $K_c > 5000$ corresponds to

AR > 1

for the open-loop transfer function, at the frequency 43 rad/min, where the open-loop phase lag is 180°.

This argument is not rigorous. We know the response B *only if* ϵ *remains constant in amplitude* because of the definition of frequency response. If, however, the change in K_c is very small, so that ϵ is amplified infinitesimally, then B will closely approximate the frequency response. While this does not *prove* anything, it shows that we are justified in suspecting instability and that closer investigation is warranted.

In Chap. 21, a rigorous proof is given that such control systems are unstable. This requires application of the theory of complex variables to establish the Nyquist stability criterion. For our present purposes it is sufficient to proceed with heuristic arguments.

The Bode Stability Criterion It is tempting to generalize the results of the analysis of the tank-temperature control system to:

A control system is unstable if the open-loop frequency response exhibits an AR exceeding unity at the frequency for which the phase lag is 180°. This frequency is called the *crossover frequency*. The rule is called the *Bode stability criterion*.

Actually, since the discussion of the previous section was based upon heuristic arguments, this rule is not quite general. It applies readily to systems for which the gain and phase curves decrease continuously with frequency. However, if the phase curve appears as in Fig. 19.4, the more

Fig. 19.4 **Phase behavior of complex system for which Bode criterion is not applicable.**

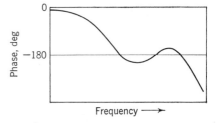

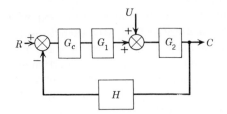

Fig. 19.5 Block diagram for general control system.

general Nyquist criterion which will be presented in Chap. 21 must usually be used to determine stability. Other exceptions may occur. Fortunately, most process control systems can be analyzed with the simple Bode criterion, and it therefore finds wide application.

Application of the criterion requires nothing more than plotting the open-loop frequency response. This may be based on the theoretical transfer function, if it is available, as we have done for the tank-temperature system. If the theoretical system dynamics are not known, the frequency response may be obtained experimentally. To do this, the open-loop system is disturbed with a sine-wave input at several frequencies. At each frequency, records of the input and output waves are compared to establish AR and phase lag. The results are plotted as a Bode diagram. This experimental technique will be illustrated in more detail in Chap. 24.

For the remainder of this chapter, we accept the Bode stability criterion as valid and use it to establish control system design procedure.

Gain and Phase Margins Let us consider the general problem of selecting $G_c(s)$ for the system of Fig. 19.5. Suppose the open-loop frequency response, when a particular controller $G_c(s)$ is tried, is as shown in the Bode diagram of Fig. 19.6. The crossover frequency, at which the phase lag is 180°, is noted as ω_{co} on the Bode diagram. At this frequency, the AR is A. If A exceeds unity, we know from the Bode criterion that the system is

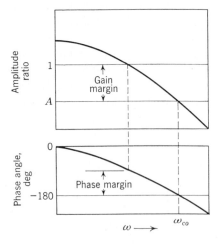

Fig. 19.6 Open-loop Bode diagram for typical control system.

unstable and that we have made a poor selection of $G_c(s)$.   In Fig. 19.6 it is assumed that A is less than unity and therefore the system is stable.

It is necessary to ascertain to what degree the system is stable.   Intuitively, if A is only slightly less than unity the system is "almost unstable" and may be expected to behave in a highly oscillatory manner even though it is theoretically stable.[1]   Furthermore, the constant A is determined by physical parameters of the system, such as time constants.   These can be only estimated and may actually change slowly with time because of wear or corrosion.   Hence, a design for which A is close to unity does not have an adequate safety factor.

To assign some quantitative measure to these considerations, the concept of gain margin is introduced.   Using the nomenclature of Fig. 19.6,

$$\text{Gain margin} = \frac{1}{A}$$

Typical specifications for design are that the gain margin should be greater than 1.7.   This means that the AR at crossover could increase by a factor of 1.7 over the design value before the system became unstable.   The design value of the gain margin is really a safety factor.   *As such, its value varies considerably with the application and designer.*   A gain margin of unity or less indicates an unstable system.   Gain margins are often given in decibels, according to the relation

$$\text{Gain margin} = 20 \log_{10} \frac{1}{A} \quad \text{db}$$

A negative gain margin (in decibels) indicates an unstable system.

Another margin frequently used for design is the phase margin.   As indicated in Fig. 19.6, it is the difference between 180° and the phase lag at the frequency for which the gain is unity.   The phase margin therefore represents the additional amount of phase lag required to destabilize the system, just as the gain margin represents the additional gain for destabilization.   *Typical design specifications are that the phase margin must be greater than* 30°.   A negative phase margin indicates an unstable system.

Example 19.1   Find a relation between relative stability and the phase margin for the control system of Fig. 19.7.   A proportional controller is to be used.

This block diagram corresponds to the stirred-tank heater system, for which the block diagram has been given in Fig. 13.6.   The particular set of constants is

$$\tau = \tau_m = 1 \qquad \frac{1}{wC} = 1$$

[1] Again, heuristic arguments are used.   This statement is self-evident to the reader who has studied Chaps. 15 to 17, where it is shown that the roots of the characteristic equation vary continuously with system parameters such as A.   Proof of the statement requires the concepts of Chap. 21.

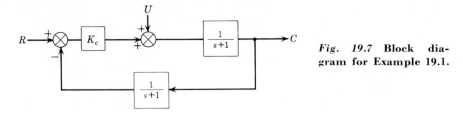

Fig. 19.7 Block diagram for Example 19.1.

These are to be regarded as fixed, while the proportional gain K_c is to be varied to give satisfactory phase margin. The *closed-loop* transfer function for this system is given by Eq. (13.17), rewritten for our particular case as

$$\frac{C}{R} = \frac{K_c}{1 + K_c} \frac{s + 1}{\tau_2{}^2 s^2 + 2\tau_2 \zeta_2 s + 1} \tag{19.2}$$

where

$$\tau_2 = \sqrt{\frac{1}{1 + K_c}}$$

$$\zeta_2 = \sqrt{\frac{1}{1 + K_c}}$$

Since the closed-loop system is second-order, it can never be *unstable*. The shape of the response of the closed-loop system to a unit step in R must resemble the curves of Fig. 8.2. The meaning of *relative stability* is illustrated by Fig. 8.2. The lower ζ_2 is made, the more oscillatory and hence the "less stable" will be the response. Therefore, a relationship between phase margin and ζ_2 will give the relation between phase margin and relative stability.

To find this relation the open-loop Bode diagram is prepared and is shown in Fig. 19.8. The simplest way to proceed from this diagram is as follows: Consider a typical frequency $\omega = 4$. If the open-loop gain were 1 at this frequency, then since the phase angle is $-152°$, the phase margin would be $28°$. To make the open-loop gain 1 at $\omega = 4$, it is required that

$$K_c = \frac{1}{0.062} = 16.1$$

Then

$$\zeta_2 = \sqrt{\frac{1}{1 + K_c}} = 0.24$$

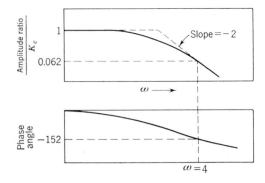

Fig. 19.8 Bode diagram for system of Example 19.1.

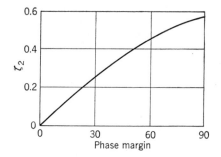

Fig. 19.9 Damping versus phase margin for system of Fig. 19.7.

Hence, a point on the curve of ζ_2 versus phase margin is

$\zeta_2 = 0.24$ phase margin $= 28°$

Other points are calculated similarly at different frequencies, and the resulting curve is shown in Fig. 19.9. From this figure it is seen that ζ_2 decreases with decreasing phase margin and that, if the phase margin is less than 30°, ζ_2 is less than 0.26. From Fig. 8.2, it can be seen that the response of this system for $\zeta_2 < 0.26$ is highly oscillatory, hence relatively unstable, compared with a response for the system with phase margin 50° and $\zeta_2 = 0.4$.

For the particular system of Example 19.1, it was shown that the response became more oscillatory as the phase margin was decreased. This result generalizes to more complex systems. Thus, the phase margin is a useful design tool for application to systems of higher complexity, where the transient response cannot be easily determined and a plot such as Fig. 19.9 cannot be made. To repeat, the rule of thumb is that the phase margin must be greater than 30°.

A similar statement can be made about the gain margin. As the gain margin is increased, the system response generally becomes less oscillatory, hence more stable. A control system designer will often try to make *both* the gain and phase margins equal to or greater than specified minimum values, typically 1.7 and 30°. Note that, for the case of Example 19.1, the gain margin is always infinite because the phase lag never quite reaches 180°. However, the phase margin requirement of 30° necessitates that $\zeta_2 > 0.26$, hence $K_c < 14$, which means that an offset of $\frac{1}{15}$ [see Eq. (19.2)] must be accepted. This illustrates the importance of considering both margins. The reader should refer to Fig. 19.6 to see that both margins exist simultaneously.

Example 19.2 Specify the proportional gain K_c for the control system of Fig. 18.11. The Bode diagram for the particular case $K_c = 10$ is presented in Fig. 18.12. The gain is to be specified for the two cases:

1. $\tau_D = 0.5$ min
2. $\tau_D = 0$ (no derivative action)

1. Consider first the gain margin. The crossover frequency for the curve with derivative action is 8.0 rad/min. At this frequency, the open-loop gain is 0.062 if the value of K_c is unity. (Including the factor of $\frac{1}{10}$ in the ordinate is actually equivalent to plotting the case $K_c = 1$.) Therefore, according to the Bode criterion, the value of K_c necessary to destabilize the loop is $1/0.062$ or 16. To achieve a gain margin of 1.7, K_c must be taken as $16/1.7$ or 9.4. To achieve proper phase margin, note that the frequency for which the phase lag is $150°$ (phase margin is $30°$) is 5.3 rad/min. At this frequency, a value for K_c of $1/0.094$ or 10.6 will cause the open-loop gain to be unity. Since this is higher than 9.4, we use 9.4 as the design value of K_c. The resulting phase margin is then $38°$.

2. Proceeding exactly as in case 1 but using the curve in Fig. 18.12 for no derivative action, it is found that $K_c = 5.3$ is needed for satisfactory gain margin and $K_c = 3.7$ for satisfactory phase margin. Hence K_c is taken as 3.7 and the resulting gain margin is 2.4.

To see the advantage of adding derivative control in this case, note from Fig. 18.11 that the final value of C for a unit-step change in U is $1/(1 + K_c)$ for any value of τ_D. The addition of the derivative action allows increase of the value of K_c from 3.7 to 9.4 while maintaining approximately the same relative stability in terms of gain and phase margins. This reduces the offset from 21 per cent of the change in U to 9.6 per cent of the change in U.

The reader is cautioned that the values of K_c selected in this way should be regarded as initial approximations to the actual values which give "optimal" control of the system of Fig. 18.11. More will be said about this matter later in this chapter in conjunction with the two-tank chemical-reactor control system of Chap. 11.

Thus far, nothing has been said about upper limits on the gain and phase margins. Referring to Example 19.1 and Fig. 8.2, it is seen that, if ζ_2 is too large, the response is sluggish. In fact, Fig. 8.2 suggests that for the system of Fig. 19.7 one should choose a value of ζ_2 low enough to give a short rise time without causing excessive response time and overshoot. In other words, one wants the most rapid response which has sufficient relative stability. The results of Example 19.1 generalize to many systems of higher complexity, in terms of margin. Hence, the designer frequently chooses the controller so that either the gain or phase margin is equal to its lowest acceptable value and the other margin is (probably) above its lowest acceptable value. This was the procedure followed in Example 19.2. In almost every situation, the designer faces this conflict between speed of response and degree of oscillation. In addition, if integral action is not used, the amount of offset must be considered.

The concepts of gain and phase margin are useful in selecting K_c for proportional action. However, for additional modes of control such as PD, these concepts are difficult to apply in practice. Consider the selection of K_c *and* τ_D in Example 19.2. For a different value of τ_D the derivative contribution is shifted to the right or left on the Bode diagram of Fig. 18.12. This means that a different value of K_c will provide the proper margins. A typical design procedure is to select the value of τ_D for which the value of K_c resulting in a $30°$ phase margin is maximized. The motivation for this choice is that the offset will be minimized. However, the procedure is clearly trial and error. In the case of three-mode control, there are two parameters, τ_I and τ_D, which must be varied by trial to meet various design criteria. Fortunately, for this case and others there are simple rules for directly establishing values of the control parameters which usually give satisfactory gain and phase margins. These are the Ziegler-Nichols[1] rules which we develop in the next section.

Ziegler-Nichols Controller Settings Consider selection of a controller G_c for the general control system of Fig. 19.5. We first plot the Bode diagram for the final control element, the process, and the measuring element in series, $G_1G_2H(j\omega)$. It should be emphasized that the controller is omitted from this plot. Suppose the diagram appears as in Fig. 19.6. As noted on the figure, the crossover frequency for these three components in series is ω_{co}. At the crossover frequency, the overall gain is A, as indicated. According to the Bode criterion, then, the gain of a proportional controller which would cause the system of Fig. 19.5 to be on the verge of instability is $1/A$. We define this quantity to be the ultimate gain K_u. Thus

$$K_u = \frac{1}{A} \tag{19.3}$$

The ultimate period P_u is defined as the period of the sustained cycling which would occur if a proportional controller with gain K_u were used. From the discussion of Fig. 19.3, we know this to be

$$P_u = \frac{2\pi}{\omega_{co}} \quad \text{time/cycle} \tag{19.3a}$$

The factor of 2π appears, so that P_u will be in units of time per cycle rather than time per radian. It should be emphasized that K_u and P_u are easily determined from the Bode diagram of Fig. 19.6.

The Ziegler-Nichols settings for controllers are determined directly from K_u and P_u according to the rules summarized in Table 19.1. Unfor-

[1] J. G. Ziegler and N. B. Nichols, Optimum Settings for Automatic Controllers, *Trans. ASME*, **64**: 759 (1942).

Table 19.1 **Ziegler-Nichols Controller Settings**

Type of control	$G_c(s)$	K_c	τ_I	τ_D
Proportional	K_c	$0.5K_u$		
Proportional-integral (PI)	$K_c\left(1 + \dfrac{1}{\tau_I s}\right)$	$0.45K_u$	$\dfrac{P_u}{1.2}$	
Proportional-integral-derivative (PID)	$K_c\left(1 + \dfrac{1}{\tau_I s} + \tau_D s\right)$	$0.6K_u$	$\dfrac{P_u}{2}$	$\dfrac{P_u}{8}$

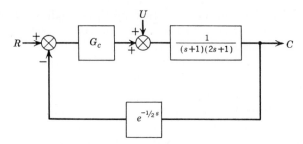

Fig. 19.10   Block diagram for two–tank chemical reactor system.

tunately, specifications of K_c and τ_D for PD control cannot be made using only K_u and P_u. In general, the values $0.6K_u$ and $P_u/8$, which correspond to the limiting case of no integral action in a three-mode controller, are too conservative. That is, the resulting system will be too stable. There exist methods for this case which are in principle no more difficult to use than the Ziegler-Nichols rules. One of these is selection of τ_D for maximum K_c at 30° phase margin, which was discussed above. Another method, which utilizes the step response and avoids trial and error, is presented in Chap. 23.

The reasoning behind the Ziegler-Nichols selection of values of K_c is relatively clear. In the case of proportional control only, a gain margin of 2 is established. The addition of integral action introduces more phase lag at all frequencies (see Fig. 18.9); hence a lower value of K_c is required to maintain roughly the same gain margin. Adding derivative action introduces phase lead. Hence, more gain may be tolerated. This was demonstrated in Example 19.2. However, by and large the Ziegler-Nichols settings are based on experience with typical processes and should be regarded as first estimates.

Example 19.3   Using the Ziegler-Nichols rules, determine controller settings for various modes of control of the two-tank chemical-reactor system of Chap. 11. The block diagram is reproduced in Fig. 19.10.

For convenience, the loop gain K_L and the controller gain K_c are combined into an overall gain K_1. The equivalent controller transfer function is regarded as

$$G_c = K_1 \left(1 + \frac{1}{\tau_I s} + \tau_D s \right)$$

where K_1 (as well as τ_I and τ_D) is to be selected by the Ziegler-Nichols rules. The required value of K_c is then easily determined as

$$K_c = \frac{K_1}{K_L}$$

where $K_L = 0.03$ for the present case (see chap. 11.)

The Bode diagram for the transfer function *without the controller*

$$\frac{e^{-(\frac{1}{2})s}}{(s+1)(2s+1)}$$

is prepared by the usual procedures and is shown in Fig. 19.11. From this figure, it is found that

$$\omega_{co} = 1.56 \text{ rad/min}$$

$$K_{1_u} = \frac{1}{0.145} = 6.9$$

$$P_u = \frac{2\pi}{1.56} = 4.0 \text{ min/cycle}$$

(19.4)

(The reader who has studied the root-locus techniques should refer for comparison to Fig. 17.15, where these constants can be determined by root locus.) Hence, the Ziegler-Nichols control constants determined from Table 19.1 and Eq. (19.4) are given in Table 19.2.

A plot comparing the open-loop frequency responses *including the controller* for the three cases, using the controller constants of Table 19.2, is given in Fig. 19.12. This figure

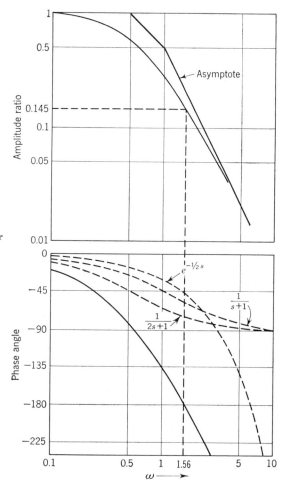

Fig. 19.11 **Bode diagram for** $e^{-\frac{1}{2}s}/[(s + 1)(2s + 1)]$.

Table 19.2

Control	K_1	τ_I	τ_D
P	3.5		
PI	3.1	3.3	
PID	4.2	2.0	0.50

shows quite clearly the effect of the phase lead due to the derivative action. The resulting gain and phase margins are listed in Table 19.3. From this table it may be seen that the margins are adequate and generally conservative.

Note that to obtain the Bode diagram for systems including the PID controller, the controller transfer function is rewritten as

$$K_c \left(1 + \frac{1}{\tau_I s} + \tau_D s\right) = K_c \frac{\tau_D \tau_I s^2 + \tau_I s + 1}{\tau_I s} \tag{19.5}$$

This is second-order in the numerator and has integral action in the denominator. In general, the numerator factors into first-order factors; hence it contributes two curves similar to that of Fig. 18.10 to the overall diagram. For the Ziegler-Nichols settings it is seen from Table 19.1 that $\tau_I = 4\tau_D$. Making this substitution into Eq. (19.5)

$$G_c = K_c \frac{4\tau_D^2 s^2 + 4\tau_D s + 1}{4\tau_D s} = \frac{K_c (2\tau_D s + 1)^2}{4\tau_D s} \tag{19.6}$$

shows that the numerator is equivalent to two PD components in series. This AR is represented by a high-frequency asymptote of slope $+2$ passing through the frequency $\omega = 1/2\tau_D$ and a low-frequency asymptote on the line AR $= 1$. It should be emphasized that these special considerations apply only to the Ziegler-Nichols settings. In the general case, the two time constants obtained by factoring the numerator of Eq. (19.5) will be different. The Bode plot of the denominator follows from

$$\frac{1}{\tau_I j\omega} = \left|\frac{1}{\omega \tau_I}\right| \angle -90°$$

The gain is a straight line of slope -1 passing through the point (AR $= 1$, $\omega = 1/\tau_I$). The phase lag is 90° at all frequencies. Plotting of the overall Bode diagram for the PID case to check the results of Fig. 19.12 is recommended as an exercise for the reader.

Table 19.3

Control	Gain margin	Phase margin
P	2.0	45°
PI	1.9	33°
PID	2.6	34°

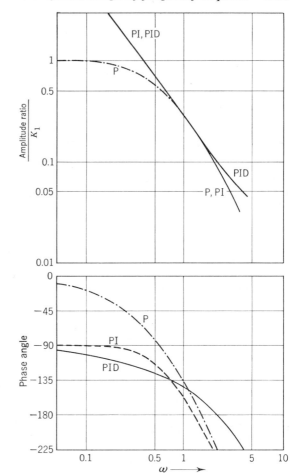

Fig. 19.12 Open-loop Bode diagrams for various controllers with system of Fig. 19.10.

Transient Responses For instructive purposes, the two-tank reactor system of Fig. 19.10 was simulated on the analog computer. Responses of $C(t)$ to a unit-step change in $R(t)$, which were obtained on the analog computer, are shown in Fig. 19.13. These responses were obtained using the Ziegler-Nichols controller settings determined in Example 19.3.

The responses to a step load change were also obtained on the analog computer. These are the curves of Fig. 10.6 which were discussed in Chap. 10 to illustrate the function of the various modes of control. A load change for this system corresponds to a change in the inlet concentration of reactant to tank 1 (refer to Fig. 11.1). As process control engineers, we would be more interested in controlling against this kind of disturbance than against a set-point change because the set point or desired product concentration is likely to remain relatively fixed. In other words, this is a regulator problem

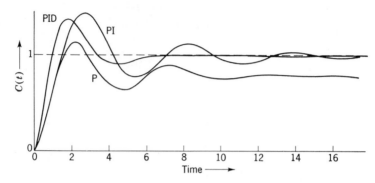

Fig. 19.13   Closed-loop responses to step change in $R(t)$ for control system of Fig. 19.10, using various control modes (obtained on analog computer).

and the curves of Fig. 10.6 are those we would use to determine the quality of control.

However, the step change in set point is frequently used to test control systems despite the fact that the system will be primarily subject to load changes during actual operation. The reason for this is the existence of well-established terminology used to describe the step response of the under-damped second-order system. This terminology, which was presented in Chap. 8, is used to assign quantitative measure to responses which are not truly second-order, such as those of Fig. 19.13. Of course, the terminology can be applied only to responses which *resemble* damped sinusoids. Values of the various parameters determined for the responses of Fig. 19.13 are summarized in Table 19.4. Offset, realized only with proportional control, is included for completeness.

It can be seen from Fig. 19.13 and Table 19.4 that addition of integral action eliminates offset at the expense of a more oscillatory response. When derivative action is also included, the response is much faster (lower rise time) and much less oscillatory (lower response time). The large over-shoots realized in all three cases are characteristic of systems with rela-tively large time delays. In this case the controller is receiving information

Table 19.4

Control	Overshoot	Decay ratio	Rise time, min	Response time, min	Period of oscillation, min	Offset
P	0.49	0.26	1.3	10.4	5.0	0.21
PI	0.46	0.29	1.5	11.8	5.5	0
PID	0.42	0.05	0.9	4.9	5.0	0

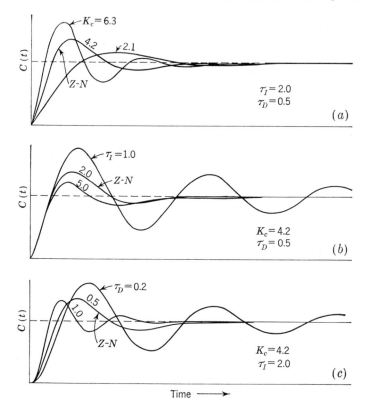

Fig. 19.14 **Effects of varying controller settings on system response.** (Z-N indicates response using Ziegler-Nichols settings.)

about the concentration in the second reactor which was true ½ min ago. This is to be compared with the reactor time constants of 1 and 2 min. Hence, it is not surprising that the system overshoots before the controller can take sufficient action.

Figure 19.14 is presented for two purposes: (1) to illustrate that the Ziegler-Nichols controller settings should be regarded as first guesses rather than fixed values and (2) to show the effects of changing the various controller settings. These figures, which were obtained on the analog computer, are transient responses to step changes in set point for the three-mode PID control. They show the effects of individually varying the three control parameters K_c, τ_I, and τ_D.

As an example of the use of these figures, suppose that it is decided that the maximum overshoot which can be tolerated is 25 percent. Figure 19.14a shows that overshoot may be reduced by decreasing K_c at the expense of a considerably more sluggish response. From Fig. 19.14b, we see that overshoot may be reduced by increasing τ_I (decreasing integral action) at a lesser expense in speed of response. Thus, for $\tau_I = 5$ min, the overshoot is reduced to 20 per cent without a serious sacrifice in speed. The overshoot cannot be significantly reduced by changing τ_D, as can be seen from Fig. 19.14c. However, the speed of response may be significantly increased by increasing the derivative action, at the expense of more oscillation before the response has settled (higher decay ratio, lower period). From this brief study of these figures, it may be concluded that, to decrease overshoot without seriously slowing the response, a combination of changes should be made. A possible combination, which should be tried, is to reduce K_c slightly and to increase τ_I and τ_D moderately. These changes would probably be tried on the actual reactor system when it is put into operation. Such adjustments from the preliminary settings are usually made by experienced control engineers, using trial procedures which are more art than science. For this reason, we leave the problem of adjustment at this point. It is recommended that, if facilities are available, the reader simulate this control system with PID control on the analog computer. Starting from the Ziegler-Nichols settings, the controllers should be adjusted until a load change response is obtained which is "optimal" to the eye. *A typical criterion for optimal load change response is minimum area under the response curve.* Only in this way can there be obtained some insight into the art of controller adjustment.

The preliminary design analysis of the previous paragraphs indicates that good control can be attained even with the presence of a ½-min delay in the feedback loop. This is evident from the PID curve of Fig. 10.6, which shows that, within 4 min after a load change occurs, the effluent concentration has returned to the desired value. Undoubtedly, this period could even be reduced by further controller adjustment. A design question which this analysis might have answered is: Is it necessary, for satisfactory control, to purchase a more costly type of composition analyzer in order to reduce the effect of the delay in the feedback loop? Probably, the answer is negative.

It should be emphasized that the analysis of the effects of changing the control parameters was carried out using the analog computer. In a particular case, this may not be feasible for any number of reasons, some of which are: (1) The system is complex and requires considerable time and effort to simulate. (2) A computer is not readily available. (3) The only system data available are the experimental open-loop frequency response rather than the theoretical transfer function. Simulation is then a trial-and-error process, seeking a computer circuit which duplicates the process frequency response. For these reasons, it is desirable to have some means of estimating the transient response without the use of a computer. We devote the next chapter to this subject.

PROBLEMS

19.1 Calculate the value of gain K_c needed to produce continuous oscillations in the control system shown in Fig. P19.1 when

    a. n is 2.
    b. n is 3.

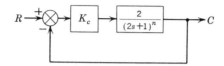

Fig. P19.1

Do not use a graph for this calculation.

  19.2 *a.* Plot the asymptotic Bode diagram $|B/\epsilon|$ versus ω for the control system shown in Fig. P19.2.

      b. The gain K_c is increased until the system oscillates continuously at a frequency of 3 rad/min. From this information, calculate the transportation lag parameter τ_d.

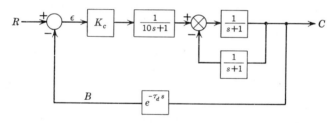

Fig. P19.2

19.3 The frequency response for the block G_p is given in the following table:

f, cycles/min	Gain	Phase angle, degr
0.06	1.60	−68
0.08	1.40	−88
0.10	1.20	−105
0.15	0.84	−145
0.20	0.61	−177
0.30	0.35	−235
0.40	0.22	
0.60	0.11	
0.80	0.066	

G_p contains a distance velocity lag $e^{-\tau s}$ with $\tau = 1$ (this transfer function is included in the data given in the table).

    a. Find the value of K_c needed to produce a phase margin of 30° for the system if $\tau_I = 0.2$.

    b. Using the value of K_c found in part **a** and using $\tau_I = 0.2$, find the percentage change in the parameter τ to cause the system to oscillate continuously with constant amplitude.

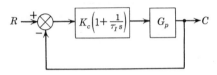

R $\xrightarrow{+}\bigotimes \xrightarrow{\;\;-\;\;}$ $\boxed{K_c\left(1+\dfrac{1}{\tau_I s}\right)}$ $\longrightarrow$ $\boxed{G_p}$ $\longrightarrow$ C

Fig. P19.3

19.4   The system shown in Fig. P19.4 is controlled by a proportional controller. The concentration of salt in the solution leaving the tank is controlled by adding a concentrated solution through a control valve.   The following data apply:

1. Concentration of concentrated salt solution $C_1 = 25$ lb salt/ft³ solution.
2. Controlled concentration $C = 0.1$ lb salt/ft³ solution.
3. Transducer: The pen on the controller moves full scale when the concentration varies from 0.08 to 0.12 lb/ft³.   This relationship is linear.   The pen moves 4.25 in during full-scale travel.
4. Control valve: The flow through the control valve varies from 0.002 to 0.006 cfm with a change of valve-top pressure from 3 to 15 psi.   This relationship is linear.
5. Distance velocity lag: It takes 1 min for the solution leaving the tank to reach the concentration-measuring element at the end of the pipe.
6. Neglect lags in the valve and transducer.

    a. Draw a block diagram of the control system.   Place in each block the appropriate transfer function.   Calculate all the constants and give the units.

    b. Using a frequency-response diagram and the Ziegler-Nichols rules, determine the settings of the controller.

    c. Using the controller settings of part **b**, calculate the offset when the set point is changed by 0.02 unit of concentration.

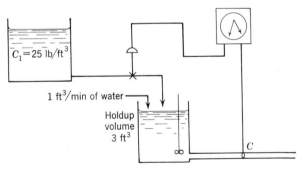

Fig. P19.4

19.5 The stirred-tank heater system shown in Fig. P19.5 is controlled by a PI controller. The following data apply:

w, flow rate of liquid through the tanks: 250 lb/min
Holdup volume of each tank: 10 ft³
Density of liquid: 50 lb/ft³
Transducer: A change of 1°F causes the controller pen to move 0.25 in.
Final control element: A change of 1 psi from the controller changes the heat input q by 400 Btu/min. The final control element is linear.

    a. Draw a block diagram of the control system. Show in detail such things as units and numerical values of the parameters.
    b. Determine the controller settings by the Ziegler-Nichols rules.
    c. If the control system is operated with *proportional mode only*, using the value of K_c found in part **b**, determine the flow rate w at which the system will be on the verge of instability and oscillate continuously. What is the frequency of this oscillation?

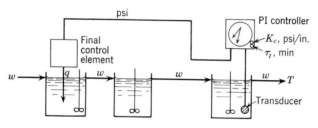

Fig. P19.5

19.6 Rework Prob. 17.2 using frequency-response methods. Compare Ziegler-Nichols settings with those obtained on the root-locus plot.

19.7 The transfer function of a process and measurement element connected in series is given by

$$\frac{e^{-0.4s}}{(2s + 1)^2}$$

    a. Sketch the open-loop Bode diagram (gain and phase) for a control system involving this process and measurement lag.
    b. Specify the gain of a proportional controller to be used in this control system.

20 Closed-loop Response by Frequency-response Methods

In order to evaluate the effects of the different modes of control and of changing controller parameters, we used an analog computer simulation of a two-tank chemical-reactor system in Chap. 19. We noted that, for several reasons, it is desirable to have means for predicting the closed-loop response without resorting to a computer simulation. The closed-loop *frequency response* of a control system can be readily obtained from the open-loop frequency response. This will be demonstrated below. However, the closed-loop transient response to a step or similar input, such as those shown in Fig. 19.13, are of more direct use in control system design. The closed-loop transient response is much more difficult to obtain than the frequency response. Hence, we shall try to find qualitative relations between step and frequency response which will provide us with approximate design guides.

Closed-loop Frequency Response Consider the unity-feedback control system shown in Fig. 20.1. The overall transfer function, from R to

C, is

$$\frac{C}{R} = \frac{G}{1 + G} \tag{20.1}$$

where $G = G_cG_1G_2$. Assume that the open-loop frequency response $G(j\omega)$ is available as a Bode diagram. It is desired to prepare the Bode diagram for the closed-loop response C/R. From Eq. (20.1)

$$\frac{C}{R}(j\omega) = \frac{G(j\omega)}{1 + G(j\omega)} \tag{20.2}$$

Let the magnitude and phase angle of G be denoted by A and ϕ, respectively. Then

$$G(j\omega) = A \angle \phi = A(\cos \phi + j \sin \phi) \tag{20.3}$$

Substituting Eq. (20.3) into Eq. (20.2) yields

$$\frac{C}{R}(j\omega) = \frac{A(\cos \phi + j \sin \phi)}{1 + A \cos \phi + jA \sin \phi} \tag{20.4}$$

Let the magnitude ratio and phase angle of C/R be M and θ, respectively. Then, from Eq. (20.4)

$$M = \frac{A}{|1 + A \cos \phi + jA \sin \phi|} \tag{20.5}$$
$$\theta = \phi - \angle(1 + A \cos \phi + jA \sin \phi)$$

After some elementary manipulations, Eq. (20.5) becomes

$$M = \frac{1}{\sqrt{[1 + (\cos \phi/A)]^2 + (\sin \phi/A)^2}} \tag{20.6}$$
$$\theta = \tan^{-1}\left[\frac{\sin \phi/A}{1 + (\cos\phi/A)}\right] \tag{20.7}$$

Since A and ϕ are available on the open-loop Bode diagram, Eqs. (20.6) and (20.7) provide a means of directly calculating the magnitude and phase angle of the closed-loop frequency response. This procedure is illustrated by the following example.

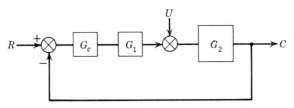

Fig. 20.1   Unity-feedback control system.

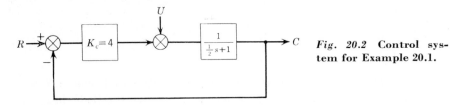

Fig. 20.2 Control system for Example 20.1.

Example 20.1   Calculate the magnitude ratio and phase angle of C/R for the control system of Fig. 20.2 when the input frequency is 2 rad/min.   The time constant of the first-order component is $\frac{1}{2}$ min.

The open-loop response can be quickly calculated.   Since the input frequency is the corner frequency of the first-order component, its magnitude ratio and phase angle are 0.707 and $-45°$, respectively.   Hence, $A = 4(0.707) = 2.828$ and $\phi = -45°$.   Substituting these values into Eqs. (20.6) and (20.7) yields

$$M = \frac{1}{\sqrt{(1 + 0.707/2.828)^2 + (-0.707/2.828)^2}} = 0.785$$

$$\theta = \tan^{-1}\left(\frac{-0.707/2.828}{1 + 0.707/2.828}\right) = -11.3° = -0.197 \text{ rad}$$

which is the desired answer.   The significance is that, if we apply a signal R of the form

$$R = \sin 2t$$

the ultimate response of C is

$$C = 0.785 \sin (2t - 11.3°)$$

In practice, it would be too cumbersome to perform the calculation of Example 20.1 at each frequency.   Fortunately, this calculation is not necessary.   The Nichols[1] chart of Fig. 20.3 is a graphical equivalent of

[1] H. M. James et al., "Theory of Servomechanisms," pp. 180–182, McGraw-Hill Book Company, New York, 1947.

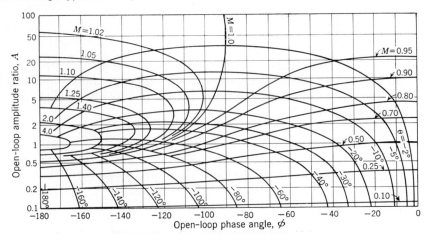

Fig. 20.3 Nichols chart.

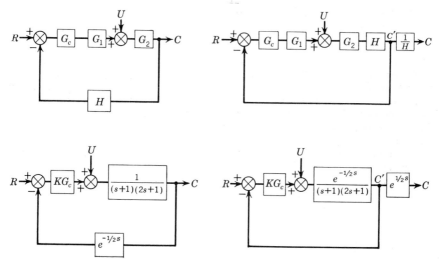

Fig. 20.4 **Manipulation of block diagram to equivalent unity-feedback system.**

Eqs. (20.6) and (20.7). These equations have been plotted as lines of constant M and θ (contours) on coordinates of A and ϕ. Thus, to do Example 20.1, one would enter the chart at the ordinate $A = 2.828$, move horizontally to the intersection with the abscissa $\phi = -45°$, and read off the values of M and θ by interpolating between the nearest contours. To prepare the closed-loop Bode diagram, this operation is repeated at each frequency, using the open-loop diagram to obtain values of A and ϕ at selected frequencies. The resulting values of M and θ are plotted against frequency to yield the closed-loop Bode diagram.

If the control system does not have unity feedback, the simplest procedure is to manipulate the block diagram so that *the closed-loop portion does have unity feedback*. Figure 20.4 is an example of this manipulation. The frequency response of the closed-loop portion C'/R is found as above. If desired, it can then be combined in series with the Bode diagram for $1/H$ to obtain C. However, since the *characteristic equations* of the two closed loops are the same, i.e.,

$$1 + G_cG_1G_2H = 0$$

the step responses of C and C' will be similar in form, differing only in the constants of the partial-fraction expansion. Hence, the frequency response of C'/R will probably give us the same qualitative information we can expect from C/R, and the inclusion of $1/H$ may be unnecessary.

Example 20.2 Prepare closed-loop Bode diagrams for the two-tank reactor control system for which step responses are shown in Fig. 19.13. The block diagram and its unity-feedback equivalent are presented in Fig. 20.4.

The Bode diagrams for the open-loop transfer function

$$\frac{KG_c e^{-0.5s}}{(s+1)(2s+1)}$$

were prepared in Chap. 19 for the three cases of proportional, PI, and PID control, using the Ziegler-Nichols controller settings. These are shown in Fig. 19.12. The closed-loop responses are prepared from these Bode diagrams using the Nichols chart and are shown in Fig. 20.5. To obtain C/R, note that we must add the factor $e^{0.5s}$ in series with the Bode diagram of Fig. 20.5. However, this does not affect the AR and serves only to add $\omega/2$ rad of phase lead at each frequency. The nature of the transient response corresponding to Fig. 20.5 is the same as that of the actual response C; it merely lags C by ½ min. We shall return to Fig. 20.5 for further discussion after the succeeding development.

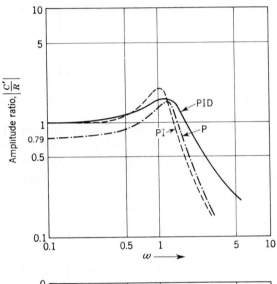

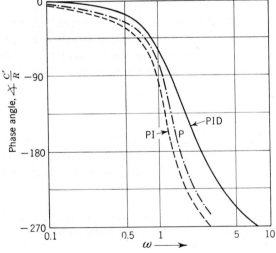

Fig. 20.5 Closed-loop Bode diagrams for control system of Fig. 20.4.

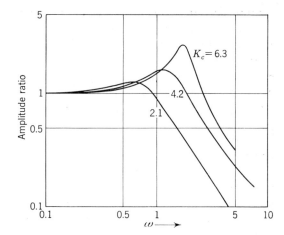

Fig. 20.6 Effect of K_c on closed-loop magnitude.

Qualitative Relations between Step and Frequency Response To develop this subject we have prepared in Fig. 20.6 closed-loop Bode diagrams (AR only) for the systems whose step responses are shown in Fig. 19.14a. Recall that these systems are the result of PID control of the two-tank reactor system. The proportional gain K_c is varied from one system to the other. A definite trend in the AR curves can be seen. To assign quantitative measure to this trend, the following quantities are defined *in conjunction with the AR curve of Fig. 20.7, which is typical of* the closed-loop responses of Fig. 20.6:

1. G_0 is the gain at zero frequency. According to the final-value theorem, the steady-state value of the response to a unit-step input is obtained by setting s to zero in the closed-loop transfer function. Hence, G_0 is this steady-state value and $(1 - G_0)$ is the offset. With integral action, G_0 is always unity (see Fig. 20.5).

2. M_p is defined as the ratio G_p/G_0, where G_p is the resonant peak

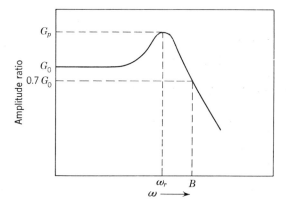

Fig. 20.7 Definitions of terms describing closed-loop response.

exhibited by the system.   In general, a higher value of M_p indicates a less damped, more oscillatory step response.   Compare Figs. 19.14a and 20.6 to verify this relation.   The basis for this generalization is the behavior of the underdamped second-order system obtained by comparison of Figs. 8.2 and 18.6.   A low value of M_p generally indicates a system which is damped and sluggish in response.

3. ω_r, the resonant frequency, is the frequency at which the resonant peak occurs.   In general, a higher value of ω_r indicates a smaller period of oscillation.   A comparison of Figs. 19.14a and 20.6 verifies this relation, which is also based on the second-order system.

4. B, the bandwidth, is the frequency at which the AR is $0.7G_0$.   Physically, it is the range of frequencies for which input signals will be attenuated by less than 30 percent.   Generalizing again, a higher value of B indicates a faster response (smaller rise time).   This is also verified by Figs. 19.4a and 20.6.

At this point, it is necessary to caution the reader that the generalizations indicated in the preceding paragraphs are just that.   There is every possibility that they may lead to erroneous conclusions in a particular case, and they should be used only as design guides and with caution.

Returning now to Fig. 20.5, we see by comparison with Fig. 19.13 and Table 19.4 that the generalizations lead to correct qualitative conclusions about the step responses in this case.   It is also seen that Fig. 20.5 correctly indicates the offset for proportional control to be 0.21.

To summarize, the generalizations can be used to predict the qualitative effect of controller settings on the responses.   Much of the preliminary design work can be done by this method, which involves use of the Nichols chart to obtain the closed-loop frequency response.   When this preliminary phase is passed, the "final" adjustments in the controller settings may be made by trial on the actual system.

Quantitative Relations between Step and Frequency Response
A possible basis for assignment of quantitative values to the predicted step response is comparison of the actual closed-loop frequency response with the frequency response of a second-order system.   It is tempting to assume that the system will exhibit a step response similar to that of the second-order system whose frequency response shows the same value of M_p and ω_r as does that of the actual system.   An illustration of this technique, together with an indication of its limitations, is provided by Example 20.3.

Example 20.3   Use the standard second-order system to estimate the response of the control system of Fig. 20.8 to a unit-step change in set point.   This is the system discussed in Chap. 14, for which the constants have been deliberately selected so that the true response can be easily evaluated.   Compare the true and estimated responses.

For review we first obtain the true step response.   The overall transfer function

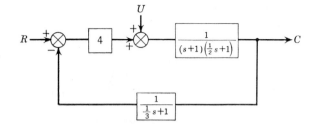

Fig. 20.8 Control system for Example 20.3.

C/R is given by

$$\frac{C}{R} = \frac{\dfrac{4}{(s+1)(s/2+1)}}{1 + \dfrac{4}{(s+1)(s/2+1)(s/3+1)}}$$

which, after some manipulation, reduces to

$$\frac{C}{R} = \frac{8(s+3)}{s^3 + 6s^2 + 11s + 30}$$

Letting $R = 1/s$ and factoring the denominator yield

$$C(s) = \frac{8(s+3)}{s(s+5)(s+\frac{1}{2}+j\sqrt{23}/2)(s+\frac{1}{2}-j\sqrt{23}/2)}$$

This may be inverted by standard partial-fraction techniques to give

$$C(t) = \frac{4}{5} + \frac{8}{65}e^{-5t} + \frac{e^{-0.5t}}{13}\left(\frac{4}{\sqrt{23}}\sin\frac{\sqrt{23}}{2}t - 12\cos\frac{\sqrt{23}}{2}t\right)$$

Reverting to decimals and using a trigonometric identity, the final expression for the true step response is

$$C(t) = 0.8 + 0.123e^{-5t} + 0.923e^{-0.5t}\sin(2.4t - 1.50) \tag{20.8}$$

It is clear that the term involving e^{-5t} becomes negligible quickly, and hence the response is approximately second-order.

To *estimate* the response by comparison of frequency response, we first form the equivalent unity-feedback system, as in Fig. 20.9. The open-loop Bode diagram for

$$\frac{1}{(s+1)(\frac{1}{2}s+1)(\frac{1}{3}s+1)}$$

is shown in Fig. 20.10 together with the closed-loop gain curve which was prepared using

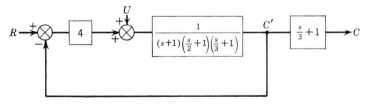

Fig. 20.9 Equivalent unity-feedback system for Fig. 20.8.

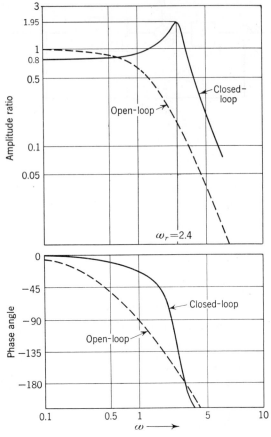

Fig. 20.10 **Bode diagrams for Example 20.3. (The open-loop AR curve is for $K = 1$.)**

the open-loop diagram and the Nichols chart. Using the definition of M_p,

$$M_p = \frac{1.95}{0.8} = 2.44$$

From Fig. 18.7, the second-order system which exhibits a response for which M_p is 2.44 has a damping coefficient ζ of 0.20. From Fig. 20.10, the resonant frequency ω_r is 2.4 rad/min. Then utilizing Eq. (18.25)

$$\omega_r = \frac{1}{\tau}\sqrt{1 - 2\zeta^2} \tag{18.25}$$

(which actually applies only to a second-order system) together with the value 0.20 for ζ, we have

$$\tau = \frac{\sqrt{1 - 2(0.20)^2}}{2.4} = 0.40$$

On this basis, the true system response to a unit-step input may be approximated by that of a second-order system with parameters

$$\zeta = 0.20$$
$$\tau = 0.40$$

Using Eq. (8.17) for these values of ζ and τ and noting that the final value (because of offset) must be 0.8 instead of unity,

$$C'(t) = 0.8[1 - 1.02e^{-0.50t} \sin{(2.45t - 4.90)}] \tag{20.9}$$

It is clear from Fig. 20.9 that

$$C(t) = \frac{1}{3}\frac{dC'(t)}{dt} + C'(t) \tag{20.10}$$

Substituting Eq. (20.9) for $C'(t)$ into Eq. (20.10) yields the final result

$$C(t) = 0.8 + 0.952e^{-0.50t} \sin{(2.45t - 0.996)} \tag{20.11}$$

Equation (20.11) for the approximate response is to be compared with the true response, Eq. (20.8). The term $0.123e^{-5t}$ in the true response becomes less than 5 percent of the final value 0.8 after 0.2 min has elapsed. Hence, it is evident from Eqs. (20.8) and (20.11) that the two responses have essentially the same damping and period of oscillation. However, the approximate response leads the true response by $\frac{1}{2}$ rad and has a greater amplitude at corresponding peaks by a factor of approximately

$$\frac{0.952e^{-0.50t}}{0.923e^{-0.50(t+0.5/2.45)}} = 1.14$$

These comparisons are depicted in Fig. 20.11.

The conclusion to be reached is that the approximation may be useful for predicting the *character* of the step response but should not be used to estimate actual values of the response. Thus, in this example, we successfully predicted the damping and period. However, we could not use Eq. (20.11) to predict accurately the value of $C(t)$ at a particular instant of time, as is evident from Fig. 20.11.

It should also be pointed out that the system of Example 20.3 was actually almost second-order. The character of the response of a system which is not close to second-order, such as the PID response of Fig. 19.13, can be only crudely approximated by that of a second-order system. However, the second-order approximation is simple, often gives reasonable results, and hence is a useful tool.

Fig. 20.11 Comparison of approximate and actual step responses for system of Fig. 20.8.

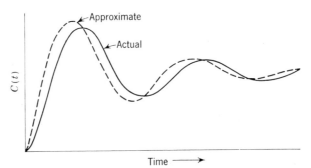

Rigorous Conversion of Step to Frequency Response   Just as a formula for the Laplace transform of a function $f(t)$ may be written,

$$f(s) = \int_0^\infty e^{-st} f(t) \, dt$$

a rigorous formula for the inverse transform of a function $f(s)$ is available. This formula is a complex line integral[1]

$$f(t) = \frac{1}{2\pi j} \int_{\sigma - j\infty}^{\sigma + j\infty} e^{st} f(s) \, ds \tag{20.12}$$

The integration is to be carried over all complex values of s lying on the line

$$\text{Re}\,(s) = \sigma$$

where σ is any real constant which guarantees that

$$\int_0^\infty |f(t)| e^{-\sigma t} < \infty$$

We have not used this formula because, for all transforms we wished to invert, it was more convenient to use partial fractions. The reader may be interested to know, however, that evaluation of the integral of Eq. (20.12) by residue theory leads directly to the partial-fraction technique.

Using Eq. (20.12), it can be shown that the response of a control loop with overall transfer function $G(s)$ to an impulse function input is

$$g(t) = \frac{1}{2\pi} \int_{-\infty}^{\infty} G(j\omega) e^{j\omega t} \, d\omega \tag{20.13}$$

The step response is calculated as the integral of the impulse response. Hence, Eq. (20.13) provides a way of calculating the step response directly from the closed-loop frequency response $G(j\omega)$.

A graphical procedure for carrying out this calculation is described and illustrated by Truxal.[2]   Suffice it to say that the calculation is tedious and not overly accurate. In fact, if the calculation is to be repeated several times, as would be the case if the effects of varying one or more control parameters are of interest, a digital computer would probably be required. Under these circumstances, it may be advantageous to simulate the system on the analog computer and obtain the response directly, as was done in Chap. 19.

Summary   To recapitulate, several methods are available for estimating the transient response (or its character) from frequency response.

[1] R. V. Churchill, "Operational Mathematics," 2d ed., McGraw-Hill Book Company, New York, 1958.

[2] J. G. Truxal, "Automatic Feedback Control System Synthesis," pp. 382–388, McGraw-Hill Book Company, New York, 1955.

A few of these have been discussed here. A more thorough discussion will be found in most servomechanism texts, such as those referred to in this chapter. The state of this art is far from satisfactory. It is easy to obtain preliminary control designs in the frequency domain but difficult to return to the time domain for improvements in the design. However, useful *approximations* to the effects of varying control parameters can be made directly in the frequency domain without undue labor.

PROBLEMS

20.1 The closed-loop Bode diagram for a control system appears as shown in Fig. P20.1 (gain only). Estimate the transfer function of this closed-loop system.

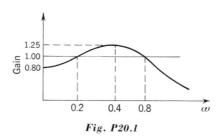

Fig. P20.1

20.2 Consider the process control systems described in Probs. 19.4 and 19.5. For these situations, plot the closed-loop Bode diagrams and use the results to characterize each loop as a second-order system. Estimate the response of the loop to a step change in set point, and compare with the results obtained:

    a. Using root-locus techniques
    b. Using the analog computer

21 *The Nyquist Stability Criterion*

We have thus far used the Bode stability criterion without formal proof. This criterion is limited to systems which have simple Bode diagrams, as was discussed in Chap. 19. Fortunately, most systems of the process industries meet this restriction.

We now develop the Nyquist stability criterion for system analysis. It will be seen that it leads rigorously to the Bode criterion. Furthermore, it may be applied to linear systems of arbitrary complexity. To develop this criterion, it is necessary to use some mathematical tools from complex variable theory. No prior knowledge of this subject is presumed. However, some readers may wish to skip this chapter depending on their particular interests. We remark that none of the later chapters of the book depend upon the material of the present chapter.

Complex Mapping The general feedback control system has an overall transfer function

$$\frac{C(s)}{R(s)} = \frac{G(s)}{1 + G(s)H(s)}$$

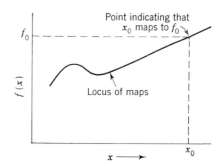

Point indicating that
x_0 maps to f_0

f_0

Locus of maps

$x \longrightarrow$

x_0

Fig. 21.1 Graph of a typical function.

where $G(s)$ and $H(s)$ are the forward and feedback transfer functions, respectively. We are interested in the roots of the denominator $1 + G(s)H(s)$. In particular, it is necessary to establish whether or not there are any roots to the right of the imaginary axis. If the answer is affirmative, the control system is unstable according to the discussion of Chap. 14.

To answer this question, it will be convenient to regard the product $G(s)H(s)$ as a *mapping* from the s plane to the GH plane. The process of mapping a function of a complex variable s is analogous to that of graphing a function of a real variable, say $f(x)$. In the latter case, for each value of x, say x_0, we calculate the value of $f(x)$, f_0. We place a point on the graph to indicate the fact that this value of x maps to this value of f. The locus of such points, for various values of x, is called the graph (or *map*) of the function $f(x)$ as illustrated in Fig. 21.1. However, if we wish to plot $G(s)H(s)$, as a function of s, this procedure must be altered, because we are interested not only in real values of s but also in complex values. For each complex number s, there is a complex value of $G(s)H(s)$. These complex numbers cannot be represented as points on a line; rather they must be represented by points on a plane. The complex numbers s and $G(s)H(s)$ are mapped on planes called the s plane and the GH plane. These are analogous to the abscissa and ordinate, respectively.

As an example, consider the function

$$GH(s) = \frac{1}{s+1}$$

where $GH(s)$ is used as an abbreviation for $G(s)H(s)$. In Fig. 21.2, the point E in the s plane is located at

$$s = -1 + j$$

Corresponding to this, the point E in the GH plane is located at

$$GH = \frac{1}{s+1} = \frac{1}{-1+j+1} = -j$$

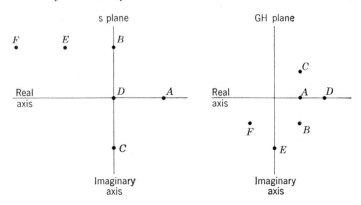

Fig. 21.2 **Mapping of typical points under** $GH(s) = 1/(s + 1)$.

In Fig. 21.2 are plotted some other typical points in the s plane. Next to this is plotted the GH plane, with the corresponding points located. This is called mapping the function $GH(s)$. In Table 21.1 are indicated the actual values of s and $GH(s)$ at the various points of Fig. 21.2.

While this is a valid way to plot a function of a complex variable, it is clear that the diagram would become very confusing if we tried to plot a large number of points. This is in contrast to the case of the function of a real variable $f(x)$ where all corresponding values of x and $f(x)$ may be indicated by a single line.

Nonetheless, this process of mapping is useful because we are usually interested in mapping only the points on some specified line in the s plane. For example, suppose it is desired to map all points on the unit circle of the s plane to the GH plane for the case $GH(s) = 1/(s + 1)$. Note that the

Table 21.1 **Actual Values for Mapping of Fig. 21.2**

Point	s	$GH(s) = \dfrac{1}{s + 1}$
A	1	$\dfrac{1}{2}$
B	j	$\dfrac{1}{2} - \dfrac{j}{2}$
C	$-j$	$\dfrac{1}{2} + \dfrac{j}{2}$
D	0	1
E	$-1 + j$	$-j$
F	$-2 + j$	$-\dfrac{1}{2} - \dfrac{j}{2}$

points A, B, and C of Fig. 21.2 are on the unit circle in the s plane. From the positions of the corresponding points in the GH plane, it might be guessed that the map of the entire circle is merely the line

$$\operatorname{Re}\left[GH(s)\right] = \tfrac{1}{2} \tag{21.1}$$

That this is actually the case can be demonstrated as follows: If we let x be the real part of s and y be the imaginary part,

$$s = x + jy$$

With this notation the points on the unit circle are characterized by

$$x^2 + y^2 = 1 \tag{21.2}$$

Substituting $s = x + jy$ into $1/(s+1)$ gives

$$GH(s) = \frac{1}{x + 1 + jy} = \frac{x + 1 - jy}{x^2 + y^2 + 2x + 1}$$

and, making use of Eq. (21.2),

$$GH(s) = \frac{1}{2} - j\,\frac{y}{2(x+1)}$$

From this it is seen that the real part of GH is $\tfrac{1}{2}$ independently of x and y for all x and y satisfying Eq. (21.2), i.e., for all points on the unit circle in the s plane. We say that the unit circle maps into the line of Eq. (21.1) under the transformation

$$GH(s) = \frac{1}{s+1} \tag{21.3}$$

As another example, the imaginary axis maps into a circle under this transformation. From Fig. 21.2 (points B, C, D), the circle has center at $GH(s) = \tfrac{1}{2}$ and radius $\tfrac{1}{2}$. The equation of the circle must then be

$$\{\operatorname{Re}\left[GH(s)\right] - \tfrac{1}{2}\}^2 + \{\operatorname{Im}\left[GH(s)\right]\}^2 = \tfrac{1}{4} \tag{21.4}$$

To prove this mapping, note that all points on the imaginary axis in the s plane can be represented by the equation

$$s = j\omega$$

where ω takes all values from minus infinity to infinity. Then, we have from Eq. (21.3)

$$GH(j\omega) = \frac{1}{1 + j\omega} = \frac{1 - j\omega}{1 + \omega^2}$$

We verify Eq. (21.4) by direct substitution:

$$\left(\frac{1}{1 + \omega^2} - \frac{1}{2}\right)^2 + \left(\frac{-\omega}{1 + \omega^2}\right)^2 = \frac{1}{(1 + \omega^2)^2} - \frac{1}{1 + \omega^2} + \frac{1}{4} + \frac{\omega^2}{(1 + \omega^2)^2}$$
$$= \tfrac{1}{4} \quad \text{Q.E.D.}$$

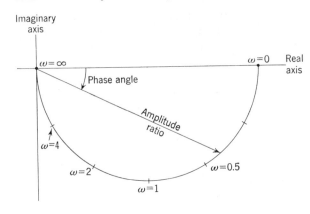

Fig. 21.3 **Polar plot of frequency response for** $GH(s) = 1/(s + 1)$.

This gives us a new way to represent frequency response. Figure 21.3 depicts the map of the positive imaginary axis under the transformation of $GH(s) = 1/(s + 1)$. Since $GH(j\omega)$ is the frequency response and $s = j\omega$ is the equation of the positive imaginary axis, the semicircle centered at $GH = \frac{1}{2}$ with radius $\frac{1}{2}$ is merely a polar plot of the frequency response of $1/(s + 1)$. At various points along the circle, we mark the values of ω. Since the AR and phase angle are the magnitude and angle of $GH(j\omega)$, they are obtained as indicated on Fig. 21.3. The reader will find it instructive to compare Figs. 21.3 and 18.4 carefully.

Nyquist's Stability Criterion These preliminary mapping concepts will now be used to discuss the following theorem:

Let $GH(s)$ be a function of the complex variable s satisfying certain mild restrictions. Let C be a closed curve in the s plane which encircles m points at which $GH(s)$ is equal to some constant GH_0. Then the map of C in the GH plane under the transformation $GH(s)$ encircles the point GH_0 m times. The only restriction on $GH(s)$ of possible significance to us is that C must not encircle any points for which $GH(s)$ becomes infinite. If it does, the number of encirclements of GH_0 will be reduced by the number of points inside C for which GH is infinite. This and all other restrictions are usually satisfied by functions of interest to us. The reader with some knowledge of complex variable theory may refer to the text by Wilts[1] for a rigorous statement and proof of this theorem.

We can immediately find use for such a theorem in our work. To find out whether or not the characteristic equation for some control system

$$1 + GH(s) = 0$$

has any roots in the right half of the complex plane, the following procedure is used: (1) Draw a curve C which encircles the right half of the s plane.

---

[1] C. H. Wilts, "Principles of Feedback Control," pp. 100–101, Addison-Wesley Publishing Company, Inc., Reading, Mass., 1960.

(2) Map C on the GH plane. (3) If the map of C encircles the point $GH = -1$, there are points inside C which are roots of the characteristic equation and the system is unstable. The number of such points is determined by the number of encirclements of the point $GH = -1$ by the map of C. (4) If there are no encirclements, the system is stable.

A closed curve encircling the *entire* right half of the s plane would be difficult to construct. However, the curve of Fig. 21.4 is suitable. The curve is composed of a semicircle of very large radius R, and the imaginary axis $s = j\omega$ for $-R < \omega < R$. It encloses enough of the right half of the s plane to determine stability, since if R is taken sufficiently large, there can be no points in the right half plane lying outside C which satisfy $GH = -1$. This is because if s becomes large (as it does on and outside the semicircle), all functions $GH(s)$ of interest to us become very small in magnitude because the denominators of these transfer functions are of higher degree in s than the numerators. [In fact, by choosing R sufficiently large we can make $GH(s)$ as close to the origin as desired for all s in the right half plane outside C.] The list of steps of the previous paragraph, when used with the curve C of the present paragraph, is referred to as Nyquist's stability criterion.

The curve C of Fig. 21.4 is not only suitable but also convenient for mapping under the transformation $GH(s)$. The portion along the positive imaginary axis is easily mapped by a polar plot of the open-loop frequency

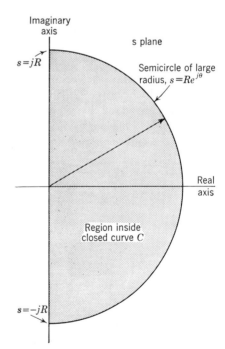

Fig. 21.4 Closed curve for Nyquist's criterion.

response $GH(j\omega)$, which can be readily prepared from the Bode diagram. The portion of C along the negative imaginary axis is simply the reflection of the polar frequency response in the real axis, because $GH(-j\omega)$ is the complex conjugate of $GH(j\omega)$.[1] The map of the large semicircle cannot affect the question of encirclement of the point $GH = -1$ because it remains very close to the origin $GH = 0$.

At this point, it is profitable to consider an example of the application of this criterion.

Example 21.1 Determine the values of K_c for which the control system of Fig. 14.1 is stable.

The open-loop frequency response is determined by

$$GH(s) = \frac{K_c}{(s+1)(\frac{1}{2}s+1)(\frac{1}{3}s+1)}$$

The Bode diagram is plotted for $K_c = 1$ in Fig. 20.10. The Nyquist plot of Fig. 21.5 has been constructed from this Bode diagram.

As can be seen, for $K_c = 1$ the -1 point is not encircled[2] and the system is stable. The Nyquist line crosses the negative real axis at $GH = 0.1$. Hence, instability will first occur if K_c is raised to $1/0.1$ or 10. This is true because raising K_c to 10 is equivalent to multiplying all magnitudes by 10, which shifts the left portion of the Nyquist curve to the dashed position on Fig. 21.5, at which the -1 point is just encircled. We conclude that stability requires that

$$K_c < 10$$

as was also found from the Routh criterion in Example 14.3.

The Nyquist criterion verifies mathematically the Bode stability criterion of Chap. 19. Proof of this statement for the class of transfer functions which have continuously decreasing AR and phase curves is left as an exercise for the reader. The construction of the Nyquist diagram is usually accomplished by first preparing the Bode diagram and then transferring the results to a polar plot.

In Example 21.1, either the Bode or Nyquist criterion could be used for determination of stability. The next example is one for which the Bode

---

[1] To see this write GH in polar form

$$GH(j\omega) = Me^{j\theta}$$

then

$$GH(-j\omega) = Me^{-j\theta} = GH^*(j\omega)$$

[2] Encirclement of the -1 point is most easily determined by counting the net number of rotations made by a vector from the -1 point to the Nyquist contour as the head of the vector traces out the entire contour. The number of encirclements, which equals the number of roots of the characteristic equation in the right half plane, is equal to the net number of rotations of this vector.

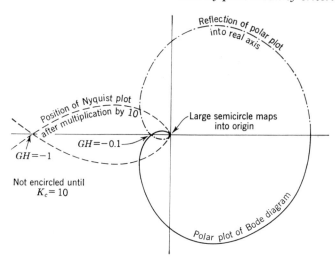

Fig. 21.5 **Nyquist diagram for Example 21.1.**

criterion cannot be used directly. It also illustrates the method for constructing the Nyquist diagram when there is an integration in $GH(s)$.

Example 21.2 A two-tank level system is to be controlled with a PID controller. The time constants of the tanks are 20 and 10 min, while that of the first-order level-measuring system is 30 sec. The integral time is 3 min, and the derivative time is 40 sec. Find the allowable values of the overall loop gain K for stability.

The open-loop transfer function is

$$GH(s) = K\left(1 + \frac{2}{3}s + \frac{1}{3s}\right)\frac{1}{(20s + 1)(10s + 1)(0.5s + 1)} \tag{21.5}$$

The gain K may be expressed as K_pK_c, where K_c is the proportional gain of the three-mode controller and K_p arises from *fixed* process constants such as valve resistances, etc. Since K_p is presumed to be known, the allowable values for K give us the allowable values for K_c.

Equation (21.5) can be rearranged to

$$GH(s) = K\frac{(2s + 1)(s + 1)}{3s(20s + 1)(10s + 1)(0.5s + 1)}$$

As a first step, the Bode diagram for this transfer function is prepared. The numerator is obtained as the reciprocal of two first-order systems, in accord with the discussion of the Bode diagram for a PD controller in Chap. 18. The Bode diagram for $3s$ in the denominator can be obtained as in Example 19.3. This factor contributes 90° of phase lag at all frequencies and has an AR curve with a slope of -1 which crosses unity at $\omega = \frac{1}{3}$ rad/min.

From this discussion, it can be seen that the Bode diagram appears as in Fig. 21.6. The phase lag is 180° at *two* frequencies, 0.096 and 1.1 rad/min. The ARs, divided by K, at these frequencies are 1.7 and 0.0028, respectively. Application of the Bode stability criterion of Chap. 19 is not possible.

The Nyquist diagram must be constructed. However, a restriction placed upon the curve C in the s plane is that it must not pass through any points for which $GH(s)$ becomes

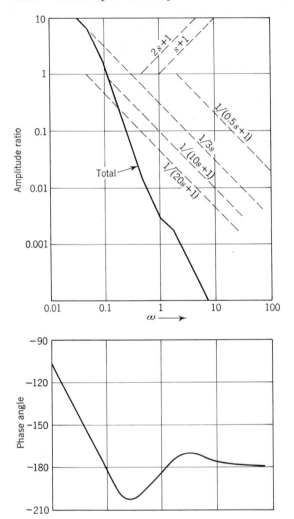

Fig. 21.6   **Bode diagram for Example 21.2.**

infinite.   The curve C of Fig. 21.4 is unsatisfactory because it passes through the point $s = 0$, where $GH(s)$ for this example is infinite.   We therefore use the curve C of Fig. 21.7. This is identical with the curve C of Fig. 21.4, with the important exception that the origin is bypassed by the semicircle of radius δ, where δ is very small.   The region of the right half plane inside the semicircle δ is excluded from curve C.   However, we know that $GH(s)$ is very large inside this semicircle and therefore cannot be equal to -1.   Hence, we have not excluded any possible roots of the characteristic equation from the inside of curve C.

The map of the small semicircle can be deduced as follows: The equation of the semicircle is

$$s = \delta e^{j\theta}$$

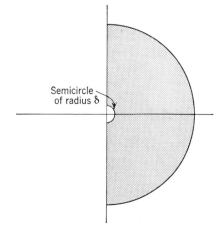

Semicircle of radius δ

Fig. 21.7 **Curve C for use with transfer functions becoming infinite at origin.**

where θ ranges between $-\pi/2$ and $\pi/2$. The function $GH(s)$ on this curve is

$$K \frac{(2\delta e^{i\theta} + 1)(\delta e^{i\theta} + 1)}{3\delta e^{i\theta}(20\delta e^{i\theta} + 1)(10\delta e^{i\theta} + 1)(0.5\delta e^{i\theta} + 1)}$$

Now if δ is very small,

$$(2\delta e^{i\theta} + 1) \cong 1$$

and similar approximations are valid for the other factors in $GH(s)$. Hence $GH(s)$ is approximated closely by

$$GH(s) = \frac{K}{3\delta e^{i\theta}} = \frac{K}{3\delta} e^{-i\theta} \tag{21.6}$$

on the semicircle. Equation (21.6) describes a semicircle of large radius with angle ranging from $\pi/2$ to $-\pi/2$ as θ ranges from $-\pi/2$ to $\pi/2$.

Making use of these facts and the Bode diagram, the Nyquist diagram, which is shown (*not to scale*) in Fig. 21.8, is prepared. From Fig. 21.8, it is seen that the -1 point is encircled for $K = 1$ and that there are two encirclements, indicating two roots in the right-half plane. This is confirmed by the root-locus diagram for this system which was presented in Fig. 15.6. Making further use of Fig. 21.8, together with the Bode diagram for actual numerical values, it is seen that, if K does not exceed $1/1.7$, or 0.59, the -1 point is not encircled and the system is stable.

Further study of the Nyquist diagram of Fig. 21.8 shows that, if K is made very large, the diagram will appear as in Fig. 21.9. Under these conditions, when K exceeds $1/0.0028$ or 360, the -1 point is again not encircled and the system is stable. Again, it must be emphasized that the diagram has not been drawn to scale. The scale has been distorted to bring out the essential behavior.

The conclusion reached is that the system is unstable for $0.59 < K < 360$ and stable for all other values of K.

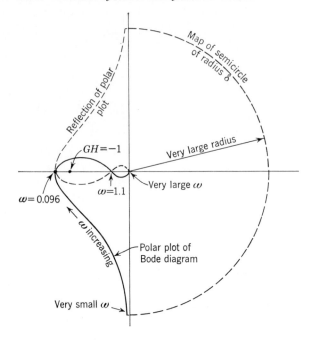

Fig. 21.8 **Nyquist plot for** $GH(s) = [K(2s + 1)(s + 1)]/[3s(20s + 1)(10s + 1)(0.5s+1)]$ **when** $K = 1.$ **(Not to scale.)**

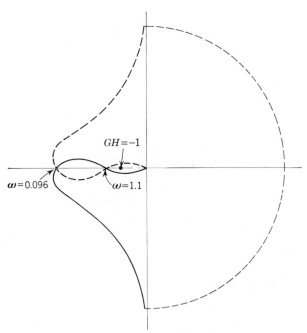

Fig. 21.9 **Nyquist plot for** $GH(s) = [K(2s+1)(s+1)]/[3s(20s+1)(10s+1)(0.5s+1)]$ **when** $K = 400.$ **(Not to scale.)**

Summary This last example points up the main advantage of Nyquist analysis over Bode analysis. That is, the actual mathematical stability test is based upon Nyquist's criterion which, in turn, rests upon complex variable theory. The Bode criterion is merely a simplification of the Nyquist criterion and is applicable to most systems of interest but may not be applicable in certain cases. Of course, the Bode diagram is always prepared first because of its graphical facility. The Nyquist plot is then used only if analysis cannot be made directly on the Bode diagram, as in Example 21.2.

All the design methods which are applicable to the Bode diagram are also applicable to the Nyquist diagram. Development of these methods, such as determination of the gain and phase margins, is left for the problems following the chapter.

PROBLEMS

21.1 For the systems described in Fig. P21.1, certain portions of the Nyquist diagram have been sketched for your convenience. Sketch the remainder of these

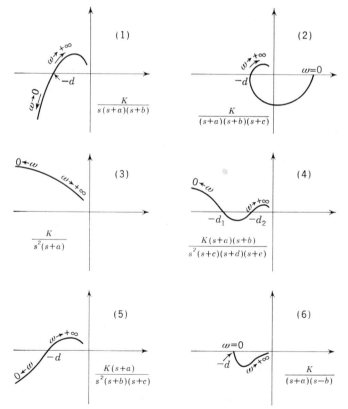

Fig. P21.1

diagrams and determine if the system is stable, unstable, or conditionally stable. State the requirement for stability for those that are conditionally stable in terms of K and d.

All constants which appear in the transfer functions (for example, a, b, c, K, etc.) are positive.

21.2   Prove that the Nyquist diagram for the open-loop transfer function:

$$GH(s) = \frac{K(\tau_a s + 1)(\tau_b s + 1)(\tau_c s + 1) \cdots}{s^k(\tau_1 s + 1)(\tau_2 s + 1)(\tau_3 s + 1) \cdots}$$

a. is asymptotic to the line

$$\mathrm{Re}\ [GH(j\omega)] = K(\tau_a + \tau_b + \tau_c + \cdots - \tau_1 - \tau_2 - \tau_3 - \cdots)$$

as ω goes to zero, for the case $k = 1$

b. is asymptotic to the parabola

$$\{\mathrm{Im}\ [GH(j\omega)]\}^2 = -K(\tau_a + \tau_b + \tau_c + \cdots - \tau_1 - \tau_2 - \tau_3 - \cdots)^2\{\mathrm{Re}\ [GH(j\omega)]\}$$

as ω goes to zero for the case $k = 2$.

21.3   Deduce an extension of the Nyquist criterion which will detect the presence of roots of the closed-loop transfer function whose real parts are greater than $-\sigma$ for a specified $\sigma > 0$.

21.4   Deduce an extension of the Nyquist criterion which will detect roots with per-unit critical damping factors less than some specified ζ. Compare the advantages and disadvantages of the results of this and the previous problem.

21.5   Show that, for a unity-feedback system, contours of constant M_p, the height of the resonant peak, appear on the Nyquist diagram as circles of radii

$$\frac{M_p}{|1 - M_p^2|}$$

with centers on the real axis at

$$\frac{M_p^2}{1 - M_p^2}$$

Sketch these circles for $M_p = 4, 2, 1.5, 1.1, 1.0, 0.9, 0.7, 0.5,$ and 0.25. Suggest a design criterion, based on Chap. 20, which uses these M_p circles.

21.6   Rework Prob. 20.2 using M_p circles on the Nyquist diagram.

21.7   Show how to calculate gain and phase margins from the Nyquist diagram.

part VI Process Applications

22 *Controller Mechanisms*

In this section of the book, Chaps. 22 to 25, we study some of the process applications of control theory. In general, these chapters can be taken in any order that satisfies the needs and interests of the reader. The material presented is highly selective from a very broad field of control applications; the emphasis is placed on applications from the process industries.

In this chapter, we shall discuss some of the mechanisms which are used in controllers to approach the ideal controller action described in Chap. 10. Historically, the pneumatic controller has been the predominant type used in the process industries, and a great variety of models is available from a large number of manufacturers. However, during the past few years, the electronic controller has found increasing application, especially in new plants.

Some of the advantages of pneumatic controllers over electronic controllers, which led to their extensive use in the process industries, are as follows:

1. The pneumatic controller is very rugged and almost free of main-

tenance.   There are no vacuum tubes or transistors to fail as in the electronic controller.

2. The pneumatic controller appears to be safer in a potentially
explosive atmosphere, which is often present in the chemical industry.

3. If, as is usually the case, the final control element is a valve, the
output pressure signal can be used directly to operate a pneumatic diaphragm control valve.   In contrast to this simplicity, the output signal
from an electronic controller (current or voltage) must be changed to pressure by means of an appropriate transducer in order to operate a pneumatic
valve.   Alternatively, it may be applied directly to a motorized valve, but
such valves are more expensive.

In recent years, most of these advantages of pneumatic controllers
have been nearly eliminated by new technology in controller design, and
now nearly all principal manufacturers of controllers offer both types.   The
electronic controller has an advantage over the pneumatic controller in
that the knob calibrations for the various controller parameters (gain,
integral time, and derivative time) are more permanent.   Also, it is easier
to obtain a desired transfer function with the electronic circuits.

A great number of improvements has occurred recently in industrial
controllers.   In the past, a controller was procured for the measurement
and control of a specific variable.   For example, if one bought a temperature
controller, the unit consisted of a controller case that housed the controller
mechanism and the recorder.   The primary measuring element was attached
more or less permanently to the controller case.   A recent trend is to furnish
the controller, the recorder, and the measuring element as separate units.
This unitized construction, which offers greater flexibility, is represented
for pneumatic systems by the block diagram of Fig. 22.1.   In the newer
system, the input signal to the controller consists of a pressure which usually
varies from 3 to 15 psig.   This means that a transducer must be available
for converting a measured variable to a pneumatic signal.   Controller
manufacturers offer a large selection of such transducers for the common
measurements of pressure, temperature, etc.   As an example of the flexibility of the unitized system, a temperature-pressure transducer of one manu-

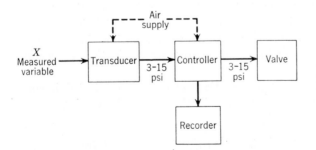

Fig. 22.1 **Pneumatic recorder-controller uni-tized assembly.**

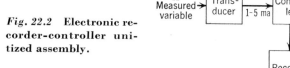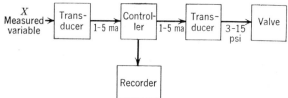

Fig. 22.2 **Electronic re-corder-controller uni-tized assembly.**

facturer can be used with the controller of another if both use the standard range of pneumatic signal of 3 to 15 psig.

The new electronic control systems are also based on the unitized concept, and a block diagram of this type of system is shown in Fig. 22.2. Notice that a transducer is needed on the output of the controller to change the current to a pneumatic signal for operating a pneumatic valve. In those cases where a valve is not the final control element, the electrical signal from the controller may be used to advantage to actuate electrically operated final control elements. For example, an electromagnetic relay may be used in an on-off, electrically heated control system. Another example of the use of an electrically operated final control element is the electromechanical servomechanism used to position the control rods in a nuclear reactor.

Space does not make it possible to examine many of the types and manufacturers of controllers and other control system hardware, such as transmitters, valves, etc. As a guide to further information about such devices, the following bibliography is offered.

Handbooks

1. Considine, D. M.: "Process Instruments and Controls Handbook," McGraw-Hill Book Company, New York, 1957. Section 9 of this book discusses pneumatic, hydraulic, and electric controllers. Considerable information is given on the mechanisms and transfer functions of pneumatic controllers.
2. Grabbe, Ramo, and Wooldridge: "Handbook of Automation, Computation and Control," vol. 3, John Wiley & Sons, Inc., New York, 1961. Chapter 7 discusses instrumentation systems, and includes under one chapter a very broad discussion of controllers, transmitters, and other control system hardware.
3. Carroll, G. C.: "Industrial Instrument Servicing Handbook," McGraw-Hill Book Company, New York, 1960. Section 11 covers the repair and calibration of a number of controllers of specific manufacturers. This book does not discuss controller mechanisms in general, and is most useful as a service manual.

Textbooks

1. Young, A. J.: "An Introduction to Process Control System Design," Longmans, Green & Co., Inc., New York, 1955. This book gives a very clear discussion of controller mechanisms and their transfer functions. The book also describes the controller mechanisms used in a number of well known controllers manufactured by companies in the United States and Europe.   Frequency-response characteristics are also given for these industrial controllers.

2. Eckman, D. P.: "Automatic Process Control," John Wiley & Sons, Inc., New York, 1958.   Chapter 6 is devoted to the mechanisms of pneumatic, hydraulic, and electric controllers.

ACTUAL VERSUS IDEAL CONTROLLERS

Before discussing the actual mechanisms used in controllers, we shall present the forms of the actual transfer functions which describe many industrial controllers and compare them with the ideal transfer functions of Chap. 10.   For the present, the actual transfer functions are presented without proof, except to say that they are verified by experiments on controllers.   Later in the chapter, these transfer functions will be justified by an analysis of the controller mechanisms.

Proportional-integral Control   The ideal transfer function for this controller was given in Chap. 10 as

$$G_c = K_c\left(1 + \frac{1}{\tau_I s}\right) \tag{22.1}$$

The Bode diagram for this transfer function was given in Fig. 18.9, from which we see that below the corner frequency $1/\tau_I$ the controller gain increases as the frequency decreases, and approaches infinity at $\omega = 0$. An actual controller cannot produce an infinite gain, and from an analysis of the controller mechanism and from frequency-response tests on an actual controller, it is found that a more realistic transfer function for a typical PI controller is given by the equation

$$G_c = K_c\frac{1 + 1/\tau_I s}{1 + 1/\alpha\tau_I s} \tag{22.2}$$

For many controllers α is very large and the transfer function reduces to the ideal transfer function.   The asymptotic Bode diagram for gain corresponding to Eq. (22.2) is shown in Fig. 22.3.   A typical phase-angle

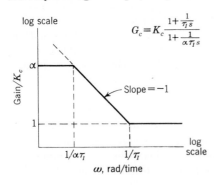

Fig. 22.3 Asymptotic Bode diagram for gain of an actual PI controller.

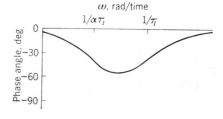

Fig. 22.4 Typical phase-angle plot for an actual PI controller.

frequency curve corresponding to Eq. (22.2) is shown in Fig. 22.4. Notice that the gain is limited at low frequency and the phase angle does not approach $-90°$ at low frequency as for the ideal controller. Actually, Eq. (22.2) does not describe the transfer function at higher frequency, because the gain decreases at high frequency. Usually the frequency at which the gain drops is much higher than those encountered in industrial processes, and we shall not modify Eq. (22.2) to account for the high-frequency drop. [If it is essential to account for the high-frequency behavior, one can often introduce a term $(1 + \tau_h s)$ in the denominator of Eq. (22.2) as a means of improving the accuracy of the transfer function. The parameter τ_h can be found experimentally.]

Proportional-derivative Control   The ideal transfer function for the PD controller, which was given by Eq. (10.8), is

$$G_c = K_c(1 + \tau_D s) \tag{22.3}$$

The Bode diagram for this transfer function was given in Fig. 18.10. An actual PD controller can be approximated by

$$G_c = K_c \frac{1 + \tau_D s}{1 + (\tau_D/\gamma)s} \tag{22.4}$$

For many controllers γ is very large and Eq. (22.4) reduces to the ideal transfer function. The asymptotic Bode diagram for gain corresponding to Eq. (22.4) is shown in Fig. 22.5, and a typical phase-angle frequency

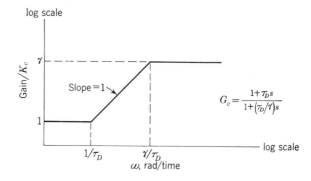

Fig. 22.5 Asymptotic Bode diagram for gain of an actual PD controller.

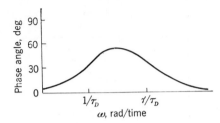

Fig. 22.6   Typical phase-angle plot for
an actual PD controller.

curve is shown in Fig. 22.6.   Notice that the gain is limited at high frequency and that the phase angle does not approach 90° at high frequency,
as for the ideal controller, but returns to zero.

Proportional-integral-derivative Control   The ideal transfer function[1] for this controller was given in Chap. 10 as

$$G_c = K_c \left(1 + \tau_D s + \frac{1}{\tau_I s} \right) \tag{22.5}$$

The asymptotic Bode diagram for this controller is shown in Fig. 22.7.

Many industrial controllers are more realistically represented by the
transfer function

$$G_c = K_c \frac{1 + \tau_D s + 1/\tau_I s}{1 + (\tau_D/\gamma)s + 1/\alpha \tau_I s} \tag{22.7}$$

---

[1] Some control engineers prefer to let the standard form of a three-mode controller be

$$G_c = K_c \left(1 + \frac{1}{\tau_I s} \right) (1 + \tau_D s) \tag{22.6}$$

The basis for this last form is that some industrial controllers are actually represented by
this equation in which the parameters K_c, τ_I, and τ_D correspond directly to knob settings.
A controller obeying the last form is often called a *cascade* controller because the three-
mode action is accomplished by connecting in series a PI controller and a PD controller.
If the right-hand side of the cascade form [Eq. (22.6)] is multiplied out, we can get the
same form as Eq. (22.5); thus

$$G_c = K_c \left(1 + \frac{\tau_D}{\tau_I} \right) \left[1 + \frac{\tau_D s}{1 + \tau_D/\tau_I} + \frac{1}{\tau_I(1 + \tau_D/\tau_I)s} \right]$$

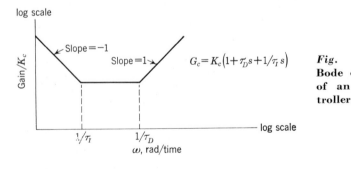

log scale

Slope $= -1$

Slope $= 1$

$G_c = K_c \left(1 + \tau_D s + 1/\tau_I s \right)$

Gain/K_c

$1/\tau_I$ $1/\tau_D$

log scale

ω, rad/time

*Fig.   22.7 Asymptotic
Bode diagram for gain
of an ideal PID con-
troller.*

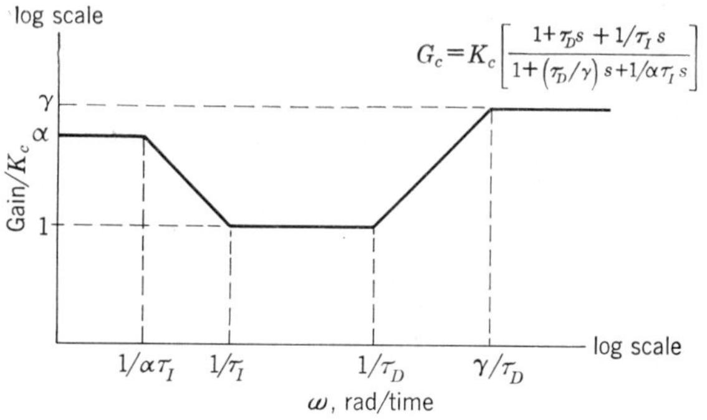

log scale

$$G_c = K_c \left[\dfrac{1 + \tau_D s + 1/\tau_I\, s}{1 + \left(\tau_D/\gamma \right) s + 1/\alpha \tau_I\, s} \right]$$

Fig. 22.8 Asymptotic Bode diagram for gain of an actual PID controller.

Gain/K_c

γ
α

1

$1/\alpha\tau_I$ $1/\tau_I$ $1/\tau_D$ γ/τ_D

log scale

ω, rad/time

If α and γ are very large, Eq. (22.7) reduces to the expression for an ideal controller given by Eq. (22.5). When $\tau_D = 0$, the transfer function reduces to Eq. (22.2). When $\tau_I = \infty$ (or reset rate = 0), the transfer function reduces to Eq. (22.4). The Bode diagram of Eq. (22.7) is shown in Fig. 22.8.

In this section, the transfer function of an ideal controller has been compared with that of an actual controller as determined by experiment. The main difference between the transfer functions for a particular mode of control is the limited gain at high or low frequencies for the actual controller. In the following sections, some of the pneumatic mechanisms which provide the various modes of control will be described and analyzed to determine their transfer functions.

PNEUMATIC CONTROLLER MECHANISMS

In this section, we shall describe the motion-balance type of pneumatic mechanism which is commonly used in controllers.

Baffle Nozzle   The basic element in this mechanism is the baffle-nozzle system shown in Fig. 22.9. In an industrial controller, the nozzle is usually supplied with air at 20 psig. When the baffle is against the nozzle ($\delta = 0$), the nozzle pressure p is at supply pressure. As the baffle is pulled away from the nozzle, the pressure in the nozzle decreases. A typical curve relating nozzle pressure to baffle-nozzle displacement is shown in Fig. 22.10. Notice that the pressure change is nearly complete when the baffle has moved only a few thousandths of an inch. This means that the

Fig. 22.9 Baffle-nozzle system.

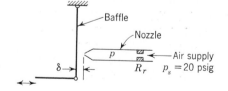

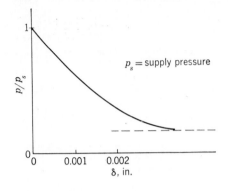

Fig. 22.10 **Baffle-nozzle character-istics.**

gain of the baffle, $m = |dp/d\delta|$, is very high.   Also notice that the curve is quite linear over a considerable range of pressure.   In order that p change over a suitable range, the restriction R_r in Fig. 22.9 is made smaller than that of the nozzle.   If the resistance R_r were zero (no restriction), the pressure p in the nozzle would not change as the baffle is pulled away from the nozzle.   When the baffle is far from the nozzle, the pressure in the nozzle will depend on the resistance R_r and the resistance of the unobstructed nozzle orifice.   We shall now show how this simple baffle-nozzle system is used to produce a desired control action.

Proportional Controller   Mechanism: The baffle-nozzle system of Fig. 22.9 is much too sensitive to be used without modification for producing proportional control.   A mechanism which reduces this sensitivity is shown in Fig. 22.11.   Notice that a spring-loaded bellows has been added to the baffle nozzle.   This bellows is often called the feedback bellows.   The right end of the bellows is fixed, and the left end is free to move horizontally as the pressure in the bellows changes.   The upper end of the baffle, which is hinged to the support attached to the bellows, moves horizontally as the bellows expands or contracts.   The pressure in the bellows is assumed to be equal to that in the nozzle at any instant.   This assumption is valid at lower frequencies, for which case there is sufficient time for the pressure to equalize throughout the nozzle and bellows.   When the mechanism is used in a controller, the lower end of the baffle is moved by an amount which is proportional to the error ϵ; that is, $x = C\epsilon$ where C is an amplification factor for the mechanism.   A simple mechanism for moving the baffle as a result of change in either set point or measured variable is also shown in Fig. 22.11.

Before writing any equations for the analysis of this mechanism, we shall first show that the sensitivity of the baffle nozzle has been reduced. Consider the system to be at steady-state conditions.   If a step change in error is introduced, the lower end of the baffle moves to the right by an amount $\Delta x = C \, \Delta\epsilon$.   This causes the displacement between baffle and nozzle to decrease with the result that the pressure in the nozzle increases by an amount Δp according to the baffle-nozzle characteristic curve of Fig. 22.10.

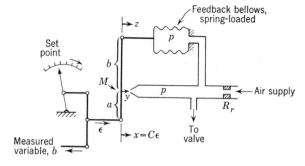

Fig. 22.11 **Proportional controller mechanism.**

However, the increase in pressure in the nozzle is transmitted to the bellows, causing the bellows to expand and to move the upper end of the baffle to the left. This motion to the left is opposite to that of the lower end of the baffle; hence, the action of the bellows is to undo, in part, the original motion caused by the step change in error. As a result of this feedback effect, the change in p is much less than would be the case if the upper end of the baffle were fixed. All this motion, which has been described sequentially, actually occurs simultaneously. Since a verbal description of such a mechanism is not precise, we shall analyze the mechanism mathematically and show that it actually does produce proportional action.

Analysis: Consider the mechanism to be at steady state with the nozzle pressure at some reference value p_0. Under these steady-state conditions, we shall take arbitrarily the positions of the ends of the baffle to be $x = 0$, $z = 0$, and the position of the baffle opposite the nozzle (point M in Fig. 22.11) to be $y = 0$. This, of course, does not mean that the baffle displacement is zero; the displacement will simply be that value required to produce p_0 in the nozzle.

If the baffle nozzle operates in the linear range of its characteristic curve, we have the relationship

$$p = p_0 + my \tag{22.8}$$

where m is the gain of the baffle nozzle, which is equal to the magnitude of the slope of the linear portion of the curve of Fig. 22.10. Notice that as y increases (the baffle moves to the right), the displacement between the baffle and nozzle decreases, causing p to increase.

A second relationship can be written from Fig. 22.11 which relates y to the positions of the ends of the baffle; thus

$$y = \frac{b}{a + b} x + \frac{a}{a + b} z \tag{22.9}$$

Replacing x by $C\epsilon$ gives

$$y = \frac{b}{a + b} C\epsilon + \frac{a}{a + b} z \tag{22.10}$$

The motion involved in an actual mechanism is so slight that the baffle remains nearly vertical throughout its range of motion. Furthermore, the baffle-nozzle displacement is so slight over the range of operation that the baffle is essentially pivoted at point M.

A third relation, which is based on a force balance for the bellows, is

$$Kz = -A(p - p_0) \qquad (22.11)$$

where A is the surface area of the bellows on which the pressure p acts and K is the spring constant of the bellows. Notice that the negative sign is needed in Eq. (22.11) because the upper end of the baffle moves to the left (negative direction) as p increases.

Equations (22.8), (22.10), and (22.11) can now be combined to find p as a function of ϵ. For convenience, define the deviation variable P as

$$P = p - p_0$$

Combining Eqs. (22.8) and (22.10) to eliminate y gives

$$P = p - p_0 = \frac{mb}{a + b} C\epsilon + \frac{ma}{a + b} z \qquad (22.12)$$

Combining the last equation with Eq. (22.11) to eliminate z gives

$$P = \frac{mb}{a + b} C\epsilon - \frac{ma}{a + b} \frac{A}{K} P$$

Rearranging this last equation gives

$$\frac{P}{\epsilon} = \frac{Cmb/(a + b)}{1 + [maA/(a + b)K]} \qquad (22.13)$$

If $maA/(a + b)K \gg 1$, Eq. (22.13) can be approximated by

$$\frac{P}{\epsilon} = C \frac{b}{a} \frac{K}{A} \qquad (22.14)$$

We see that either form of the result, Eq. (22.13) or (22.14), does have the form of the transfer function for an ideal proportional controller.

It is instructive to represent the action of the pneumatic mechanism of Fig. 22.11 by means of a block diagram, from which we can readily see the negative feedback. Such a block diagram is drawn directly from Eqs. (22.8), (22.10), and (22.11) as shown in Fig. 22.12a. The diagram shown in Fig. 22.12a may be reduced to that shown in Fig. 22.12b. From this diagram, it is apparent that there is negative feedback of the pressure signal P.

It should be clear now that the baffle nozzle acts as a high-gain amplifier in a negative feedback circuit. This means that any nonlinearity in the

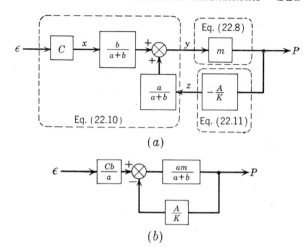

Fig. 22.12 Block diagram for proportional controller mechanism.

(a)

(b)

baffle nozzle will have very little effect on the linearity of the overall feedback system; i.e., the characteristics of the proportional control mechanism are essentially independent of the gain m. This is evidenced by Eq. (22.14), which is valid only for high $maA/[(a + b)K]$. Equation (22.14) follows by writing the closed-loop transfer function for the block diagram of Fig. 22.12b, to yield Eq. (22.13), which then gives Eq. (22.14) for high $maA/[(a + b)K]$.

Proportional-integral Control   To obtain PI control, another bellows and a variable resistance R are added to the mechanism of Fig. 22.11. This new arrangement is shown in Fig. 22.13. We shall explain how this system works by considering what happens when a step change in error is introduced into the controller. From the discussion of the ideal PI controller of Chap. 10, we know that a positive step input should produce a sudden rise in output pressure followed by a linearly increasing pressure (see Fig. 10.4).

In the mechanism of Fig. 22.13, assume that steady-state conditions prevail, for which case the pressures in the two bellows are equal; that is, $p = p_b$. At time $t = 0$, the lower end of the baffle is moved to the right by an amount Δx as a result of a step change in error $\Delta\epsilon$. Immediately

Fig. 22.13 Proportional-integral controller mechanism.

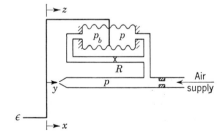

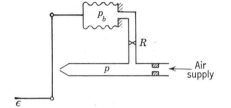

Fig. 22.14 **Proportional-derivative controller mechanism.**

following this change, the controller will respond as the proportional controller which was described earlier. Because of the resistance R, the pressure in the left-hand bellows will not change immediately after the input disturbance. However, as time proceeds, air will leak through the resistance R as a result of the difference in pressure $p - p_b$. The rising pressure in the left-hand bellows will force the upper end of the baffle to the right, thereby decreasing the baffle-nozzle displacement, which in turn will cause p to increase with time. With this argument, we have shown that the mechanism of Fig. 22.13 produces a response to a step input which agrees qualitatively with the desired response for an ideal PI controller. The transfer function for this mechanism will be derived after the controller mechanisms for PD and PID controllers have been described.

Proportional-derivative Control To obtain PD control from the mechanism of Fig. 22.11, a variable resistance is introduced between the nozzle and the feedback bellows as shown in Fig. 22.14. To explain how this mechanism operates, consider what happens when the error changes linearly with time; i.e., the lower end of the baffle moves to the right at a constant rate. From Chap. 10, we know that the response of an ideal PD controller to this input consists of a sudden rise in pressure followed by a linear change in pressure with time (see Fig. 10.5). Without any resistance in the line to the bellows, p would increase linearly with time as required by proportional control. However, with a resistance present, the pressure in the bellows p_b lags behind p, with the result that, at any instant after introducing the disturbance, the baffle-nozzle displacement is less than for the case of no resistance in the feedback line. Therefore, p will be greater when the resistance is present. Furthermore, the greater the resistance, the more the delay in the pressure reaching the bellows and the higher the pressure signal p. We see from this rather crude argument that the mechanism will produce a response which resembles that desired for PD action. A mathematical derivation of this result is given later in the chapter.

Proportional-integral-derivative Control As the reader may anticipate, the mechanism for a three-mode controller is obtained by combining the mechanisms of Figs. 22.13 and 22.14 to produce the mechanism shown in Fig. 22.15. The configuration shown in Fig. 22.15 is called a parallel arrangement. Rather than attempt a verbal explanation of how

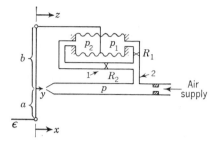

Fig. 22.15 **Proportional-integral-derivative controller mechanism—parallel arrangement.** R_1 **is located at position 2 for a series arrangement.**

this mechanism functions, we shall analyze it later in order to obtain the transfer function. The results of this analysis will also be used to obtain the transfer functions for the mechanisms for PI and PD control, Figs. 22.13 and 22.14.

Relay Before examining an actual industrial controller, we must describe the relay, which has been neglected up to this point for the sake of clarity. The pressure signal from the controller mechanisms just shown cannot be used directly to operate a large control valve. If the nozzle of Fig. 22.11 were to feed a valve directly, the valve would operate very sluggishly. This can be understood by considering what happens when the lower end of the baffle is moved to the right by a fixed amount. Without a valve attached to the nozzle, the pressure will rise suddenly to a higher value. If a valve is connected to the nozzle, the pressure will increase slowly because the air which flows into the space above the diaphragm must pass through the very small restriction R_r in the nozzle. To avoid this loading of the pneumatic mechanism by the valve, a relay is used. A common relay is shown in Fig. 22.16 in conjunction with a proportional controller mechanism.

The relay consists of a bellows A of small volume and a specially designed valve B. Since the bellows A has small volume, its pressure closely follows the nozzle pressure. As the pressure in the bellows A increases, the bellows expands and causes the ball in the cavity of the valve B to lower. The lowering of the ball decreases the leak to the atmosphere, and the pressure

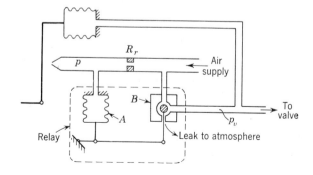

Fig. 22.16 **Proportional controller with relay (direct-action).**

p_v rises.   The relay is made in such a way that the output pressure p_v is proportional to the signal pressure p.   The flow rate of air from the relay can be much greater than the flow rate through the nozzle because the relay is supplied with air from upstream of the resistance R_r.   Hence, the valve responds very quickly to changes in nozzle pressure.   A relay valve can be direct-acting, in which case the output pressure is directly proportional to input pressure, or it can be inverse-acting, in which case the output pressure is inversely proportional to input pressure.   The relay shown in Fig. 22.16 is direct-acting.   If the relay shown in Fig. 22.16 has a gain of 1, the controller mechanism operates in the same manner as that of Fig. 22.11.

EXAMPLE OF AN INDUSTRIAL PNEUMATIC CONTROLLER

In order to illustrate how the components of a controller mechanism are arranged in a typical industrial controller, we shall examine the Taylor Fulscope proportional controller.   After studying this particular controller, the reader should be able to understand the operation of controllers of other types and manufacturers.   Most manufacturers publish literature which describes the operation of their controllers.

An overall photograph of the Taylor Fulscope controller is shown in Fig. 22.17.   The set point can be adjusted by the small knob near the top of the case.   The gain is adjusted by the calibrated knob.   The gain knob is calibrated in terms of pounds per square inch per inch of pen travel. The inches of pen travel refer to the error, or displacement, between the set pointer and the pen, measured along the arc traced by pen and pointer. In this particular controller, the pen moves about 4.25 in. as it traverses the chart, and the sensitivity or gain can be varied from 1 to 1,000 psi/in.

The diagram of the controller mechanism is shown in Fig. 22.18. The novel feature of this mechanism is that the feedback is accomplished by motion of the nozzle.   In the simple diagram of Fig. 22.11, the feedback was caused by motion of the upper end of the baffle.

To explain the action of this controller we shall show first how the baffle is moved by changes in set point and measured variable.   An enlarged view of the baffle is shown in Fig. 22.19.   As shown by the insert of Fig. 22.19, the baffle A is circular in the region where the nozzle discharge is directed.   The baffle is pivoted at M on disk C.   The left end of the baffle rests on yoke B at point N.   Disk C can be rotated about O by means of set-point gear D.   For the moment, consider the set point to be fixed, which means that disk C is fixed and that the right end of baffle A pivots about a fixed position M.   If the measured variable increases, link E is pulled down, which causes the yoke B, which pivots about O, to rotate clockwise and thereby lift the baffle which is resting on point N of the yoke.   Note that this rotation also moves the pen to indicate the change in measured variable.

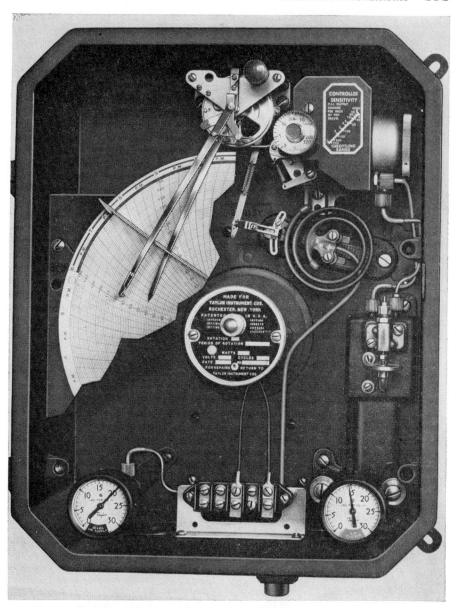

Fig. 22.17 A typical pneumatic recorder controller, the Taylor Fulscope controller.   (Courtesy of Taylor Instrument Companies.)

To understand how the baffle moves when the set point changes, consider the measured variable to remain fixed; this means that the left end of the baffle is pivoted at a fixed point N.   If the set-point gear D rotates clockwise, disk C rotates counterclockwise.   The motion of disk C moves pivot M upward, and the baffle is lifted away from the nozzle.

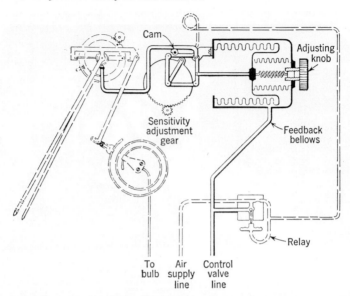

Cam

Adjusting knob

Sensitivity
adjustment
gear

Feedback
bellows

Relay

To   Air   Control
bulb   supply   valve
        line   line

Fig. 22.18   Controller mechanism.   (Courtesy of Taylor Instrument Companies.)

We shall now describe how the nozzle is moved to follow the motion of the baffle.   An enlarged view of the sensitivity adjustment mechanism is shown in Fig. 22.20.   Disk *F* is rotated by means of the sensitivity adjustment gear.   Five links pivoted at *o, y, x, w,* and *q* are used to provide nozzle motion in the following manner: Pivot *o* is fixed to the instrument case.   Pivot *q* is attached to disk *F* and rotates with the disk.   A push rod, actuated by the feedback bellows, pushes against link *oy*.   The nozzle rests on a cam which is firmly attached to link *wo*.   Let us now follow the action which occurs if the pressure in the bellows increases.   Let pivot *q*

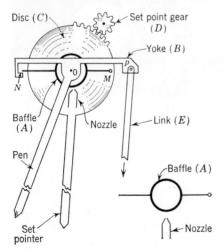

Disc (*C*)

Set point gear
(*D*)

Yoke (*B*)

N

O   *M*

Baffle
(*A*)

Nozzle

Link (*E*)

Pen

Baffle (*A*)

Set
pointer

Nozzle

Fig. 22.19   Baffle-nozzle arrangement in Taylor Fulscope controller.   (The insert in lower right-hand corner shows more clearly the baffle-nozzle system.)

be at the position shown in Fig. 22.20*b*.   For convenience, let the four links initially take the shape of the square shown by the solid lines in Fig. 22.20*b*. An increase in bellows pressure causes the push rod to move to the left with the result that the four links take the shape shown by the dashed lines. This action causes the nozzle to be lifted upward, since it rests on the cam attached to link *ow*.   The sensitivity is changed by rotating disk *F* to move pivot *q*.   Thus, to consider an extreme case, if the pivot *q* is placed directly behind pivot *w* when the four links form a square, the result is as shown in Fig. 22.20*c*.   It can be seen from the dashed lines that the motion of the push rod does not change the position of link *wo*.   This means that the nozzle does not move and there is no feedback; hence, the gain of the controller is very high.   From this discussion, it should be clear that the gain (or sensitivity) of the controller is lowered by rotating disk *F* clockwise.

The relay shown in Fig. 22.18 is an inverse acting relay with a gain of about 10:1.   Its operation was described in Fig. 22.16.

Having described the features of the controller, we shall now describe its overall action to a step change in measured variable.   Let the measured variable suddenly increase, causing link *E* (Fig. 22.19) to be pulled down and baffle *A* to rise.   The increase in baffle-nozzle displacement causes the pressure in the nozzle to fall, which in turn causes the pressure from the inverse-acting relay to rise by a proportional amount.   The pressure rise from the relay goes to the valve and the feedback bellows (see Fig. 22.18), and the expansion of the bellows pushes link *oy* (Fig. 22.20) to the left and causes the nozzle to rise and follow the motion of the baffle.

The basic mechanism which has just been described for proportional control is also used in two- and three-mode controllers.   Another bellows is placed to the left of the one shown in Fig. 22.18, and variable resistances

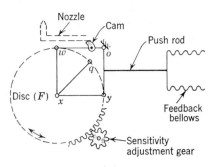

Fig. 22.20   Adjustment of sensitivity in Taylor Fulscope controller.

(*a*)

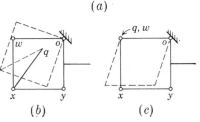

(*b*)   (*c*)

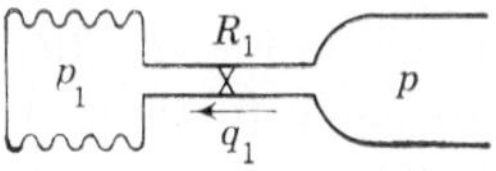

Fig. 22.21   Bellows.

are added as shown in the mechanisms of Figs. 22.13 to 22.15. These details will not be considered here, but the interested reader can consult the Taylor literature[1] for the instrument diagrams for the two- and three-mode controllers.

In the next section, the three-mode controller mechanism of Fig. 22.15 will be analyzed for the purpose of obtaining its transfer function. The result will also be used to obtain transfer functions for the PI and PD controllers shown in Figs. 22.13 and 22.14.

Transfer Function   Consider the mechanism shown in Fig. 22.15 to be at steady state with the pressure in both bellows and the nozzle equal to some reference pressure p_0; furthermore, the values of x, y, and z are taken as zero at steady state. The two relationships used in the analysis of a proportional controller [Eqs. (22.8) and (22.10)] also apply here and are given again for convenience.

$$p = p_0 + my \qquad (22.8)$$

$$y = \frac{b}{a + b} C_\epsilon + \frac{a}{a + b} z \qquad (22.10)$$

The position of the upper end of the baffle will be related to the pressures in the bellows by the expression

$$Kz = -A(p_1 - p_0) + A(p_2 - p_0) \qquad (22.15)$$

where A is the surface area common to both bellows against which the pressures act and K is the spring constant for the pair of bellows.

At this stage of the derivation, there are three equations and six variables (ϵ, y, z, p, p_1, p_2), and to solve for p as a function of ϵ requires that two more equations be written. These equations are obtained from a mass balance around each bellows as follows:

Consider only the right-hand bellows as a separate system, as shown in Fig. 22.21. The procedure will be to write a differential mass balance around this bellows so that we can obtain a differential equation involving p and p_1. Assume that the volume of the bellows is constant[2] at V. Also assume that the temperature of the system remains constant at T.

A mass balance around the bellows shown in Fig. 22.21 gives

$$q_1 = \frac{dM}{dt} \qquad (22.16)$$

---

[1] Taylor Fulscope Controllers, *Taylor Bull.* 98291, Sept., 1960.

[2] In a typical controller, the motion of the bellows is so slight that the fractional change in bellows volume is negligible.

where $q_1 =$ flow rate into bellows, moles/sec

$M =$ air in bellows, moles

Assuming the ideal gas law to apply, we have

$$M = \frac{p_1 V}{RT} \tag{22.17}$$

Combining Eqs. (22.16) and (22.17) gives

$$q_1 = C_1 \frac{dp_1}{dt} \tag{22.18}$$

where $C_1 = V/RT$. The term C_1 is called the capacity of the bellows. Assuming the mass flow rate to be proportional to the pressure drop across the resistance gives

$$q_1 = \frac{p - p_1}{R_1} \tag{22.19}$$

Combining Eqs. (22.18) and (22.19) gives

$$\tau_1 \frac{dp_1}{dt} = p - p_1 \tag{22.20}$$

where $\tau_1 = R_1 C_1 = R_1 V/RT$ is the time constant of the right-hand bellows. From this analysis, we see that the bellows is a first-order system.

In a similar manner, the following differential equation can be written for the left-hand bellows of Fig. 22.15:

$$\tau_2 \frac{dp_2}{dt} = p - p_2 \tag{22.21}$$

Equations (22.8), (22.10), (22.15), (22.20), and (22.21) can now be combined to obtain the desired relationship between p and ϵ. Introducing the deviation variables

$$P = p - p_0$$
$$P_1 = p_1 - p_0$$
$$P_2 = p_2 - p_0$$

into these five equations and taking the Laplace transform of each equation give the following algebraic equations:

$$P(s) = my(s) \tag{22.22}$$

$$y(s) = \frac{bC}{a+b} \epsilon(s) + \frac{a}{a+b} z(s) \tag{22.23}$$

$$z(s) = -\frac{A}{K} P_1(s) + \frac{A}{K} P_2(s) \tag{22.24}$$

$$\frac{P_1(s)}{P(s)} = \frac{1}{\tau_1 s + 1} \tag{22.25}$$

$$\frac{P_2(s)}{P(s)} = \frac{1}{\tau_2 s + 1} \tag{22.26}$$

Notice that ϵ, y, and z are deviation variables, for we have taken these variables to be zero at steady state. Solving Eqs. (22.22) through (22.26) simultaneously gives the following result:

$$\frac{P(s)}{\epsilon(s)} = \frac{bC/(a+b)}{\dfrac{a}{a+b}\dfrac{A}{K}\left(\dfrac{1}{\tau_1 s + 1} - \dfrac{1}{\tau_2 s + 1}\right) + \dfrac{1}{m}} \tag{22.27}$$

If m is very large, the term $1/m$ can be neglected in Eq. (22.27) and the equation can now be placed in the standard form of Eq. (22.5) to give

$$\frac{P}{\epsilon} = \frac{b}{a}\frac{K}{A} C \frac{\tau_1 + \tau_2}{\tau_2 - \tau_1}\left[1 + \frac{\tau_1 \tau_2}{\tau_1 + \tau_2} s + \frac{1}{(\tau_1 + \tau_2)s}\right] \tag{22.28}$$

Comparing the last equation with Eq. (22.5) reveals that

$$K_c = \frac{b}{a}\frac{K}{A} C \frac{\tau_1 + \tau_2}{\tau_2 - \tau_1} \tag{22.29}$$

$$\tau_I = \tau_1 + \tau_2 \tag{22.29a}$$

$$\tau_D = \frac{\tau_1 \tau_2}{\tau_1 + \tau_2} \tag{22.29b}$$

We see from this analysis that the gain of the controller, K_c, can be changed conveniently by varying the ratio b/a. Also, τ_I and τ_D can be varied by changing the values of the resistances R_1 and R_2. Notice that there is considerable *interaction*, which means that changing τ_1 or τ_2 separately affects the value of all three parameters K_c, τ_I, and τ_D. We shall return to this problem of interaction after discussing the transfer functions for two-mode controllers.

As can be seen from a comparison of Fig. 22.13 with Fig. 22.15, the three-mode controller becomes a PI controller by simply removing resistance R_1, that is, make $\tau_1 = 0$. For this case, Eq. (22.28) reduces to

$$\frac{P}{\epsilon} = \frac{b}{a}\frac{K}{A} C\left(1 + \frac{1}{\tau_2 s}\right) \tag{22.30}$$

Thus, we see that Eq. (22.30) is of the same form as the ideal transfer function given by Eq. (10.6) and that

$$K_c = \frac{b}{a}\frac{K}{A} C$$

$$\tau_I = \tau_2$$

In this case there is no interaction, for we can change τ_I without changing K_c. Notice that τ_2 is exactly the integral time τ_I.

To obtain a PD controller from the three-mode controller of Fig. 22.15, we simply close the resistance R_2 completely, which means that $\tau_2 \rightarrow \infty$.

For this case, Eq. (22.28) reduces to

$$\frac{P}{\epsilon} = \frac{b}{a}\frac{K}{A} C(1 + \tau_1 s) \tag{22.31}$$

This last equation is of the same form as the ideal transfer function given by Eq. (10.8) with

$$K_c = \frac{b}{a}\frac{K}{A} C$$

$$\tau_D = \tau_1$$

Again there is no interaction, for we can change τ_D without changing K_c. In this case, τ_1 is exactly equal to the derivative time τ_D.

Interaction in Three-mode Controllers A three-term industrial controller, which is based on the mechanism of Fig. 22.15 and which follows Eq. (22.28), is calibrated so that the markings on the knobs correspond to the value of τ_I (or τ_D) when only PI (or PD) action is used. This means that the markings on the knobs are in terms of τ_1 and τ_2 of Eq. (22.28). Furthermore, the gain knob is marked to correspond to the gain when only the PI or PD mode is used. The effect of interaction between the knob settings for a three-term controller is not always appreciated. If certain values of K_c, τ_I, and τ_D in Eq. (22.5) are wanted, then the actual settings of the controller knobs must be computed by the formulas given in Eqs. (22.29).

It should be emphasized at this time that other types of mechanisms are used in industrial controllers besides the one considered in Fig. 22.15.[1] For example, derivative action can be obtained from the mechanism of Fig. 22.15 by means of a "series arrangement" in which R_1 is located at point 2 in Fig. 22.15. For this case, one can show[2] that the controller transfer

---

[1] A. R. Aikman and C. A. Rutherford, "The Characteristics of Air-operated Controllers," Automatic and Manual Control, Papers contributed to the Conference at Cranfield, 1951, pp. 175–187, Butterworth & Co. (Publishers), Ltd., London, 1952. This discusses the transfer functions for five types of pneumatic controllers and shows the interaction terms associated with each type. The parallel and series arrangement described by Eqs. (22.28) and (22.32) are included in their article.

A. J. Young, "An Introduction to Process Control System Design," Longmans, Green & Co., Inc., New York, 1955. This also gives the transfer functions of three-mode pneumatic controllers and the interaction terms associated with each one.

[2] When the series arrangement is used, Eqs. (22.25) and (22.26) must be replaced by

$$sP_1 = \frac{P - P_1}{\tau_1} - \frac{P_1 - P_2}{\tau_2}$$

and

$$sP_2 = \frac{P_1 - P_2}{\tau_2}$$

where P, P_1, and P_2 are deviation variables. Solving these two equations and Eqs. (22.22) to (22.24) simultaneously will finally lead to Eq. (22.32).

function for large m is

$$\frac{P}{\epsilon} = \frac{b}{a}\frac{K}{A}C\left(1 + 2\frac{\tau_1}{\tau_2}\right)\left[1 + \frac{\tau_1}{1 + 2\tau_1/\tau_2}s + \frac{1}{\tau_2(1 + 2\tau_1/\tau_2)s}\right] \quad (22.32)$$

The reader can show from Eq. (22.32) that when only two-mode control is used,

$$K_c = \frac{b}{a}\frac{K}{A}C$$

$$\tau_I = \tau_2$$

or

$$\tau_D = \tau_1$$

With all three modes in use, we have from Eq. (22.32)

$$K_c = \frac{b}{a}\frac{K}{A}C\left(1 + \frac{2\tau_1}{\tau_2}\right)$$

$$\tau_D = \frac{\tau_1}{1 + 2\tau_1/\tau_2}$$

$$\tau_I = \tau_2\left(1 + \frac{2\tau_1}{\tau_2}\right)$$

The formulas given above for series arrangement are quite different from those of the parallel arrangement given in Eqs. (22.29). One disadvantage of the parallel arrangement is that $K_c \to \infty$ if $\tau_1 = \tau_2$ [see Eq. (22.29)]. For the series arrangement, K_c remains finite for $\tau_1 = \tau_2$.

Controller Transfer Function With Finite Baffle-nozzle Gain In deriving the transfer functions of the previous section, we neglected the term $1/m$ in Eq. (22.27) with the result that the transfer functions agreed with the ideal transfer functions of Chap. 10. Neglecting $1/m$ is equivalent to stating that $m \to \infty$. Actually m is finite. We shall now derive the transfer functions for the controller without dropping the term $1/m$ to show that this yields the more realistic controller transfer functions given earlier in the chapter, Eqs. (22.2) and (22.4).

Proportional-integral For this case $\tau_1 = 0$, and Eq. (22.27) becomes

$$\frac{P(s)}{\epsilon(s)} = \frac{bC/(a + b)}{\dfrac{a}{a + b}\dfrac{A}{K}\left(1 - \dfrac{1}{\tau_2 s + 1}\right) + \dfrac{1}{m}} \quad (22.33)$$

This last equation can be rearranged to give

$$\frac{P(s)}{\epsilon(s)} = \frac{\dfrac{b}{a + b}\dfrac{Cm}{maA/[(a + b)K] + 1}\left(1 + \dfrac{1}{\tau_2 s}\right)}{1 + \dfrac{1}{\{maA/[(a + b)K] + 1\}\tau_2 s}}$$

If $maA/[(a + b)K] \gg 1$, we have

$$\frac{P(s)}{\epsilon(s)} = \frac{b}{a}\frac{K}{A}C\frac{1 + 1/\tau_2 s}{1 + 1/\alpha\tau_2 s} \tag{22.34}$$

where

$$\alpha = \frac{maA}{(a + b)K} + 1 \cong \frac{maA}{(a + b)K}$$

From Eq. (22.34) the inclusion of finite baffle-nozzle gain leads to the transfer function which was given previously in Eq. (22.2). By referring to the asymptotic Bode diagram of Fig. 22.3, we see that the low-frequency gain is $K_c\alpha$, or in terms of the symbols of Eq. (22.34), we have

$$\text{Low-frequency gain} = \frac{bCK}{aA}\frac{maA}{(a + b)K} = \frac{mCb}{a + b}$$

We can show that this result is to be expected on purely physical grounds in the following manner: Imagine that the lower end of the baffle in Fig. 22.13 is moved sinusoidally at a very low frequency. Under such slowly varying conditions, the pressure in the left-hand bellows will be approximately equal to the pressure in the right-hand bellows (i.e., there will be sufficient time for equalization of the pressures in the two bellows). If there is no difference in pressure, the upper end of the baffle will not move and the mechanism will operate without feedback. Under these conditions the gain of the controller will be independent of frequency and the value will be simply the product of baffle-nozzle gain m, the ratio of distances $b/(a + b)$, and amplification factor C. (Recall that C merely relates the motion of the lower end of the baffle to the error ϵ.)

Proportional-derivative For this case, $\tau_2 \to \infty$ and Eq. (22.27) becomes

$$\frac{P(s)}{\epsilon(s)} = \frac{bC/(a + b)}{\left(\dfrac{a}{a + b}\dfrac{A}{K}\right)\left(\dfrac{1}{\tau_1 s + 1}\right) + \dfrac{1}{m}}$$

This equation can be rearranged to give

$$\frac{P(s)}{\epsilon(s)} = \frac{\dfrac{bCm}{(a + b)\{maA/[(a + b)K] + 1\}}(\tau_1 s + 1)}{\dfrac{\tau_1 s}{maA/[(a + b)K] + 1} + 1}$$

If $maA/[(a + b)K] \gg 1$, we obtain

$$\frac{P(s)}{\epsilon(s)} = \frac{b}{a}\frac{K}{A}C\frac{1 + \tau_1 s}{1 + (\tau_1/\gamma)s} \tag{22.35}$$

where

$$\gamma = \frac{maA}{(a+b)K} + 1 \cong \frac{maA}{(a+b)K}$$

This last equation agrees with the form of Eq. (22.4), and the effect of finite baffle-nozzle gain is to produce a finite high-frequency gain as shown in Fig. 22.5. From Fig. 22.5, we see that the high-frequency gain is $K_c\gamma$, or in terms of the symbols used in Eq. (22.35) this becomes

$$\text{High-frequency gain} = \frac{bCK}{aA}\frac{maA}{(a+b)K} = \frac{mCb}{a+b}$$

Again, we can give a physical explanation for the limited gain at high frequency. Imagine the lower end of the baffle of Fig. 22.14 to be moving sinusoidally at a very high frequency. Under these conditions, there will not be sufficient time for the pressure in the feedback bellows to vary, and the upper end of the baffle will remain fixed. The controller then has no feedback and the gain will be directly proportional to the product of the baffle nozzle gain m and the ratio $Cb/(a+b)$.

PROBLEMS

22.1 Rework the control system design problems given in Chaps. 17 and 19, including the nonideal characteristics of the controller. Consider typical cases such as

a. $\alpha = \gamma = 5$
b. $\alpha = \gamma = 10$

23 *Control of Complex Processes*

Thus far, we have presented and studied a number of tools for linear control system analysis and synthesis. To develop this subject, we have considered the dynamics and control of very simple systems, such as the stirred-tank heater. In practice, however, the systems we wish to control are quite complex. The transfer functions are usually unknown. There are often several inputs and outputs to each system. For example, a chemical reactor will respond to changes in the flow rate, composition, and temperature of each reactant stream; the pressure; the rate of external heat removal or addition; and perhaps some other independent input variables. These changes will be reflected in the dynamic behavior of the product composition, flow rate, and other output variables. For a system as complex as a chemical reactor, usually not one of the transfer functions between pairs of input and output variables is known. This is because the present theory is not adequate to predict these transfer functions accurately and the experiments necessary for their determination are expensive and difficult to perform, interpret, and extrapolate.

Nevertheless, it is often imperative that an engineering answer to the

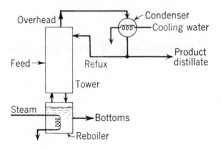

Fig. 23.1   **Typical distillation tower.**

problem of control of such processes be reached quickly and inexpensively. For this reason, the practice of process control has been built largely upon the results of past experience and simple tests and, to some extent, upon intuition.   The purpose of the present chapter is to give a brief introduction to this practice through the example of control of a distillation tower.   In the succeeding two chapters we shall discuss more elaborate experimental methods for determining transfer functions, and theoretical prediction of complex transfer functions, respectively.   In all cases, attention will be devoted to processes which are linear or which may be considered to operate sufficiently closely to steady state to exhibit only linear behavior.

Control of a Distillation Tower   The following discussion should be regarded as illustrative only; a comprehensive study of distillation control would require an entire text.   Suppose we are trying to control the distillation tower of Fig. 23.1.   This means it is desired that the column operate at or very near a specified steady-state condition despite changes in the independent variables.

Specification of a set of independent variables for a distillation tower generally requires careful application of the phase rule.   For our present purposes, assume that, if we set the feed flow rate and composition, the steam rate to the reboiler, the cooling-water flow rate, and the reflux ratio, then the steady-state distillate and bottoms flow rates and compositions and the column pressure are fixed.   Hence, the following will be considered as input variables:

1. Feed flow rate (assuming constant feed temperature)
2. Feed composition
3. Steam rate (assuming constant saturation temperature)
4. Cooling-water flow rate (assuming constant water temperature)
5. Reflux ratio

and the following as output variables:

1. Distillate composition
2. Distillate flow rate
3. Bottoms composition
4. Bottoms flow rate
5. Column pressure

In the general case, the number of input and output variables need not be equal. Furthermore, representation of composition by only one variable is strictly true only in the case of a binary separation. However, as we shall see below, for control purposes this representation will often be adequate even for multicomponent systems.

Strictly speaking, there is a transfer function between each of the input and output variables; in the present case this means 25 transfer functions. For example, a change in reflux ratio will cause a change in each of the five output variables. However, in many cases, its effect on the distillate composition will be more direct, and we may consider only this transfer function in our control scheme.

There is usually a control objective of primary importance. For illustrative purposes, assume that it is imperative that the distillate purity be maintained. It is also assumed that moderate changes in the composition of the less important bottoms product are tolerable, in our endeavor for distillate purity. The reader can easily imagine that in other cases only the bottoms purity, or possibly *both* product purities, must be carefully controlled.

It is evident that changes in any of the five input variables will affect the distillate composition. However, changes in the feed flow rate and steam rate can be almost entirely eliminated by placing flow controllers on these streams, as shown in Fig. 23.2. To see how these work, consider the block diagram of Fig. 23.3 in conjunction with Fig. 23.2. An orifice is placed in the feed line to meter the flow. The result of this measurement, the pressure drop across the orifice, is transduced to a controller pen displacement. This displacement is compared with the set point, and a resulting pneumatic signal is applied to the valve top by the controller. The flow rate through the valve is affected by this pneumatic control signal and also by the pressure of the feed stream upstream of the valve, which is the load variable. We choose to represent the form of this load variable

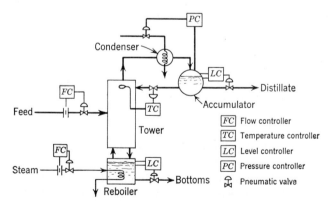

Fig. 23.2 Control of a typical distillation tower.

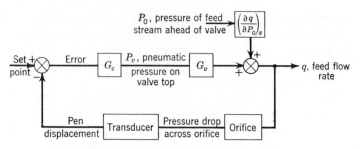

Fig. 23.3   **Control of feed stream flow rate.**

by a linearized constant $(\partial q/\partial P_0)_s$. This is the change in flow rate per unit change in upstream feed pressure when all other variables are fixed at their steady-state values. Of course, in the same sense, $G_v(s)$ is the transfer function giving the change in flow rate per unit change in pneumatic pressure applied to the valve, at steady-state conditions. Thus, for a typical first-order valve

$$G_v(s) = \frac{(\partial q/\partial P_v)_s}{\tau_v s + 1} = \frac{K_v}{\tau_v s + 1}$$

where P_v is the applied pneumatic pressure.

The orifice is an inherently nonlinear device, since the flow rate is proportional to the square root of the pressure drop. Hence, we shall have to linearize the square-root relation, as we have done previously for liquid-level control systems, in order to effect a linear analysis.

In any event, since the dynamics of all elements in the control loop of Fig. 23.3 (the valve, orifice, and transducer) are fast, little difficulty is anticipated in selecting a controller to give good control of the feed rate. Possibly, proportional-integral control will be chosen so that no offset will occur. Derivative action is probably not necessary. Similar considerations apply to control of the steam rate. Therefore, we may confidently expect to prevent potential disturbances in the feed and steam flow rates from reaching the column. This kind of control is sometimes called environmental control.

However, there is no easy way of preventing changes in the feed composition from reaching the column. Changes in this composition will certainly have an effect on distillate composition, and we shall have to look for a feedback type of control scheme for maintaining this product purity.

Before the distillate composition can be controlled, there must be some means for measuring it. Continuous composition measurements are usually expensive and sometimes impossible. However, *if the column pressure is constant*, the temperature of the liquid on the top plate, i.e., its boiling point, is a simple measure of the distillate composition. This measure is exact only in the case of a binary system; however, it is often used in multi-component systems because of its simplicity. The accuracy in the case of multicomponent distillation will depend on the type of mixture being fractionated. A more complete discussion of this subject can be found in any text on distillation.

Assuming that somehow the column pressure is maintained constant, we now plan to use measurement of the column top temperature as the basis for a distillate composition control loop. The distillate composition may be controlled by changing the reflux ratio. If the reflux ratio is increased, the distillate purity is increased and vice versa. However, in order to be able to increase and decrease the reflux ratio, there must be a reserve supply of distillate available. To accomplish this we put an accumulator tank on the overhead line leaving the condenser. A level controller on the accumulator adjusts the distillate product rate to maintain a constant level of distillate in the accumulator, as shown in Fig. 23.2.

Now if the temperature on the top tray increases, indicating a higher boiling point and hence a lower distillate purity, this information is transmitted to a temperature controller which increases the reflux rate. A control loop for this purpose is described by the block diagram of Fig. 23.4. (The only load variable indicated in Fig. 23.4 is the feed composition. Actually, we know that there will be other load variables, such as column pressure which, incidentally, also affects the measuring circuit. Discussion of these interactions is deferred to a later section.)

Selection of a controller for this system is not so simple as was selection of the controller for the flow-control system of Fig. 23.3. This is because the transfer functions between reflux rate and top temperature, and between feed composition and top temperature (G_1 and G_2), will involve long transportation lags and long time constants for commercially sized equipment. Thus, it may be a matter of hours between the time a change in the feed composition occurs and the time its full effect appears in the top temperature. Furthermore, as we have indicated previously, the transfer functions G_1 and G_2 are generally not known. In practice, for loops showing these characteristics it is often merely assumed that three-mode control, propor-

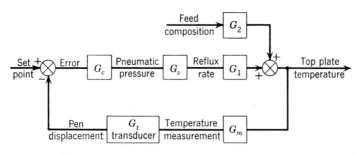

Fig. 23.4 Control of top plate temperature.

tional plus integral plus derivative, is required.   The closed-loop system is placed on stream with estimated starting values for the controller constants K_c, τ_I, and τ_D, which are then adjusted for better control as operating experience is acquired.   Methods for obtaining initial estimates of K_c, τ_I, and τ_D are discussed in later sections of this chapter.

It remains now to control the column pressure and the reboiler level. The pressure can be controlled by adjusting the flow rate of cooling water to the overhead condenser.   Increasing this flow rate decreases the pressure and vice versa.   The reboiler level can be maintained by adjusting the rate of bottoms product withdrawal.   Design of the bottoms level controller and level controller on the overhead accumulator is much simpler than design of the column pressure controller.   These level systems are the same as those we have studied in earlier chapters.   The dynamic behavior of the tanks may be approximated by first-order transfer functions, with time constants given by the volume of liquid in the tank divided by throughput, at steady state.   Control engineers often call this the *holdup time*.   In the case of the overhead accumulator, it represents the number of hours for which reflux could be supplied to the tower in the event that, owing to some upset, no overhead product was being produced.   If the tank can be made large enough to give a long holdup time, a simple proportional system may be adequate to control the level, since offset can be tolerated and speed of response is not of primary concern.   If, however, a large tank is not possible or practical, other control modes may be required for satisfactory control.

The column pressure-control system involves unknown transfer functions with long time constants.   Thus, as in the case of the top temperature-control system, design is often achieved by approximate methods, to be discussed later.   The complete system appears in Fig. 23.2.

Interactions in the Control System   Even if each of the individual control loops is designed to operate satisfactorily, there is no guarantee that, when they are placed in simultaneous operation on the distillation tower, the overall column operation will be satisfactory.   In fact, it is quite possible for the loops to be individually stable and yet lead to unstable operation of the entire column.   To see why this is so, it is necessary to recall that we have assumed control of each variable to be single-loop.   We know there is considerable interaction between the variables.   Thus, a more realistic block diagram for the pressure- and top temperature-control loops might look like Fig. 23.5.   Process control system design can seldom be based upon such complex diagrams because the transfer functions are not known.   (These are referred to as multivariable control systems, and their study is beyond the scope of this text.)   However, the existence of interactions must be recognized, as they may render the scheme of Fig. 23.2 inoperative and a new control system which minimizes the effect of the interactions may be required.   There are numerous pitfalls in this type of

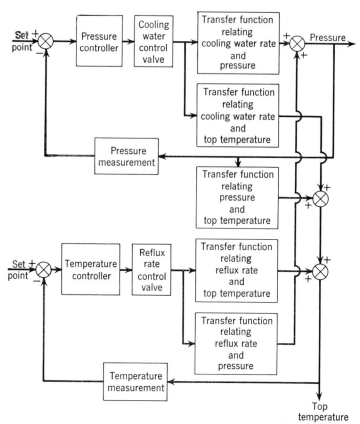

Fig. 23.5 **Block diagram for distillation control illustrating interactions between control loops.**

design. A more detailed discussion of interaction effects in the case of a distillation column is available elsewhere.[1] In practice, the pitfalls are generally avoided on the basis of past experience. However, when new and different applications arise, considerable difficulty may be involved in obtaining satisfactory process control. There is admittedly considerable room for improvement in our present state of knowledge on control system design.

Choice of Different Modes of Control After specification of types (e.g., flow, level, etc.) and location of controllers, as we have done for the hypothetical distillation column, it is necessary to specify the modes of control and the values of the controller settings (K_c, τ_I, and τ_D where applicable) which should be used to give satisfactory control. In this

[1] T. J. Williams, "Systems Engineering for the Process Industries," McGraw-Hill Book Company, New York, 1961.

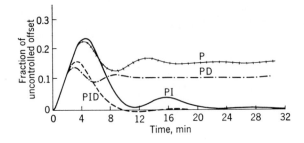

Fig. 23.6 **Load response of typical control system using various modes of control (control system shown in Fig. 23.7).**

section we summarize the criteria which generally guide the selection of control modes, and in the next sections we present simple methods for estimating satisfactory controller settings.

Figure 23.6 represents the response of the control system of Fig. 23.7 to a unit-step load change, using various modes of control. To fix the absolute scale, we remark that the offset realized by the curve for proportional control is 15 percent of the ultimate change which would have occurred in the absence of control. The open-loop transfer function of Fig. 23.7 is typical of the behavior of many process control systems (see Chap. 24). Therefore Fig. 23.6 is typical of the responses of process control systems to load changes and allows the following generalizations to be drawn: (The reader should also refer to Figs. 10.6 and 19.13 to supplement this discussion.)

1. Simple proportional control results in a response showing a high maximum deviation, a moderate period of oscillation, and, most important, a maximum offset. In addition, a significant length of time is required before the system ceases to oscillate.

2. Proportional-integral control has no offset. However, elimination of offset comes at the expense of a higher maximum deviation, a longer period of oscillation, and a longer time required for oscillations to cease, compared with proportional control.

3. Proportional-derivative control generally brings the system to steady state in the shortest time with the least oscillation and smallest maximum deviation. However, it has offset which, while reduced from the case of proportional control, is still significant.

4. Proportional-integral-derivative control is essentially a compromise

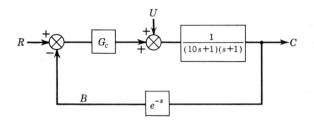

Fig. 23.7 **Block diagram of control system whose load responses are illustrated in Fig. 23.6.**

Table 23.1   Typical Costs of
Pneumatic Recorder-Controllers†

Mode	Cost
On-off	$335
P	425
PI	490
PD	475
PID	535

† This table is based on recent
cost data of a United States manufacturer for a controller which will control
and record one temperature. The cost
includes the controller, the recorder,
and the temperature-measuring element (mercury-filled bulb).

between the advantages of PD and PI control.   Offset is eliminated by the
integral action, while the derivative action serves to lower the maximum
deviation and to eliminate some of the oscillation realized with PI control.
Figure 19.13 also shows that addition of derivative mode increases the speed
of response to set-point changes.   This increase in speed of response does not
appear in the load response curves of Fig. 23.6 because of the distance-
velocity lag in the feedback loop.

It is important to emphasize the general nature of the above discussion;
exceptions may occur in some systems.   However, the basic characteristics
attributable to each mode of control (e.g., elimination of offset by integral
mode, increase of speed by derivative mode) generally persist even in complex systems.

The question of which modes to use in a specific application does not,
in general, have a definitive answer.   Ideally, the objective is to choose
the simplest controller which will give adequate control.   Unfortunately,
it is usually impossible to state in advance what is the simplest controller
unless the application is simple (such as the flow controller on the distilla-
tion-tower feed stream) or unless a considerable backlog of experience is
available for the particular application.   Frequently, a PID controller is
selected simply because it is most likely to be capable of satisfactory control.
The additional cost associated with each mode is indicated in Table 23.1.
It may be seen that the additional purchased cost of each mode is not
excessive.   Probably of more importance is the additional controller adjustment and maintenance associated with each mode.

Adjustment of Controllers   In this section we shall discuss two
methods of testing industrial processes which can be used to determine good

initial controller settings. The characteristic of these methods which distinguishes them from the methods which will be discussed in Chap. 24 is that they are primarily oriented toward setting the controller and give little or no information about the general dynamic behavior of the process. However, together with other similar methods they are quite expedient and hence find widespread application in the process industries.

The first method will be called *loop tuning*. It was first proposed by Ziegler and Nichols[1] and actually forms the basis for the Ziegler-Nichols rules presented in Chap. 19. That is, Ziegler and Nichols did not suggest that the values of K_u and P_u, the ultimate gain and period, be determined from the frequency response. Instead, they assumed that the frequency response was unknown and suggested that K_u and P_u be determined in the following manner:

The controller is operated in *closed loop* with the system to be controlled. The integral and derivative modes (if any) are rendered inoperative, and the proportional gain is slowly increased to the value at which continuous cycling of the system variables first occurs. This value of proportional gain corresponds to K_u of the Ziegler-Nichols method. In practice, it may be necessary to introduce momentary disturbances in order to cause the onset of continuous cycling, because of process sluggishness. The period of the continuous oscillations is also observed from the instrument chart record and taken as P_u. The Ziegler-Nichols rules may now be applied to determine the controller settings.

While this method is simple and fairly rapid, it does have several disadvantages:

1. Closed-loop operation in a condition of borderline stability may suddenly turn into unstable operation owing to external influences. This could cause physical damage to the equipment. Particularly dangerous in this regard would be an exothermic chemical reactor.

2. The proportional gain calibrations on pneumatic instruments are frequently inaccurate. This error in K_u, which may be of the order of 30 percent, leads to errors in the control settings, and hence good control may not be realized.

3. No information about the actual process transfer function is obtained. Hence, if a change is made in some portion of the loop, it is likely that the entire test procedure will have to be repeated.

The second method for setting controllers will be called the process reaction curve method. It was proposed by Cohen and Coon[2] and consists of applying a small step change in the manipulated variable to the opened control loop and recording the curve of measured variable versus time. The

[1] J. G. Ziegler and N. B. Nichols, Optimum Settings for Automatic Controllers, *Trans. ASME*, **64**:759 (1942).

[2] G. H. Cohen and G. A. Coon, *Trans. ASME*, **75**:827 (1953).

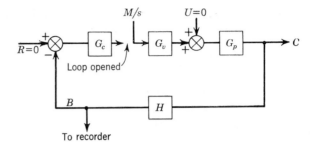

Fig. 23.8 Block diagram for measurement of process reaction curve.

changes must be small enough to ensure operation in the linear range.   The output curve is called the *process reaction curve.*   The block diagram for this test is shown by Fig. 23.8.   It must be assumed that no load changes occur during the test.   In addition, all the dynamic components of the loop other than the controller must be included between the point of application of the manipulated variable change and the point of recording the response.   In Fig. 23.8, it is suggested that the step change is applied to the valve-top pressure and that the loop be opened at the controller output.   Similar procedures apply to nonpneumatic systems.

A typical process reaction curve is given in Fig. 23.9.   We first present the method for estimating controller settings from Fig. 23.9.   In the succeeding discussion we shall present some of the rationale for this method.

First, a tangent is drawn to the curve at the point of inflection.   The intercept of this tangent on the abscissa is taken as the apparent dead time T_d.   The slope of the tangent, S, is proportional to $1/T$, the reciprocal of the apparent time constant.   In this case the slope is equal to B_u/T where B_u is the ultimate response.   Hence,

$$T = \frac{B_u}{S}$$

The steady-state gain between M and B is calculated as

$$K_p = \frac{B_u}{M}$$

Fig. 23.9 Typical process reaction curve.

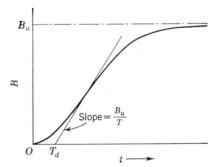

Using the values of T_d, T, and K_p determined in this manner, the following are recommended controller settings:

Proportional

$$K_c = \frac{1}{K_p} \frac{T}{T_d} \left(1 + \frac{T_d}{3T} \right)$$

Proportional-integral

$$K_c = \frac{1}{K_p} \frac{T}{T_d} \left(\frac{9}{10} + \frac{T_d}{12T} \right)$$

$$\tau_I = T_d \frac{30 + 3T_d/T}{9 + 20T_d/T}$$

Proportional-derivative

$$K_c = \frac{1}{K_p} \frac{T}{T_d} \left(\frac{5}{4} + \frac{T_d}{6T} \right)$$

$$\tau_D = T_d \frac{6 - 2T_d/T}{22 + 3T_d/T}$$

Proportional-integral-derivative

$$K_c = \frac{1}{K_p} \frac{T}{T_d} \left(\frac{4}{3} + \frac{T_d}{4T} \right)$$

$$\tau_I = T_d \frac{32 + 6T_d/T}{13 + 8T_d/T}$$

$$\tau_D = T_d \frac{4}{11 + 2T_d/T}$$

A brief rationale of this method is in order at this point. The process reaction curve method essentially assumes the open-loop system to behave as though it had a transfer function given by

$$G_p(s) = \frac{K_p e^{-T_d s}}{Ts + 1} \tag{23.1}$$

Using this form as the true transfer function, Cohen and Coon derived theoretical values of the controller settings to give responses having one-quarter decay ratios, minimum offset, minimum area under load response curve, and other favorable properties. These values are the ones given above. The quantities estimated from the process reaction curve, K_p, T, and T_d, are taken as "best" values to use for representing the process transfer function in the form of Eq. (23.1).

To illustrate the two methods of setting controllers, the control system of Fig. 23.7 was simulated on the analog computer. The settings calculated using each method are tabulated in Table 23.2. (For PD control,

Table 23.2   **Controller Settings Obtained for System of Fig. 23.7**

Control type	Parameter	Values	
		Loop tuning	Process reaction curve
P	K_c	6.0	7.3
PI	K_c	5.0	6.4
	τ_I	6.4	4.6
PD	K_c	10.0†	8.9
	τ_D	0.75†	0.46
PID	K_c	6.7	9.5
	τ_I	3.9	4.2
	τ_D	1.0	0.64

† Obtained by design for 30° phase margin at maximum K_c.

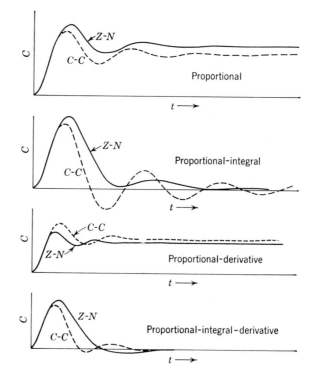

Fig. 23.10 **Comparison of load responses for system of Fig. 23.7 using controller settings obtained by the loop-tuning method of Ziegler-Nichols (Z-N) and the process reaction curve method of Cohen-Coon (C-C).**

the setting indicated under the loop-tuning method was obtained by using the theoretical frequency response and designing for a 30° phase margin and maximum K_c, as described in Chap. 19.) The responses using the two different methods are compared for each kind of control in Fig. 23.10. *It is important to emphasize that no general conclusions about the relative merits of the two methods can be drawn from Fig.* 23.10. The only inference that can be made is that both methods give reasonable first guesses for the control parameters.

There are many variants of these two methods. Often, controller adjustment instructions are supplied by the instrument manufacturer. These methods are similar to the ones presented here in basis, accuracy, and ease of application.

Summary We have considered only simple, illustrative examples of the controller selection and adjustment problem. Furthermore, even here we cannot say with assurance that the control scheme of Fig. 23.2 will work on a particular column. It is often necessary to resort to trial and error. The reader should be aware, however, that processes of all kinds have been controlled for many years, and a considerable backlog of experience and know-how is available to most manufacturing concerns. This information on past performance is frequently used for designing control schemes for new processes which are still in the design stage. The work of the control engineer is largely to bring together the control fundamentals we have developed in this text and the experience gained from the previous operation of control loops and apply them to new situations. This combination of science and art is typical of many engineering functions.

As more research is accomplished in process control, more of the art is replaced by science. In addition, although the discussion presented in this chapter is pertinent to present technology, tremendous inroads have been made in the use of more sophisticated controllers, such as computers. It is not unlikely that a few more years will bring significant revisions in the typical methodology of process control engineering. Hence, the student is urged to keep abreast of the literature in this changing field.

24 *Experimental Dynamics of Complex Processes*

We have devoted considerable effort to the study of *linear* control theory in the previous chapters. Given the transfer function of the component whose output is to be controlled, we now have at our disposal tools for specifying a control system. With reference to Fig. 24.1, this problem usually involves specification of $G_c(s)$, the controller transfer function, so that the process output C is maintained near a desired level despite fluctuations in the process load input U.

The processes considered so far in this text have been of sufficient simplicity that a transfer function can be obtained by application of basic

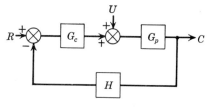

Fig. 24.1 Typical control system.

principles such as mass and energy balances. However, there are many components used in manufacturing processes for which transfer functions are neither known nor readily derived. Examples of such components are distillation towers, heat exchangers, chemical reactors, etc., all of which occur frequently in the process industries and which normally must be controlled for smooth operation. The transfer functions usually cannot be easily derived because the processes occurring in the components are complex and incompletely understood.

The major objective of this chapter is to present the reader with a brief study of *experimental* methods for determination of the process transfer function $G_p(s)$. In Chap. 25, we shall consider examples of the *theoretical* determination of $G_p(s)$ for the class of *complex* components which most easily lends itself to theoretical analysis, namely, heat- and mass-transfer systems. Even here it will be seen that the transfer functions are quite unwieldy.

Distinguishing Characteristics of Process Dynamics In this section, we list and describe those features of the dynamic behavior of process industry equipment which distinguish them from those of typical electrical or mechanical systems.

1. Long time constants The time constants in the transfer functions will be of the order of minutes for a moderately sized heat exchanger and may be of the order of hours for a large fractionating column. Suppose, for example, that a step change occurs in the feed composition to a fractionating column. What we are saying is that it may require an hour for the compositions of the overhead and bottoms to complete 63 percent of their ultimate total change.

2. Long distance-velocity lags Lags on the order of minutes are not uncommon. Thus, with reference to the fractionator discussed in 1, it may be several minutes after the occurrence of the step change in feed composition before any change at all is observed in the overhead and bottoms compositions. The disturbance must first work its way through the column, plate by plate. This long dead-time phenomenon has no simple analog in electrical systems. In fact, the simulation of the long dead time on the analog computer is a relatively complicated operation (see Chap. 32).

3. Absence of underdamped systems Few underdamped systems have been reported in the literature on process dynamics. In fact, the approximate response of many process components, such as heat exchangers, fractionators, etc., can be adequately represented by a transfer function of the form

$$G(s) = \frac{e^{-T_d s}}{(T_1 s + 1)(T_2 s + 1)} \tag{24.1}$$

i.e., overdamped second-order, with a distance-velocity lag. Experimental measurements of process dynamics are sometimes directed toward evaluation of the dead time T_d and time constants T_1 and T_2, so that Eq. (24.1) best represents the response of the particular process component.

4. Nonlinearity   Linear dynamics are the exception, rather than the rule, in process dynamics.   While the dynamics of an electrical component such as a capacitor can be adequately described by a linear equation

$$\frac{de}{dt} = \frac{i(t)}{C}$$

over a wide range of operating conditions, the equations describing process dynamics are usually nonlinear.   Even in a simple liquid-level system, the outflow from the tank is usually proportional to the square root of the tank level.   The rate of a chemical reaction

$$A + B \rightarrow C$$

is frequently expressed as

$$r = ke^{-E/RT}C_A C_B \tag{24.2}$$

which is nonlinear in the temperature and concentrations.   As we have done in previous chapters, these relations may be linearized by expansion in Taylor's series around the steady-state operating point.   However, transfer functions derived from these linearized equations are valid only in the region near the operating point.   If the region of validity of the linearized equations is small, which may well occur for a highly nonlinear relation such as Eq. (24.2), this restriction may be quite troublesome and necessitate direct analysis of the nonlinearities.   In subsequent chapters, we shall find that nonlinear analysis is difficult to perform.

5. Distributed parameters   In a simple RC electrical circuit the response of the output voltage to changes in input voltage may be accurately described by considering all the resistance of the circuit to be "lumped" in the resistor and all its capacitance to be lumped in the capacitor.   However, consider a simple steam-jacketed pipe heat exchanger.   Suppose we are interested in the dynamic response of the outlet fluid temperature to a change in the inlet fluid temperature, assuming the steam temperature to remain constant.   It is evident that the heat-transfer resistance and capacitance of the fluid flowing in the exchanger pipe are distributed along the length of the exchanger.   This is manifested in the fact that the temperature is a function of both distance and time and the pertinent differential equation is partial rather than ordinary.   In the analysis of the stirred-tank heat exchanger of Chap. 1, we assumed sufficient mixing so that the temperature in the tank was uniform.   We know physically, however, that even for very good mixing, the temperature in the immediate vicinity of the heater will be higher than elsewhere in the tank.   Thus, we have approximated the

Fig. 24.2 **Frequency testing of a process.**

distributed system by the equivalent lumped system in our derivation of the transfer function.   In this case, the accuracy of the approximation is good.   However, approximation of the steam-jacketed heat exchanger by a lumped-parameter system will not be so accurate.   Despite this loss of accuracy, such approximations are often made because no other analysis can be made without unreasonable labor.   We shall discuss a detailed example of this type of approximation in Chap. 25.

The remainder of this chapter is devoted to a discussion of various techniques for measurement and utilization of process dynamics.   Some references to the literature are made, and the reader is encouraged to read the original works.

Frequency Testing   We have already demonstrated that, for a linear process with transfer function $G_p(s)$, the output response to a sinusoidal input is determined simply from $G_p(j\omega)$.   The inverse process, determination of $G_p(s)$ from $G_p(j\omega)$, immediately suggests itself as a means for experimental determination of $G_p(s)$.   Thus, with reference to Fig. 24.2, suppose we apply an input sine wave to a process of unknown transfer function.   As we discussed in the previous section, few processes are truly linear.   Hence, we are required to use a sufficiently small amplitude A that the process will respond approximately linearly and the output will be essentially a sine wave of the same frequency as the input, ω.   Observation of the output yields directly the AR and phase lag of the process at frequency ω.   Repetition of the experiment at different frequencies yields the frequency response.

This procedure was applied to a model-sized steam-water heat exchanger by Lees and Hougen.[1]   Their apparatus is illustrated schematically in Fig. 24.3.   A sinusoidal variation in pressure was applied to the control-valve diaphragm, thus causing a similar variation in the valve-stem position and flow of water through the exchanger.   The steam pressure and inlet water temperature were maintained constant.   The effluent water temperature and control-valve-stem position, both varying sinusoidally, were recorded as functions of time.   A number of such experiments, at different

[1] S. Lees and J. O. Hougen, *Ind. Eng. Chem.*, **48**:1064 (1956).

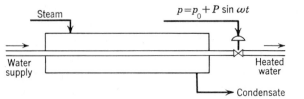

Fig. 24.3 **Schematic for frequency testing of heat exchanger.**

frequencies, produced a Bode diagram relating effluent temperatures to valve-stem position, which could be accurately represented by the transfer function

$$G_p(s) = K_p \frac{1 - 2\zeta\tau s + \tau^2 s^2}{(T_1 s + 1)(T_2 s + 1)} e^{-T_d s} \tag{24.3}$$

where K_p = steady-state change in effluent temperature with change in valve-stem position, 404°F/in. for heat exchanger tested

T_1, T_2 = process time constants, 7.6 and 1.41 sec

T_d = dead time, 1.75 sec

τ, ζ = parameters associated with resonance effects, 0.8 sec and 0.3 respectively

The latter parameters τ and ζ are associated with resonance effects, which were observed experimentally and may be predicted theoretically, as shown in Chap. 25. Since τ is an order of magnitude less than T_1, the most significant time constant, the quadratic factor in the numerator of Eq. (24.3) is unimportant except at frequencies which are ten times higher than normal operating frequencies. We shall, therefore, ignore this factor and point out that the observed behavior is that suggested by Eq. (24.1). This is true despite the fact that the exchanger is a distributed-parameter system. Hence, we have an example of the valid approximation of a distributed system by a lumped-parameter transfer function. The presence of an apparent dead time is the result of the distributed parameters. (It should be observed that the heat exchanger was much smaller than a commercial unit, for experimental convenience. This leads to the relatively small time constants.)

Once the Bode diagram associated with Eq. (24.3) is available to the control engineer, a control system for the heat exchanger may be easily and accurately designed by the methods of Chap. 19. This control system would be based on the block diagram of Fig. 24.4, indicating control of the effluent temperature by regulation of flow rate by adjustment of valve-stem position. The transfer functions of the other components in the loop are

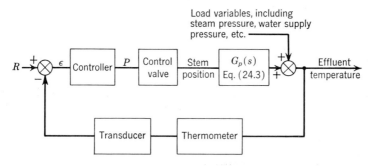

Fig. 24.4   Control of steam-jacketed heat exchanger.

generally known or easily obtained. For example, it is usually known that a variation of pressure P, from minimum to maximum controller output, causes a certain change in valve-stem displacement, and the relationship may be assumed linear.

This discussion indicates the great utility of the results of frequency testing. The approach of obtaining the Bode diagram for the process and designing a control system by frequency-response methods is a standard one for control engineers. However, there were major experimental difficulties involved in the establishment of Eq. (24.3) by direct frequency testing. It is quite difficult to generate an exactly sinusoidal variation in valve-stem position. Lees and Hougen observed that small deviations from true sinusoidal inputs caused significant changes in the observed frequency response. Furthermore, the experiments had to be performed over a range of frequencies from 0.001 to 3.0 rad/sec to establish Eq. (24.3). This 3,000-fold variation in input frequency is quite difficult to achieve, especially with any degree of accuracy.

From a commercial viewpoint, frequency testing involves not only experimental difficulties but also extensive testing time. The experiment must be repeated at several frequencies and must be sustained for a sufficiently long time so that all transients die out and only the frequency response is observed at the output. This results in excessive tie-up of equipment and manpower.

For these reasons, the major purpose of Lees's and Hougen's work was to demonstrate that virtually the same results can be obtained by pulse testing, as will be discussed in the next section.

Pulse Testing Suppose an arbitrary pulse is introduced into the valve-stem position, again with reference to the heat exchanger of the previous section. A pulse disturbance is one for which the stem position is changed from its steady-state position momentarily and then returned to the steady-state position. An arbitrary pulse is one for which the form of the displacement and return of the valve-stem position is not specified. A typical arbitrary pulse, resembling the one actually used by Lees and Hougen, is shown in Fig. 24.5, together with the resulting behavior of the

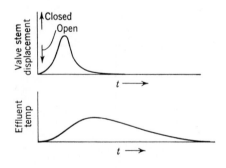

Fig. 24.5 **Typical pulse functions for process testing of heat exchanger.**

effluent temperature. It is essential that the height of the pulse be sufficiently small that operation in the linear range is assured. Experimentally, this disturbance is much easier to produce than a sine wave, since all that is required is a fairly rapid, manual variation in the pressure on the control-valve diaphragm.

To determine the process transfer function from test results such as those of Fig. 24.5, note that

$$G_p(s) = \frac{Y(s)}{X(s)} \tag{24.4}$$

where $X(s)$ = Laplace transform of function representing valve-stem displacement versus time.

$\quad\quad\quad Y(s)$ = Laplace transform of function representing effluent temperature versus time.

From Eq. (24.4)

$$
\begin{aligned}
G_p(j\omega) &= \frac{Y(j\omega)}{X(j\omega)} = \frac{\int_0^\infty Y(t)e^{-j\omega t}\,dt}{\int_0^\infty X(t)e^{-j\omega t}\,dt} \\
&= \frac{\int_0^\infty Y(t)\cos\omega t\,dt - j\int_0^\infty Y(t)\sin\omega t\,dt}{\int_0^\infty X(t)\cos\omega t\,dt - j\int_0^\infty X(t)\sin\omega t\,dt}
\end{aligned}
\tag{24.5}
$$

It is evident from Fig. 24.5 that the improper integrals in Eq. (24.5) exist, because practically speaking the functions $X(t)$ and $Y(t)$ [which must necessarily be taken as deviation variables because of Eq. (24.4)] both return to zero after a finite time.

Using the data of Fig. 24.5, the integrals in Eq. (24.5) can be evaluated numerically. This must be done at each frequency of interest, and a digital computer is warranted for the computations. At each frequency, the calculations described yield $G_p(j\omega)$ as a complex number. Conversion of this number to polar form yields the Bode diagram and frequency response. Thus, the pulse technique is experimentally more convenient but requires considerable computational effort.

Lees and Hougen found that the results of the pulse test were in good agreement with the direct frequency tests on the heat exchanger. They concluded that economically the experimental advantages of the pulse test outweigh the computational disadvantages.

Step Testing Another disturbance which is easy to generate experimentally is a step function. As an illustration of this method, we consider the work of Laspe,[1] in which the dynamic behavior of an oil cracking furnace was determined. The furnace is illustrated schematically in Fig. 24.6. The

[1] C. G. Laspe, *ISA J.*, **3**:134 (1956).

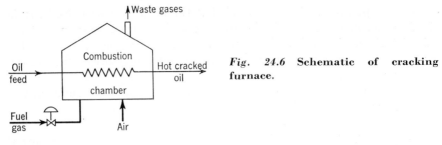

Fig. 24.6 Schematic of cracking furnace.

objective is to control the fuel gas feed rate so as to maintain a constant temperature of the hot oil effluent. Theoretical determination of the dynamic response of the oil temperature to changes in the fuel gas rate is impossible with the present state of our knowledge. This is because the overall process is made up of three major, complex processes in series: (1) the combustion of the fuel gas, (2) the heat transfer, and (3) the thermal cracking reaction of the oil to lower molecular weight hydrocarbons. The dynamics of these processes, especially (1) and (3), are essentially unknown.

Laspe's procedure was to bring the furnace to steady state and introduce a small step change in the pressure on the fuel gas control-valve diaphragm. Values of the oil effluent temperature were recorded as a function of time. The resulting time curve was found to be accurately represented by the step response of a system having the transfer function

$$G_p(s) = \frac{Ke^{-1.06s}}{(6.90s + 1)(1.03s + 1)} \tag{24.6}$$

where the numerical values are in minutes. Equation (24.6) was taken as the process transfer function, and the Bode diagram was prepared from it. The methods of Chap. 19 were used to design a control system which performed satisfactorily.

Although the constants of Eq. (24.6) were fitted by machine computation, it is possible to obtain estimates of these constants from step-test results by rapid graphical procedures. One of these, which ignores the smaller time constant, is the process reaction curve method discussed in Chap. 23. However, if a value of the smaller time constant (assuming that the process actually has a second time constant) is desired, other graphical approximations must be used. These are presented in other texts.[1]

The step method is approximately equivalent to the pulse method in experimental ease and computational difficulty. The major source of computational difficulty is in accurately finding constants T_1, T_2, and T_d, so that Eq. (24.1) will represent the response. This is not necessary in frequency testing, because the control designer requires only the Bode

---

[1] E. M. Grabbe, et al., "Handbook of Automation, Computation, and Control," vol. III, pp. 9–05 to 9–09, John Wiley & Sons, Inc., New York, 1958.

diagram, which is obtained directly. Another disadvantage of step testing is that additional time constants smaller than T_1 and T_2 may go undetected. For example, if the actual transfer function of the cracking furnace were

$$G_p(s) = \frac{Ke^{-1.06s}}{(6.90s + 1)(1.03s + 1)(0.2s + 1)} \tag{24.7}$$

instead of Eq. (24.6), there would be essentially no difference in the step response, as the reader can verify. On the other hand, at sufficiently high frequencies in a frequency response, the third time constant will cause additional attenuation and hence can be observed. It can be shown that a sufficiently narrow pulse will also yield a response containing observable effects of small time constants.

It is pertinent to compare the step test of Chap. 23 with that of the present chapter. In Chap. 23, the results of a step test were called a process reaction curve. It was used only to estimate controller settings for the particular *loop* to be controlled. In the present context, step testing is an experimental attempt to determine the process dynamics so as to increase our knowledge of the process behavior, for both present and future control problems. For example, it may be of interest to determine how the time constants in Eq. (24.6) vary as a function of hydrocarbon flow rate. More of this type of work is appearing in current literature.

Other methods exist for determination of process dynamics. These include ramp tests, impulse tests, statistical methods, and others. The ramp and impulse tests are simply extensions of the material we have already covered. For a discussion of statistical methods the reader is referred to texts on servomechanisms.[1]

There is considerable need for generalizable data on process dynamics. An extensive bibliography on available information is given by Williams.[2] This field is one of current research interest.

PROBLEMS

24.1 It is desired to evaluate the dynamic parameters in the transfer function

$$\frac{Ke^{-T_d s}}{(T_1 s + 1)(T_2 s + 1)}$$

so that it best represents the experimentally measured step response of a process.

    a. Referring to Prob. 8.11, show that estimation of T_1 and T_2 can be made using the method established there, independently of the presence of T_d.

[1] See, for example, J. G. Truxal, "Automatic Feedback Control System Synthesis," McGraw-Hill Book Company, New York, 1955.
[2] T. J. Williams, "Systems Engineering for the Process Industries," pp. 67–69, McGraw-Hill Book Company, New York, 1961.

b. Show that, for $T_d = 0$, the tangent to the step response at the inflection point intersects the abscissa at a time

$$t_x = t_i - \frac{S(t_i)}{S'(t_i)}$$

c. Using the information deduced in part **b**, present a method for graphically estimating T_d.

d. Simulate a system with two time constants and dead time on the analog computer, and for various values of the time parameters, use the technique developed in this problem to estimate the parameters from a step response.

e. Repeat part **d** introducing a third small time constant into the process and, still assuming the system to be second-order, estimating the second-order time constants as before. What is the effect of the third time constant on the determined parameters?

f. Design and test control systems, using various controller modes and various design methods, for the system of part **e**, using the fitted two-time-constant transfer function.

24.2   The results of a step change test on a chemical reactor are listed below. These data were obtained by suddenly increasing the flow rate of a stream of pure reactant from its normal value of 6 to 6.5 lb mole/min.

It is desired to design a pneumatic control system for this reactor. The system is to adjust the flow rate of the pure reactant stream so as to maintain the concentration of product in the exit stream at a mole fraction of 0.60. The flow rate of the pure reactant stream is small compared to the total react r feed.

There are available:

1. *A control valve.* Its discharge rate varies from 0 to 15 lb mole/hr of reactant a ; the pressure on the diaphragm changes from 15 to 3 psi.

2. *A concentration-measuring device.* Its electrical output varies from 0.15 to 2.0 mv, as the mole fraction of product varies from 0.5 to 0.8 in the exit stream.

3. *A transducer.* It moves the controller measuring pen 4 in. as the electrical input varies from 0 to 3 mv.

a. Design a pneumatic controller for this system.
   In your design considerations, you should:

   1. Investigate various modes of control.
   2. Prepare typical graphs showing the expected step response of the controlled system.
   3. Indicate, preferably by graphical means, the effect of change of controller settings.
   4. List any assumptions you must make to complete this design.

b. You should also recommend other dynamic information about the concentration versus flow-rate response which you would want to obtain in order to improve your design. For example, would you recommend a frequency response, impulse test, or ramp test as the next experiment? Discuss the advantages and disadvantages of each *relative to the particular system at hand.* Indicate graphically how you expect the reactor to respond to these and other forcing functions of interest. (Specify these forcing functions carefully. Thus, if you want a ramp test indicate the slope of the ramp and your reasons for this choice.)

c. Give some quantitative measure of the improvement in controlled response which could be achieved by redesigning the reactor to have different time constants.

Time since introduction of step change, min	Mole fraction of product in exit stream
−20	0.599
−10	0.600
0	0.601
5	0.600
10	0.601
15	0.602
20	0.605
25	0.607
30	0.609
35	0.610
40	0.612
45	0.613
50	0.612
55	0.615
60	0.617
75	0.620
90	0.622
105	0.623
120	0.625
135	0.626
150	0.627
180	0.629
210	0.629
240	0.630
300	0.629
360	0.631
420	0.630

25 *Theoretical Analysis of Complex Processes*

In order to investigate theoretically the control of a process, it is necessary first to know the dynamic characteristics of the process which is being controlled. In the previous chapters, the processes have been very simple for the purpose of illustrating the control theory. Many physical processes are extremely complicated, and it requires considerable effort to construct a mathematical model that will adequately simulate the dynamics of the actual system. In this chapter, we shall analyze several complex systems to indicate some of the types of problems which are encountered. In these examples, the technique of linearization, first presented in Chap. 6, will be applied to a function of several variables. One example will lead to a multiloop control system. In the last section, distributed-parameter systems will be discussed.

CONTROL OF A STEAM - JACKETED KETTLE

The dynamic response and control of the steam-jacketed kettle shown in Fig. 25.1 are to be considered. The system consists of a kettle through

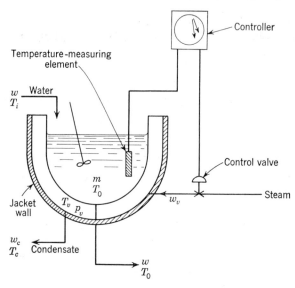

Fig. 25.1 Control of a steam-jacketed kettle.

which water flows at a variable rate w lb/time. The entering water is at temperature T_i, which may vary with time. The kettle water, which is well agitated, is heated by steam condensing in the jacket at temperature T_v and pressure p_v. The temperature of the water in the kettle is measured and transmitted to the controller. The output signal from the controller is used to change the stem position of the valve, which adjusts the flow of steam to the jacket. The major problem in this example is to determine the dynamic characteristics of the kettle. The kettle is actually a nonlinear system, and in order to obtain a linear model a number of simplifying assumptions are needed.

Analysis of Kettle The following assumptions are made for the kettle:

1. The heat loss to the atmosphere is negligible.

2. The holdup volume of water in the kettle is constant.

3. The thermal capacity of the kettle wall, which separates steam from water, is negligible compared with that of the water in the kettle.

4. The thermal capacity of the outer jacket wall, adjacent to the surroundings, is finite, and the temperature of this jacket wall is uniform and equal to the steam temperature at any instant.

5. The kettle water is sufficiently agitated to result in a uniform temperature.

6. The flow of heat from the steam to the water in the kettle is described by the expression

$$q = U(T_v - T_o)$$

where q = flow rate of heat, Btu/(hr)(ft^2)
U = overall heat-transfer coefficient, Btu/(hr)(ft^2)(°F)
T_v = steam temperature, °F
T_o = water temperature, °F

The overall heat-transfer coefficient U is constant.

7. The heat capacities of water and the metal wall are constant.
8. The density of water is constant.
9. The steam in the jacket is saturated.

The assumptions listed here are more or less arbitrary. For a specific kettle operating under a particular set of conditions, some of these assumptions may require modification.

The approach to this problem is to make an energy balance on the water side and another energy balance on the steam side. In order to aid the development of the transfer functions, a schematic diagram of the kettle is shown in Fig. 25.2. The symbols used throughout this analysis are defined as follows:

T_i = temperature of inlet water, °F
T_o = temperature of outlet water, °F
T_v = temperature of jacket steam, °F
T_c = temperature of condensate, °F
w = flow rate of inlet water, lb/time
w_v = flow rate of steam, lb/time
w_c = flow rate of condensate from kettle, lb/time
m = mass of water in kettle, lb
m_1 = mass of jacket wall, lb
V = volume of jacket steam space, ft^3
C = heat capacity of water, Btu/(lb)(°F)
C_1 = heat capacity of metal in jacket wall, Btu/(lb)(°F)
A = cross-sectional area for heat exchange, ft^2
t = time
H_v = specific enthalpy of steam entering, Btu/lb
H_c = specific enthalpy of condensate leaving, Btu/lb
U_v = specific internal energy of steam in jacket, Btu/lb
ρ_v = density of steam in jacket, lb/ft^3

An energy balance on the water side gives

$$wC(T_i - T_o) + UA(T_v - T_o) = mC\frac{dT_o}{dt} \tag{25.1}$$

In Eq. (25.1), the terms C, U, A, and m are constants. The first term in Eq. (25.1) is nonlinear, since it contains the product of flow rate and temperature, that is, wT_i and wT_o. In order to obtain a transfer function from Eq. (25.1), these nonlinear terms must be linearized. Before continuing

Fig. 25.2   **Schematic diagram of kettle.**

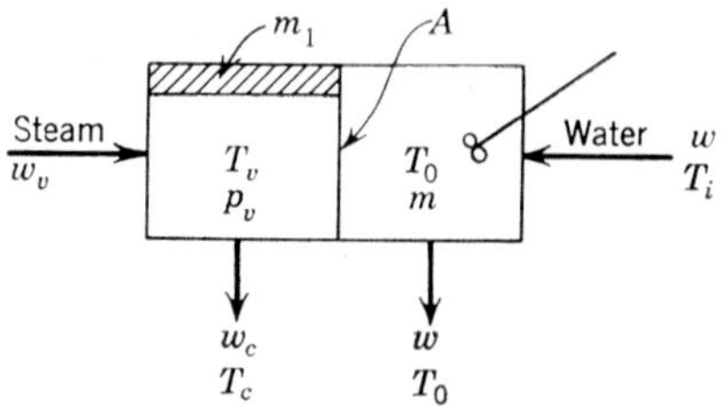

the analysis, we shall digress briefly to discuss the general problem of linearization of a function of several variables.

Consider a function of two variables, $z(x,y)$. By means of a Taylor series expansion, the function can be expanded[1] around an operating point (x_s, y_s) as follows:

$$z = z(x_s, y_s) + \frac{\partial z}{\partial x}\Big|_{x_s, y_s} (x - x_s) + \frac{\partial z}{\partial y}\Big|_{x_s, y_s} (y - y_s)$$
$$+ \text{ higher-order terms in } (x - x_s) \text{ and } (y - y_s) \quad (25.2)$$

The subscript s stands for steady state.

In control problems, the operating point (x_s, y_s), around which the expansion is to be made, is selected at steady-state values of the variables before any disturbance occurs. Linearization of the function z consists of retaining only the linear terms, on the basis that the deviations $(x - x_s)$, etc., will be small. Thus,

$$z \cong z_s + z_{x_s}(x - x_s) + z_{y_s}(y - y_s) \quad (25.3)$$

where z_{x_s} and z_{y_s} are the partial derivatives in Eq. (25.2). If z is a function of three or more variables, the linearized form is the same as that of Eq. (25.3) with an additional term for each variable.

The linearization expressed by Eq. (25.3) may be applied to the terms wT_i and wT_o in Eq. (25.1) to obtain

$$wT_i = w_s T_{i_s} + w_s(T_i - T_{i_s}) + T_{i_s}(w - w_s) \quad (25.4)$$

and

$$wT_o = w_s T_{o_s} + w_s(T_o - T_{o_s}) + T_{o_s}(w - w_s) \quad (25.5)$$

Notice that for these cases the nonlinear terms are wT_i and wT_o. The first partial derivatives, evaluated at the operating point, are

$$\frac{\partial(wT_i)}{\partial w}\Big|_{w_s, T_{i_s}} = T_{i_s}$$

$$\frac{\partial(wT_i)}{\partial T_i}\Big|_{w_s, T_{i_s}} = w_s$$

etc.

---

[1] The reader may refer to I. S. Sokolnikoff and R. M. Redheffer, "Mathematics of Physics and Modern Engineering," p. 257, McGraw-Hill Book Company, 1958, for further discussion of this expansion.

Introducing Eq. (25.4) and (25.5) into (25.1) gives the following linearized equation:

$$[(T_{i_s} - T_{o_s})(w - w_s) + w_s(T_i - T_o)]C + UA(T_v - T_o) = mC \frac{dT_o}{dt}$$
(25.6)

At steady state, $dT_o/dt = 0$, and Eq. (25.1) can be written

$$w_s C(T_{i_s} - T_{o_s}) + UA(T_{v_s} - T_{o_s}) = 0 \qquad (25.7)$$

Subtracting Eq. (25.7) from (25.6) and introducing the deviation variables

$$\begin{aligned} T'_i &= T_i - T_{i_s} \\ T'_o &= T_o - T_{o_s} \\ T'_v &= T_v - T_{v_s} \\ W &= w - w_s \end{aligned}$$

and rearranging give the result

$$C[(T_{i_s} - T_{o_s})W + w_s(T'_i - T'_o)] + UA(T'_v - T'_o) = mC \frac{dT'_o}{dt} \qquad (25.8)$$

Taking the transform of Eq. (25.8) and solving for $T'_o(s)$ give

$$T'_o(s) = \frac{K_1}{\tau_w s + 1} T'_i(s) + \frac{K_2}{\tau_w s + 1} T'_v(s) - \frac{K_3}{\tau_w s + 1} W(s) \qquad (25.9)$$

where

$$K_1 = \frac{w_s C}{UA + w_s C}$$

$$K_2 = \frac{UA}{UA + w_s C}$$

$$K_3 = \frac{C(T_{o_s} - T_{i_s})}{UA + w_s C}$$

$$\tau_w = \frac{mC}{UA + w_s C}$$

From Eq. (25.9), we see that the response of T'_o to T'_i, T'_v, or W is first-order with a time constant τ_w. The steady-state gains (K's) in Eq. (25.9) are all positive.

The following energy balance can be written for the steam side of the kettle:

$$w_v H_v - w_c H_c = UA(T_v - T_o) + \frac{Vd(\rho_v U_v)}{dt} + m_1 C_1 \frac{dT_v}{dt} \qquad (25.10)$$

Notice that we have made use of assumption 4 in writing the last term of Eq. (25.10), which implies that the metal in the outer jacket wall is always at the steam temperature.

A mass balance on the steam side of the kettle yields

$$w_v - w_c = V \frac{d\rho_v}{dt} \tag{25.11}$$

Combining Eqs. (25.10) and (25.11) to eliminate w_c gives

$$w_v(H_v - H_c) = (U_v - H_c)V \frac{d\rho_v}{dt} + m_1 C_1 \frac{dT_v}{dt} + UA(T_v - T_o)$$

$$+ V\rho_v \frac{dU_v}{dt} \tag{25.12}$$

The variables ρ_v, U_v, H_v, and H_c are functions of the steam and condensate temperatures and can be approximated by expansion in Taylor series and linearization as follows:

$$\begin{aligned}
\rho_v &= \rho_{v_s} + \alpha(T_v - T_{v_s}) \\
U_v &= U_{v_s} + \phi(T_v - T_{v_s}) \\
H_v &= H_{v_s} + \gamma(T_v - T_{v_s}) \\
H_c &= H_{c_s} + \sigma(T_c - T_{c_s})
\end{aligned} \tag{25.13}$$

where

$$\alpha = \frac{d\rho_v}{dT_v}\bigg|_s$$

$$\phi = \frac{dU_v}{dT_v}\bigg|_s$$

$$\gamma = \frac{dH_v}{dT_v}\bigg|_s$$

$$\sigma = \frac{dH_c}{dT_c}\bigg|_s$$

The parameters α, ϕ, γ, and σ in these relationships can be obtained from the steam tables once the operating point is selected.[1]

[1] For example, if the operating point is at 212°F and the deviation in steam temperature is 10°F, we obtain the following estimate of γ from the steam tables:

$$T_{v_s} = 212°F$$
$$H_{v_s} = 1150.4 \text{ Btu/lb}$$

At $T_v = 222°F$,

$$H_v = 1154.1$$

At $T_v = 202°F$,

$$H_v = 1146.6$$
$$\gamma \approx \frac{1154.1 - 1146.6}{222 - 202} = 0.375$$

and

$$H_v = 1150.4 + 0.375(T_v - 212)$$

In a similar manner, the properties of saturated steam can be used to evaluate α, ϕ, and σ.

Introducing the relationships of Eq. (25.13) into Eq. (25.12) and assuming the condensate temperature T_c to be the same as the steam temperature T_v give the following result:

$$[H_{v_s} - H_{c_s} + (\gamma - \sigma)(T_v - T_{v_s})]w_v$$
$$= \left[(U_{v_s} - H_{c_s}) + (2\phi - \sigma)(T_v - T_{v_s}) + \frac{\phi}{\alpha}\rho_{v_s} + \frac{m_1 C_1}{\alpha V} \right] \alpha V \frac{dT_v}{dt}$$
$$+ UA(T_v - T_o) \quad (25.14)$$

Some of the terms in Eq. (25.14) can be neglected. The term

$$(\gamma - \sigma)(T_v - T_{v_s})$$

can be dropped because it is negligible compared with $(H_{v_s} - H_{c_s})$. For example, for steam at atmospheric pressure, a change of $10°F$ gives a value of $(\gamma - \sigma)(T_v - T_{v_s})$ of about 7 Btu/lb while $(H_{v_s} - H_{c_s})$ is 970 Btu/lb. Similarly, the term $(2\phi - \sigma)(T_v - T_{v_s})$ can be neglected. For example, this term is about -4 Btu/lb for a change in steam temperature of $10°F$ for steam at about 1 atm pressure; the term $(U_{v_s} - H_{c_s})$ is 897 Btu/lb under these conditions. Also, the term $\phi\rho_{v_s}/\alpha$ is about 15 Btu/lb and can be neglected. Discarding these terms, writing the remaining terms in deviation variables, and transforming yield

$$T_v'(s) = \frac{1}{\tau_v s + 1} T_o'(s) + \frac{K_5}{\tau_v s + 1} W_v(s) \quad (25.15)$$

where

$$T_v' = T_v - T_{v_s}$$
$$W_v = w_v - w_{v_s}$$
$$K_5 = \frac{H_{v_s} - H_{c_s}}{UA}$$
$$\tau_v = \frac{(U_{v_s} - H_{c_s})\alpha V + m_1 C_1}{UA}$$

From Eq. (25.15), we see that the steam temperature T_v' depends on the steam flow rate W_v and the water temperature T_o'. The combination of Eqs. (25.9) and (25.15) give the dynamic response of the water temperature to changes in water flow rate, inlet water temperature, and steam flow rate. These equations are represented by a portion of the block diagram of Fig. 25.4. Before completing the analysis of the control system, we must consider the effect of valve-stem position on the steam flow rate.

Analysis of Valve   The flow of steam through the valve depends upon three variables: steam supply pressure, steam pressure in the jacket, and the valve-stem position, which we shall assume to be proportional to the pneumatic valve-top pressure p. For simplicity, assume the steam supply pressure to be constant with the result that the steam flow rate is a function

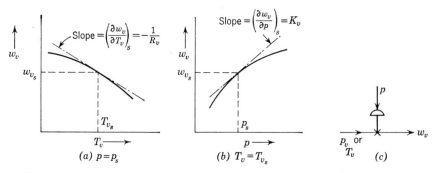

Fig. 25.3 **Linearization of valve characteristics from experimental tests.**

of only the two remaining variables; thus

$$w_v = f(p, p_v) \tag{25.16}$$

Because of the assumption that the steam in the jacket is always saturated, we know that p_v is a function of T_v; thus

$$p_v = g(T_v) \tag{25.17}$$

This functional relation can be obtained from the saturated steam tables. Equations (25.16) and (25.17) can be combined to give

$$w_v = f[p, g(T_v)] = f_1(p, T_v)$$

The function $f_1(p, T_v)$ is in general nonlinear, and if an analytic expression[1] is available, the function can be linearized as described previously. In this example, we shall assume that an analytic expression is not available. The linearized form of $f_1(p, T_v)$ can be obtained by making some experimental tests on the valve. If the valve-top pressure is fixed at its steady-state (or average) value and w_v is measured for several values of T_v (or p_v), a curve such as the one shown in Fig. 25.3a can be obtained. If the steam temperature T_v (or p_v) is held constant and the flow rate is measured at several values of valve-top pressure, a curve such as that shown in Fig. 25.3b

---

[1] The flow of steam through a control valve can often be represented by the relationship

$$w_v = A_0 C_v \sqrt{p_s - p_v} \tag{25.18}$$

where p_s = supply pressure of steam
$\quad p_v$ = pressure downstream of valve
$\quad A_0$ = cross-sectional area for flow of steam through valve
$\quad C_v$ = constant of the valve

For a linear valve, A_0 is proportional to stem position and the stem position is proportional to valve-top pressure p; under these conditions, Eq. (25.18) takes the form

$$w_v = C'_v p \sqrt{p_s - p_v} \tag{25.19}$$

can be obtained. These two curves can now be used to evaluate the partial derivatives in the linear expansion of $f_1(p, T_v)$ as we shall now demonstrate.

Expanding w_v about the operating point p_s, T_{v_s} and retaining only the linear terms give

$$w_v = w_{v_s} + \frac{\partial w_v}{\partial p}\bigg|_{p_s, T_{v_s}} (p - p_s) + \frac{\partial w_v}{\partial T_v}\bigg|_{p_s, T_{v_s}} (T_v - T_{v_s})$$

This equation can be written in the form

$$W_v = K_v P - \frac{1}{R_v} T_v' \tag{25.20}$$

where

$$W_v = w_v - w_{v_s}$$
$$P = p - p_s$$
$$T_v' = T_v - T_{v_s}$$
$$K_v = \frac{\partial w_v}{\partial p}\bigg|_{p_s, T_{v_s}}$$
$$\frac{1}{R_v} = -\frac{\partial w_v}{\partial T_v}\bigg|_{p_s, T_{v_s}}$$

The coefficients K_v and $-1/R_v$ in Eq. (25.20) are the slopes of the curves of Fig. 25.3 at the operating point p_s, T_{v_s}. This follows from the definition of a partial derivative. Notice that $1/R_v$ has been defined as the negative of the slope so that R_v is a positive quantity. The experimental approach described here for obtaining a linear form for the flow characteristics of a valve is always possible in principle. However, it must be emphasized that the linear form is useful only for small deviations from the operating point. If the operating point is changed considerably, the coefficients K_v and $1/R_v$ must be reevaluated. Notice that, in writing Eq. (25.20), we have assumed the valve to have no dynamic lag between p and stem position. This assumption is valid for a system having large time constants, such as a steam-jacketed kettle, as was demonstrated in Chap. 10.

Block Diagram of Control System We have now completed the analysis of the kettle and valve. A block diagram of the control system, based on Eqs. (25.9), (25.15), and (25.20), is shown in Fig. 25.4.

The controller action is not specified but merely denoted by G_c in the block diagram. Also, the feedback element is denoted as H. From Fig. 25.4, we see that the steam-jacketed kettle is a multiloop control system. Furthermore, the loops overlap. The block diagram can be used to obtain the overall transfer function between any two variables by applying the methods of Chap. 12. After considerable algebraic manipulation, the following result is obtained:

$$T_o' = \frac{G_c G_2 G_5 K_v}{D(s)} R + \frac{G_1(1 + G_5/R_v)}{D(s)} T_i' - \frac{G_3(1 + G_5/R_v)}{D(s)} W \tag{25.21}$$

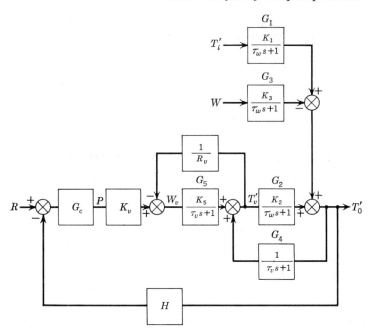

Fig. 25.4   Block diagram for control of steam-jacketed kettle.

where $D(s) = 1 + G_5/R_v + G_cG_2G_5K_vH - G_2G_4$.   The terms G_1, G_2, G_3, G_4, G_5, G_c, and H are defined in Fig. 25.4.   For example, if $G_c = K_c$ and $H = 1$, one obtains from Eq. (25.21) the transfer function

$$\frac{T'_o}{R} = \frac{K}{\tau^2 s^2 + 2\zeta\tau s + 1} \tag{25.22}$$

where

$$K = \frac{K_cK_vK_2K_5}{D_1}$$

$$\tau^2 = \frac{\tau_v\tau_w}{D_1}$$

$$2\zeta\tau = \frac{\tau_v + \tau_w + K_5\tau_w/R_v}{D_1}$$

$$D_1 = 1 + \frac{K_5}{R_v} + K_cK_vK_2K_5 - K_2$$

It is seen that the response of the control system is second-order when proportional control is used and the measuring element does not have dynamic lag.   Notice that the parameters K, τ^2, and $2\zeta\tau$ in Eq. (25.22) are positive.   This follows from the fact that the parameters K_c, **K_v**, K_2,

K_5, R_v, τ_v, and τ_w are all positive and that $K_2 < 1$. When a block diagram of a control system becomes very complicated, such as the one in this example, it is convenient to simulate the control system with an analog computer. When computer simulation is selected as the means of studying the transient response of the control system, the block diagram can be translated directly into a computer circuit. This computer-simulation technique will be covered in detail in Chap. 32.

DYNAMIC RESPONSE OF A GAS ABSORBER

Another example of a complex system is the plate absorber[1] shown in Fig. 25.5. In this process, air containing a soluble gas such as ammonia is contacted with fresh water in a two-plate column in order to remove part of the ammonia from the gas. The action of gas bubbling through the liquid causes thorough mixing of the two phases on each plate. During the mixing process, ammonia diffuses from the bubbles into the liquid. In an industrial operation, many plates may be used; however, for simplicity, we consider only two plates in this example, since the basic principles are unaffected by the number of plates.

Our problem is to analyze the system for its dynamic response. In other words, we want to know how the concentrations of liquid and gas change as a result of change in inlet composition or flow rate.

[1] The reader who has not studied gas absorption may find this subject presented in any textbook on chemical engineering unit operations. For example, see C. O. Bennett and J. E. Myers, "Momentum, Heat, and Mass Transfer," Chap. 39, McGraw-Hill Book Company, New York, 1962.

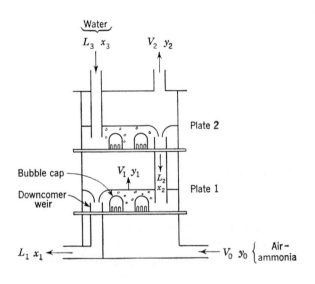

Fig. 25.5 Bubble-cap gas absorber.

Throughout the analysis, the following symbols are used:

L_n = flow of liquid leaving nth plate, moles/min
V_n = flow of gas leaving nth plate, moles/min
x_n = concentration of liquid leaving nth plate, mole fraction NH_3
y_n = concentration of gas leaving nth plate, mole fraction NH_3
H_n = holdup (or storage) of liquid on nth plate, moles

In order to avoid too many complicating details, the following assumptions will be used:

1. The temperature and total pressure throughout the column are uniform and do not vary with changes in flow rates.
2. The entering gas stream is dilute (say 5 mole percent NH_3) with the consequence that we can neglect the decrease in total molar flow rate of gas as ammonia is removed.   Likewise, we can assume that the molar flow rate of liquid does not increase as ammonia is added.
3. The plate efficiency is 100 percent,[1] which means that the vapor and liquid streams leaving a plate are in equilibrium.   Such a plate is called an *ideal* equilibrium stage.
4. The equilibrium relationship is linear and is given by the expression

$$y_n = mx_n^* + b \tag{25.23}$$

where m and b are constants which depend on the temperature and total pressure of the system and x_n^* is the concentration of liquid in equilibrium with gas of concentration y_n.   For an ideal plate

$$x_n = x_n^*$$

5. The holdup of liquid H_n on each plate is constant and independent of flow rate.   Furthermore, the holdup is the same for each plate, that is, $H_1 = H_2 = H$.
6. The holdup of gas between plates is negligible.   As a consequence of this assumption and assumption 2, the flow rate of gas from each plate

---

[1] If the efficiency of the plate is not 100 percent, we can introduce an individual tray efficiency of the Murphree type, defined as

$$E_n = \frac{x_n - x_{n+1}}{x_n^* - x_{n+1}}$$

where x_n^* is the concentration of the liquid in equilibrium with gas of composition y_n. Notice that for an ideal plate $E_n = 1$ and $x_n = x_n^*$.   In general the efficiency of a plate depends on the design of the plate, the properties of the gas and liquid streams, and the flow rates.   We could include efficiency in our mathematical model; however, to do so would greatly increase the complexity of the problem.   To account properly for the variation in efficiency with flow rates would require empirical relationships for a specific plate design.

is the same and equal to the entering gas flow rate; that is,

$$V_0 = V_1 = V_2 = V$$

In this list of assumptions, the one which is most likely to be invalid for a practical process is that the plate is an ideal equilibrium stage.

Analysis   We begin the analysis of this process by writing an ammonia balance around each plate.   A mass balance on ammonia around plate 1 gives

$$H \frac{dx_1}{dt} = L_2x_2 + Vy_0 - L_1x_1 - Vy_1 \tag{25.24}$$

This last equation states that the accumulation of NH_3 on plate 1 is equal to the flow of NH_3 into the plate minus the flow of NH_3 out of the plate. Notice that V and H do not have subscripts because of assumptions 5 and 6.

A mass balance on ammonia around plate 2 gives

$$H \frac{dx_2}{dt} = Vy_1 - L_2x_2 - Vy_2 \tag{25.25}$$

The last equation does not contain a term L_3x_3, since we have assumed that $x_3 = 0$.

For an ideal plate $x_n = x_n^*$, and the equilibrium relation of Eq. (25.23) becomes

$$y_n = mx_n + b$$

Substituting the equilibrium relationship into (25.24) and (25.25) gives

$$H \frac{dx_1}{dt} = L_2x_2 - L_1x_1 + Vm(x_0 - x_1)$$

and

$$H \frac{dx_2}{dt} = Vm(x_1 - x_2) - L_2x_2$$

where $x_0 = (y_0 - b)/m$ is the composition of liquid which would be in equilibrium with the entering gas of composition y_0.   Solving these last two equations for the derivatives gives

$$\frac{dx_1}{dt} = \frac{1}{H}(L_2x_2 - L_1x_1) + \frac{Vm}{H}(x_0 - x_1) \tag{25.26}$$

$$\frac{dx_2}{dt} = \frac{Vm}{H}(x_1 - x_2) - \frac{1}{H}L_2x_2 \tag{25.27}$$

Thus far the analysis has resulted in two nonlinear first-order differential equations.   The nonlinear terms in Eqs. (25.26) and (25.27) are L_2x_2 and L_1x_1.   The forcing functions in this process, which must be specified as functions of t, are the inlet gas concentration $[x_0 = (y_0 - b)/m]$ and the inlet liquid flow rate L_3. In order to solve for $x_1(t)$ and $x_2(t)$, we must have two more equations, which can be obtained by considering the liquid-flow dynamics on each plate. Assume that each plate can be considered as a first-order system for which the following equations hold:[1]

$$\tau_2 \frac{dL_2}{dt} = L_3 - L_2 \tag{25.28}$$

and

$$\tau_1 \frac{dL_1}{dt} = L_2 - L_1 \tag{25.29}$$

The time constants in these equations (τ_1 and τ_2) can be determined experimentally by the methods of Chap. 24. The first-order representation for liquid dynamics was found to be adequate by Nobbe.[2] We now have four differential equations [Eqs. (25.26) to (25.29)], and six variables (x_1, x_2, x_0, L_1, L_2, L_3). Since x_0 and L_3 are the forcing functions, which are specified functions of time, these four equations can be solved for $x_1(t)$, $x_2(t)$, $L_1(t)$, and $L_2(t)$ in terms of x_0 and L_3.

We shall now divide the problem into two cases. The first case requires that we find the response of y_2 to a change in the inlet gas concentration only, the liquid flow rate remaining constant. In this case, the problem is linear and only Eqs. (25.26) and (25.27) are needed.

In the second case, it is assumed that we want to know the change in outlet concentration y_2 for a change in both inlet flow and inlet gas concentration. For this case, four simultaneous differential equations must be solved, two of which contain nonlinear terms. One approach to this problem is to linearize the nonlinear terms as was done in the case of the steam-jacketed kettle of the previous example; however, since this technique

---

[1] The assumption that the plate behaves as a first-order system with respect to liquid-flow dynamics would have to be justified experimentally. For the common bubble-cap plate, liquid builds up on the plate and flows over a weir, which may consist of a circular pipe or a vertical plate. The resistance to flow from the plate is therefore a weir, for which flow-head relationships are known (see footnote in Chap. 6). However, these flow-head relationships for weirs have been developed for the flow of liquids which are not aerated. In the case of flow of liquid over a bubble-cap plate, the liquid is very turbulent as a result of the agitation of the bubbles rising through the liquid. For this reason, one cannot expect the flow-head relations developed for quiescent flow to apply to the turbulent conditions present in the liquid on a plate. The true flow-head relation should be determined experimentally.

The fact that the flow rate is assumed to vary without change in holdup on the plate (assumption 5) appears to be contradictory. Actually, to increase the flow rate, a slight increase in level (and therefore holdup volume) above the crest of the weir is required. However, for the example under consideration, it will be assumed that the change in level needed to produce a substantial increase in flow is so small that the change in the amount of liquid on the plate is a small fraction of the total liquid holdup.

[2] L. B. Nobbe, "Transient Response of a Bubble-cap Plate Absorber," M. S. Thesis, Purdue University, January, 1961.

has already been illustrated, we shall not repeat it here. Rather, we shall solve this second case by analog computer simulation in Chap. 32, and no further discussion of this case will be given in this chapter.

For the first case where the inlet liquid flow rate remains constant $(L_1 = L_2 = L)$, Eqs. (25.26) and (25.27) can be written

$$\frac{dx_1}{dt} = -ax_1 + bx_2 + cx_0 \tag{25.30}$$

$$\frac{dx_2}{dt} = cx_1 - ax_2 \tag{25.31}$$

where

$$a = \frac{L}{H} + \frac{Vm}{H}$$

$$b = \frac{L}{H}$$

$$c = \frac{Vm}{H}$$

At steady state, $dx_1/dt = dx_2/dt = 0$, and Eqs. (25.30) and (25.31) can be written

$$0 = -ax_{1_s} + bx_{2_s} + cx_{0_s}$$
$$0 = cx_{1_s} - ax_{2_s}$$

Subtracting these steady-state equations from Eqs. (25.30) and (25.31) and introducing the deviation variables $X_1 = x_1 - x_{1_s}$, $X_2 = x_2 - x_{2_s}$, and $X_0 = x_0 - x_{0_s}$ give

$$\frac{dX_1}{dt} = -aX_1 + bX_2 + cX_0 \tag{25.32}$$

$$\frac{dX_2}{dt} = cX_1 - aX_2 \tag{25.33}$$

Notice that $X_0 = Y_0/m$ because

$$X_0 = x_0 - x_{0_s}$$
$$X_0 = \frac{y_0 - b}{m} - \frac{y_{0_s} - b}{m} = \frac{y_0 - y_{0_s}}{m} = \frac{Y_0}{m}$$

Equations (25.32) and (25.33) can be transformed to give

$$sX_1 = -aX_1 + bX_2 + cX_0$$
$$sX_2 = cX_1 - aX_2$$

We now have two algebraic equations and three unknowns (X_1, X_2, and X_0). Solving this pair of equations to eliminate X_1 and replacing X_2 by Y_2/m

and X_0 by Y_0/m give the transfer function

$$\frac{Y_2(s)}{Y_0(s)} = \frac{c^2/(a^2 - bc)}{[1/(a^2 - bc)]s^2 + [2a/(a^2 - bc)]s + 1} \tag{25.34}$$

This result shows that the response of outlet gas concentration to a change in inlet gas concentration is second-order. One can show[1] that ζ for this system is greater than 1, meaning that the response is overdamped. If the analysis is repeated for a gas absorber containing n plates, it will be found that the response between inlet gas concentration and outlet gas concentration is nth-order.

DISTRIBUTED - PARAMETER SYSTEMS

Heat Conduction into a Solid   In Chap. 5, the analysis of the mercury thermometer was based on a "lumped-parameter" model. At that time, reference was made to a distributed-parameter model of the thermometer. To illustrate the difference between a lumped-parameter system and a distributed-parameter system, consider a slab of solid conducting material of infinite thickness, as shown in Fig. 25.6. Let the input to this system be the temperature at the left face ($x = 0$), which is some arbitrary function of time. The output will be the temperature at the position $x = L$. For convenience, we may consider this system to represent the response of a bare thermocouple embedded in a thick wall, as the surface of the wall experiences a variation in temperature. The conductivity k, heat capacity C, and density ρ of the conducting material are constant,

---

[1] Equation (25.34) is of the standard second-order form, $K/(\tau^2 s^2 + 2\zeta\tau s + 1)$, with the parameters

$$\tau^2 = \frac{1}{a^2 - bc} \quad \text{and} \quad 2\zeta\tau = \frac{2a}{a^2 - bc}$$

Solving these two equations to eliminate τ gives

$$\zeta = \frac{1}{\sqrt{1 - bc/a^2}}$$

Writing a and b in terms of the original system parameters (L, H, V, m) gives

$$\zeta = \left[1 - \frac{(L/H)(Vm/H)}{(L/H + Vm/H)^2}\right]^{-\frac{1}{2}}$$

Simplifying this expression gives

$$\zeta = \left[1 - \frac{Vm/L}{(1 + Vm/L)^2}\right]^{-\frac{1}{2}}$$

Since $Vm/L > 0$, we see that $\zeta > 1$.

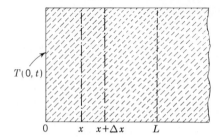

$T(0, t)$

0   x $x+\Delta x$ L

Fig. 25.6 Heat conduction in a solid.

independent of temperature.   Initially $(t < 0)$, the slab is at a uniform steady-state temperature.   Therefore in deviation variables, which will be used henceforth, the initial temperature is zero.   The cross-sectional area of the slab is A.

   Analysis: In this problem the temperature in the slab is a function of position and time and is indicated by $T(x,t)$.   The temperature at the surface is indicated by $T(0,t)$, and that at $x = L$ by $T(L,t)$.   To derive a differential equation that describes the heat conduction in the slab, we first write an energy balance over a differential length Δx of the slab.   This energy balance can be written

$$\begin{Bmatrix} \text{Flow of heat into} \\ \text{left face by} \\ \text{conduction} \end{Bmatrix} - \begin{Bmatrix} \text{flow of heat out of} \\ \text{right face by} \\ \text{conduction} \end{Bmatrix} = \begin{Bmatrix} \text{rate of accumulation} \\ \text{of internal energy} \\ \text{in the volume element} \end{Bmatrix}$$

$$(25.35)$$

   The flow of heat by conduction follows Fourier's law:

$$q = -k\,\frac{\partial T}{\partial x} \tag{25.36}$$

where q = heat flux by conduction
   $\partial T/\partial x$ = temperature gradient
      k = thermal conductivity
Applying Eq. (25.36) to Eq. (25.35) gives

$$-Ak\,\frac{\partial T}{\partial x}\bigg|_x - \left(-Ak\,\frac{\partial T}{\partial x}\bigg|_{x+\Delta x}\right) = \frac{\partial}{\partial t}\,[C\rho A\,\Delta x(T - T_r)] \tag{25.37}$$

where T_r is the reference temperature used to evaluate internal energy. The term $\partial T/\partial x\big|_{x+\Delta x}$ can be written

$$\frac{\partial T}{\partial x}\bigg|_{x+\Delta x} = \frac{\partial T}{\partial x}\bigg|_x + \frac{\partial}{\partial x}\frac{\partial T}{\partial x}\,\Delta x \tag{25.38}$$

Substituting Eq. (25.38) into (25.37) and simplifying give the fundamental equation describing conduction in a solid

$$k\frac{\partial^2 T}{\partial x^2} = \rho C\frac{\partial T}{\partial t}$$

This is often written as

$$K\frac{\partial^2 T}{\partial x^2} = \frac{\partial T}{\partial t} \tag{25.39}$$

where $K = k/\rho C =$ thermal diffusivity.

Several points are worth noticing at this time. In this analysis, we have allowed the capacity for storing heat ($\rho C A$ per unit length of x) and the resistance to heat conduction ($1/kA$ per unit length of x) to be "spread out" or distributed uniformly throughout the medium. This distribution of capacitance and resistance is the basis for the term *distributed parameter*. The analysis has also led to a partial differential equation, which in general is more difficult to solve than the ordinary differential equation that results from a lumped-parameter model.

Transfer function: We are now in a position to derive a transfer function from Eq. (25.39). First notice that, since T is a function of both time t and position x, a transfer function may be written for an arbitrary value of x. In this problem, the temperature is to be observed at $x = L$; hence the transfer function will relate $T(L,t)$ to the temperature at the left surface $T(0,t)$, which is taken as the forcing function.

Equation (25.39) will be solved by the method of Laplace transforms. Taking the Laplace transform of both sides of Eq. (25.39) with respect to t gives

$$K\int_0^\infty \frac{\partial^2 T}{\partial x^2}(x,t)e^{-st}\,dt = \int_0^\infty \frac{\partial T}{\partial t}(x,t)e^{-st}\,dt \tag{25.40}$$

Consider first the integral on the left side of Eq. (25.40). Interchanging the order of integration and differentiation[1] results in

$$\int_0^\infty \frac{\partial^2 T}{\partial x^2}(x,t)e^{-st}\,dt = \frac{\partial^2}{\partial x^2}\int_0^\infty T(x,t)e^{-st}\,dt = \frac{d^2\bar{T}(x,s)}{dx^2} \tag{25.41}$$

where $\bar{T}(x,s)$ is the Laplace transform of $T(x,t)$.[2] It should be noted that the presence of x has no effect on the second integral of Eq. (25.41) because the integration is with respect to t. Also note that the derivative on the right side of Eq. (25.41) is taken as an ordinary derivative because $T(x,s)$

---

[1] This interchange is allowed for most functions of engineering interest. See R. V. Churchill, "Operational Mathematics," 2d ed., McGraw-Hill Book Company, New York, 1958.

[2] In this chapter the overbar will often be used to indicate the Laplace transform of a function of two variables.

will later be seen to be a function of only one independent variable x and a parameter s. Next consider the integral on the right side of Eq. (25.40). Again, the presence of x has no effect on the integration with respect to t, and the rule for the transform of a derivative may be applied directly to yield

$$\int_0^\infty \frac{\partial T}{\partial t}(x,t)e^{-st}\,dt = s\bar{T}(x,s) - T(x,0) \tag{25.42}$$

where $T(x,0)$ is the initial temperature distribution in the solid. Introducing the results of the transformation into Eq. (25.39) gives

$$K\frac{d^2\bar{T}(x,s)}{dx^2} = s\bar{T}(x,s) - T(x,0) \tag{25.43}$$

The partial differential equation has now been reduced to an ordinary differential equation, which can usually be solved without difficulty. It should be clear that s in Eq. (25.43) is merely a parameter, with the result that this equation is an *ordinary* second-order differential equation in the independent variable x. This follows because there are no derivatives with respect to s in Eq. (25.43). Since we have taken $T(x,0) = 0$ for the example under consideration, Eq. (25.43) becomes

$$\frac{d^2\bar{T}}{dx^2} - \frac{s}{K}\bar{T} = 0 \tag{25.44}$$

Equation (25.44) is a linear differential equation and can be solved to give

$$\bar{T} = A_1 e^{-\sqrt{s/K}x} + A_2 e^{\sqrt{s/K}x} \tag{25.45}$$

The arbitrary coefficients A_1 and A_2 may be evaluated as follows: In order that $\bar{T}$ may be finite as $x \to \infty$, it is necessary that $A_2 = 0$. Equation (25.45) then becomes

$$\bar{T} = A_1 e^{-\sqrt{s/K}x} \tag{25.45a}$$

The transformed forcing function at $x = 0$ is $\bar{T}(0,s)$, which can be substituted into Eq. (25.45a) to determine A_1; then

$$\bar{T}(0,s) = A_1 e^0$$

or

$$A_1 = \bar{T}(0,s)$$

Substituting A_1 into Eq. (25.45a) gives

$$\frac{\bar{T}(x,s)}{\bar{T}(0,s)} = e^{-\sqrt{s/K}x} \tag{25.46}$$

By specifying a particular value of x, say $x = L$, the transfer function is

$$\frac{\bar{T}(L,s)}{\bar{T}(0,s)} = e^{-\sqrt{s/K}L} \tag{25.47}$$

Step response: To illustrate the use of this transfer function, consider a forcing function which is the unit-step function; thus

$$T(0,t) = u(t)$$

for which case $\bar{T}(0,s) = 1/s$. Substituting this into Eq. (25.47) gives

$$\bar{T}(L,s) = \frac{1}{s} e^{-\sqrt{s/K}L} \tag{25.48}$$

To obtain the response in the time domain, we must invert Eq. (25.48). A table of transforms[1] gives the following transform pair:

$$L\left\{\frac{1}{s} e^{-\sqrt{s/K}x}\right\} = \text{erfc} \frac{x}{\sqrt{4Kt}} \tag{25.49}$$

where erfc x is the error-function complement of x defined as

$$\text{erfc } x = 1 - \frac{2}{\sqrt{\pi}} \int_0^x e^{-u^2} du$$

This function is tabulated in many textbooks[2] and mathematical tables. Using this transform pair, Eq. (25.48) becomes

$$T(L,t) = \text{erfc} \frac{L}{\sqrt{4Kt}} = \text{erfc} \left[\frac{1}{2}\left(\frac{Kt}{L^2}\right)^{-\frac{1}{2}}\right] \tag{25.50}$$

A plot of T versus the dimensionless group Kt/L^2 is shown in Fig. 25.7.

Sinusoidal response: It is instructive to consider the response in temperature at $x = L$ for the case where the forcing function is a sinusoidal variation; thus

$$T(0,t) = A \sin \omega t$$

---

[1] Tables of transforms which include transform pairs that are frequently encountered in the solution of partial differential equations may be found in many textbooks on heat conduction and applied mathematics. For example, see H. S. Mickley, T. K. Sherwood, and C. E. Reed, "Applied Mathematics in Chemical Engineering," 2d ed., pp. 311–319, McGraw-Hill Book Company, New York, 1957.

Inversion of complicated transforms such as that of Eq. (25.48) can be achieved systematically by the method of complex residues, which is also discussed in the above reference.

[2] See Carslaw and Jaeger, "Conduction of Heat in Solids," 2d ed., p. 485, Oxford University Press, New York, 1959.

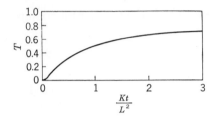

Fig. 25.7 Response of temperature in the interior of a solid to a unit-step change in temperature at the surface.

Using the substitution rule of Chap. 18, in which s is replaced by $j\omega$, Eq. (25.47) becomes

$$\frac{\bar{T}(L,j\omega)}{\bar{T}(0,j\omega)} = e^{-\sqrt{j\omega/K}\,L} \tag{25.51}$$

To obtain the AR and phase angle requires that the magnitude and argument of the right-hand side of Eq. (25.51) be evaluated. This can be done as follows: First write j in polar form; thus

$$j = e^{j\pi/2}$$

from which we get

$$\sqrt{j} = [e^{j(\pi/2)}]^{\frac{1}{2}} = \pm e^{j\pi/4} = \pm \frac{1}{\sqrt{2}}(1+j)$$

Substituting the positive form[1] of $\sqrt{j}$ into Eq. (25.51) gives

$$\frac{\bar{T}(L,j\omega)}{\bar{T}(0,j\omega)} = e^{-\sqrt{\omega/2K}\,L}e^{-j\sqrt{\omega/2K}\,L}$$

From this form, we can write by inspection

$$AR = \left|\frac{\bar{T}(L,j\omega)}{\bar{T}(0,j\omega)}\right| = e^{-\sqrt{\omega/2K}\,L} \tag{25.52}$$

$$\text{Phase angle} = \measuredangle\,\frac{\bar{T}(L,j\omega)}{\bar{T}(0,j\omega)} = -\sqrt{\frac{\omega}{2K}}\,L \qquad \text{rad} \tag{25.53}$$

From these results, it is seen that the AR approaches zero as $\omega \to \infty$ and the phase angle decreases without limit as $\omega \to \infty$. Such a system is said to have *nonminimum* phase lag characteristics. With the exception of the distance-velocity lag, all the systems that have been considered up to now have given a limited value of phase angle as $\omega \to \infty$. These are called minimum phase systems and always occur for lumped-parameter systems. The nonminimum phase behavior is typical of distributed-parameter systems.

[1] Notice that the substitution of $-(1+j)/\sqrt{2}$ into Eq. (25.51) leads to a result in which the AR is greater than 1 and the phase angle leads. This is contrary to the response of the physical system and is not admitted as a useful solution.

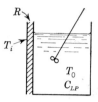

Fig. 25.8 **Lumped-parameter model of conduction in a** solid.

Lumped-parameter Models: We shall now further illustrate the relation between a distributed-parameter model and a lumped-parameter model, again using the temperature variation in the slab of Fig. 25.6 at $x = L$. We do this by trying to construct a lumped-parameter model which approximates the dynamics of the system. A simple first-order model may be conceived by considering an equivalent system (Fig. 25.8) in which all the resistance is placed in a thin layer (or film) adjacent to a well-stirred fluid while the fluid acts as a pure capacitance.

The capacitance C_{LP} in this model will be equivalent to that of the material between $x = 0$ and $x = L$ in the slab. The capacitance can be calculated as follows:

$$C_{LP} = C\rho L A \tag{25.54}$$

The resistance R will be taken as equivalent to that offered to conduction (at steady state) by the material between $x = 0$ and $x = L$. This can be calculated as follows: At steady state, Fourier's law takes the form

$$\frac{Q}{A} = \frac{k\,\Delta T}{L} \tag{25.55}$$

where Q = heat flow, Btu/hr
ΔT = temperature difference between $x = 0$ and $x = L$
Since $R = \Delta T/Q$, we have from Eq. (25.55)

$$R = \frac{L}{Ak} \tag{25.56}$$

From our experience with first-order systems, the time constant can be written

$$\tau = RC_{LP} = \frac{L}{Ak}\,(C\rho L A) = \frac{C\rho L^2}{k}$$

Using the definition of thermal diffusivity, $K = k/\rho C$, the time constant is

$$\tau = \frac{L^2}{K}$$

The transfer function relating T_o to T_i for the lumped-parameter model is

$$G(s) = \frac{T_o(s)}{T_i(s)} = \frac{1}{\tau s + 1}$$

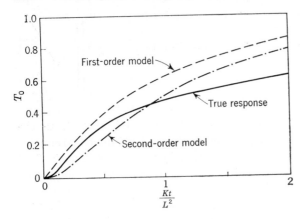

Fig. 25.9 Comparison of lumped-parameter models with distributed-parameter system.

and the corresponding unit-step response is

$$T_o(t) = 1 - e^{-t/\tau} = 1 - e^{-(K/L^2)t} \tag{25.57}$$

A comparison between the true response of Eq. (25.50) and the first-order lumped-parameter response of Eq. (25.57) is shown in Fig. 25.9. As one might expect, there is considerable discrepancy between the two response curves. The curve for the first-order lumped model is considerably above the true response curve, although it does show the same general trends.

Other lumped-parameter models can be constructed which more closely resemble the distributed-parameter system. For example, intuition suggests that the slab between $x = 0$ and $x = \frac{4}{3}L$ be divided into two zones of equal thickness as shown in Fig. 25.10. The choice of the thickness of each zone being $\frac{2}{3}L$ places the midpoint of zone 2 at $x = L$. In this case, each zone of length $\frac{2}{3}L$ can be considered as a first-order lumped-parameter system having a time constant

$$\tau_1 = \left(\frac{2}{3}L\right)^2 \frac{1}{K} = \frac{4}{9}\frac{L^2}{K}$$

The reader can show that the system reduces to two *interacting* first-order

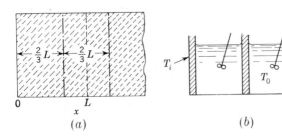

Fig. 25.10 Second-order lumped model.

systems in series with equal time constants (τ_1). The transfer function[1] relating the temperature in the second tank to the temperature at the surface ($x = 0$) is

$$\frac{T_o(s)}{T_i(s)} = \frac{1}{\tau_1{}^2 s^2 + 3\tau_1 s + 1} \tag{25.58}$$

This transfer function can be used to give the following response of T_o to a unit-step change in T_i:

$$T_o = 1 + 0.17 e^{-2.62 t/\tau_1} - 1.17 e^{-0.38 t/\tau_1} \tag{25.59}$$

The response given by Eq. (25.59) is also plotted on Fig. 25.9 for comparison with the first-order model and the true response. It can be seen that some improvement is achieved. In particular, the second-order model fits the true response curve near the origin in that the slope at the origin is zero. The first-order model gives a response with nonzero slope at the origin.

Other higher-order, lumped-parameter models can be proposed to approximate the distributed parameter system. The reader may want to create one of his own models and compare its unit-step response with the true response.

Distance-velocity Lag as a Distributed-parameter System We can demonstrate that the distance-velocity lag is, in fact, a distributed-parameter system as follows: Consider the flow of an incompressible fluid through an insulated pipe of uniform cross-sectional area A and length L, as shown in Fig. 25.11. The fluid flows at velocity v, and the velocity profile is flat. We know from Chap. 8 that the transfer function relating outlet temperature T_o to the inlet temperature T_i is

$$\frac{T_o(s)}{T_i(s)} = e^{-(L/v)s}$$

Let the pipe be divided into n zones as shown in Fig. 25.11*b*. If each zone of length L/n is considered to be a well-stirred tank, then the pipe is equivalent to n noninteracting first-order systems in series, each having a time

---

[1] It can be seen that Eq. (25.58) is equivalent to Eq. (7.27), which represented the dynamic response of a two-tank liquid-level system.

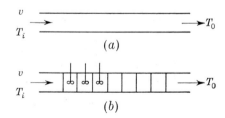

Fig. 25.11 Obtaining the transfer function of a distance-velocity lag from a lumped-parameter model.

constant[1]

$$\tau = \frac{L}{n}\frac{1}{v}$$

(Note that taking each zone to be a well-stirred tank is the step associated with lumping of parameters.)   The overall transfer function for this lumped-parameter model is therefore

$$\frac{T_o(s)}{T_i(s)} = \left(\frac{1}{\tau s + 1}\right)^n = \left[\frac{1}{(L/v)s/n + 1}\right]^n$$

To "distribute" the parameters, we let the size of the individual lumps go to zero by letting $n \to \infty$.

$$\frac{T_o(s)}{T_i(s)} = \lim_{n \to \infty}\left[\frac{1}{(L/v)s/n + 1}\right]^n$$

The thermal capacitance is now distributed over the tube length.   It can be shown by use of the calculus that the limit is

$$e^{-(L/v)s}$$

which is the transfer function derived previously.   This demonstration should provide some initial insight into the relationship between a distributed-parameter system and a lumped-parameter system and indicates that a distance-velocity lag is a distributed system.

Heat Exchanger[2]   As our last example of a distributed-parameter system, we consider the double-pipe heat exchanger shown in Fig. 25.12. The fluid which flows through the inner pipe at constant velocity v is heated by steam condensing outside the pipe.   The temperature of the fluid entering the pipe and the steam temperature vary according to some arbitrary

[1] This expression for τ is equivalent to that appearing in Eq. (9.10).   Since the transfer function for flow through a tank was developed in Chap. 9, the analysis will not be repeated here.

[2] The analysis presented here essentially follows that of W. C. Cohen and E. F. Johnson, *IEC*, **48** (1956).   These authors also present the experimental results of frequency response tests on a double-pipe, steam-to-water heat exchanger.

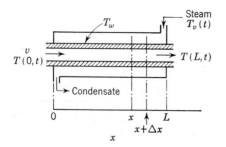

Fig.   25.12   Double-pipe   heat   exchanger.

functions of time.   The steam temperature varies with time, but not with position in the exchanger.   The metal wall separating steam from fluid is assumed to have significant thermal capacity which must be accounted for in the analysis.   The heat transfer from the steam to the fluid depends on the heat-transfer coefficient on the steam side (h_o) and the convective transfer coefficient on the water side (h_i).   The resistance of the metal wall is neglected.   The goal of the analysis will be to find transfer functions relating the exit fluid temperature $T(L,t)$ to the entering fluid temperature $T(0,t)$ and the steam temperature $T_v(t)$.

The following symbols will be used in this analysis:

$$T(x,t) = \text{fluid temperature}$$
$$T_w(x,t) = \text{wall temperature}$$
$$T_v(t) = \text{steam temperature}$$
$$T_r = \text{reference temperature for evaluating enthalpy}$$
$$\rho = \text{density of fluid}$$
$$C = \text{heat capacity of fluid}$$
$$\rho_w = \text{density of metal in wall}$$
$$C_w = \text{heat capacity of metal in wall}$$
$$A_i = \text{cross-sectional area for flow inside pipe}$$
$$A_w = \text{cross-sectional area of metal wall}$$
$$D_i = \text{inside diameter of inner pipe}$$
$$D_o = \text{outside diameter of inner pipe}$$
$$h_i = \text{convective heat-transfer coefficient inside pipe}$$
$$h_o = \text{heat-transfer coefficient for condensing steam}$$
$$v = \text{fluid velocity}$$

Analysis: We begin the analysis by writing a differential energy balance for the fluid inside the pipe over the volume element of length Δx (see Fig. 25.12).   This balance can be stated

$$\begin{Bmatrix} \text{Flow of} \\ \text{enthalpy in} \end{Bmatrix} - \begin{Bmatrix} \text{flow of} \\ \text{enthalpy out} \end{Bmatrix} + \begin{Bmatrix} \text{heat transferred} \\ \text{through film on} \\ \text{inside wall} \end{Bmatrix}$$
$$= \begin{Bmatrix} \text{rate of accum-} \\ \text{ulation of} \\ \text{internal energy} \end{Bmatrix} \quad (25.60)$$

The terms in this balance can be evaluated as follows:

$$\text{Flow of enthalpy in at } x = vA_i\rho C(T - T_r)$$
$$\text{Flow of enthalpy out at } x + \Delta x = vA_i\rho C\left[\left(T + \frac{\partial T}{\partial x}\Delta x\right) - T_r\right]$$
$$\text{Heat transfer through film} = \pi D_i h_i \Delta x(T_w - T)$$
$$\text{Accumulation of internal energy} = \frac{\partial}{\partial t}[A_i\rho \Delta x\, C(T - T_r)]$$

Introducing these terms into Eq. (25.60) gives, after simplification,

$$\frac{\partial T}{\partial t} = -v \frac{\partial T}{\partial x} + \frac{1}{\tau_1}(T_w - T) \tag{25.61}$$

where

$$\frac{1}{\tau_1} = \frac{\pi D_i h_i}{A_i \rho C}$$

An energy balance is next written for the metal in the wall, over the volume element of length Δx.   This can be stated as follows:

$$\left\{\begin{matrix} \text{Heat transfer in} \\ \text{through steam} \\ \text{condensate film} \end{matrix}\right\} - \left\{\begin{matrix} \text{heat transfer out} \\ \text{through fluid film} \end{matrix}\right\} = \left\{\begin{matrix} \text{accumulation of} \\ \text{energy in wall} \end{matrix}\right\}$$

Expressing each term in this balance by symbols gives

$$\pi D_o h_o \, \Delta x \, (T_v - T_w) - \pi D_i h_i \, \Delta x \, (T_w - T) = A_w \, \Delta x \, \rho_w C_w \frac{\partial T_w}{\partial t} \tag{25.62}$$

Simplifying this expression gives

$$\frac{\partial T_w}{\partial t} = \frac{1}{\tau_{22}}(T_v - T_w) - \frac{1}{\tau_{12}}(T_w - T) \tag{25.63}$$

where

$$\frac{1}{\tau_{12}} = \frac{\pi D_i h_i}{A_w \rho_w C_w} \qquad \frac{1}{\tau_{22}} = \frac{\pi D_o h_o}{A_w \rho_w C_w}$$

We now have obtained the differential equations which describe the dynamics of the system.   As in previous problems, the dependent variables will be transformed to deviation variables.   At steady state, the time derivatives in Eqs. (25.61) and (25.63) are zero, and it follows that

$$0 = -v \frac{dT_s}{dx} + \frac{1}{\tau_1}(T_{w_s} - T_s) \tag{25.64}$$

and

$$0 = \frac{1}{\tau_{22}}(T_{v_s} - T_{w_s}) - \frac{1}{\tau_{12}}(T_{w_s} - T_s) \tag{25.65}$$

where the subscript s is used to denote the steady-state value.   Note that to determine the steady-state values of the temperature requires the solution of two simultaneous equations, the first of which is an ordinary differential equation.   Thus, the steady-state temperature T_s is a function of x and may be obtained by solution of Eqs. (25.64) and (25.65) as

$$T_s = T_{v_s} + (T_{s_0} - T_{v_s}) \exp\left[-\frac{x}{v\tau_1} \Big/ \left(1 + \frac{\tau_{22}}{\tau_{12}}\right)\right]$$

where T_{s_0} is the normal entrance temperature. All equations for T' to be derived below should be recognized as deviations from this expression.

Subtracting Eq. (25.64) from (25.61) and Eq. (25.65) from (25.63) and introducing deviation variables give

$$\frac{\partial T'}{\partial t} = -v \frac{\partial T'}{\partial x} + \frac{1}{\tau_1} (T'_w - T') \tag{25.66}$$

and

$$\frac{\partial T'_w}{\partial t} = \frac{1}{\tau_{22}} (T'_v - T'_w) - \frac{1}{\tau_{12}} (T'_w - T') \tag{25.67}$$

where

$$T' = T - T_s$$
$$T'_w = T_w - T_{w_s}$$
$$T'_v = T_v - T_{v_s}$$

Equations (25.66) and (25.67) may be transformed with respect to t to yield

$$s\bar{T}' = -v \frac{d\bar{T}'}{dx} + \frac{1}{\tau_1} (\bar{T}'_w - \bar{T}') \tag{25.68}$$

and

$$s\bar{T}'_w = \frac{1}{\tau_{22}} (\bar{T}'_v - \bar{T}'_w) - \frac{1}{\tau_{12}} (\bar{T}'_w - \bar{T}') \tag{25.69}$$

where

$$\bar{T}' = \bar{T}'(x,s)$$
$$\bar{T}'_w = \bar{T}'_w(x,s)$$
$$\bar{T}'_v = \bar{T}'_v(s)$$

In Eqs. (25.68) and (25.69) it has been assumed that the exchanger is initially at steady state, so that $T(x,0) = T_s$, $T_w(x,0) = T_{w_s}$, and $T_v(0) = T_{v_s}$.

Eliminating $\bar{T}'_w$ from Eqs. (25.68) and (25.69) gives after considerable simplification

$$\frac{d\bar{T}'}{dx} + \frac{a}{v} \bar{T}' = \frac{b}{v} \bar{T}'_v \tag{25.70}$$

where

$$a(s) = s + \frac{1}{\tau_1} - \frac{\tau_{22}}{\tau_1(\tau_{12}\tau_{22}s + \tau_{12} + \tau_{22})}$$

$$b(s) = \frac{\tau_{12}}{\tau_1(\tau_{12}\tau_{22}s + \tau_{12} + \tau_{22})}$$

Equation (25.70) is an ordinary first-order differential equation with boundary condition $\bar{T}'(x,s) = \bar{T}'(0,s)$ at $x = 0$.

It can be readily solved to yield

$$\bar{T}'(x,s) = \bar{T}'(0,s) + [1 - e^{-(a/v)x}]\left[\frac{b}{a}\bar{T}'_v(s) - \bar{T}'(0,s)\right] \tag{25.71}$$

where $\bar{T}'(0,s)$ is the transform of the fluid temperature at the entrance to the pipe and $\bar{T}'_v(s)$ is the transform of the steam temperature. From Eq. (25.71), the transfer functions can be obtained as follows:

If the steam temperature does not vary, $T'_v(s) = 0$; the transfer function relating temperature at the end of the pipe ($x = L$) to temperature at the entrance is

$$\frac{\bar{T}'(L,s)}{\bar{T}'(0,s)} = e^{-(a/v)L} \tag{25.72}$$

Setting $1/\tau_1$ to zero in the expression for $a(s)$ [Eq. (25.70)] shows that $a(s) = s$ and hence the response is simply that of a distance-velocity lag. This is in agreement with the physical situation where h_i approaches zero [Eq. (25.61)], for which case the wall separating cold fluid from hot fluid acts as a perfect insulator. We saw in Chap. 8 that this situation is represented by a distance-velocity lag.

If the temperature of the fluid entering the pipe does not vary, the transfer function relating the exit fluid temperature to the steam temperature is

$$\frac{\bar{T}'(L,s)}{\bar{T}'_v(s)} = \frac{b}{a}[1 - e^{-(a/v)L}] \tag{25.73}$$

In principle, the response in the temperature of the fluid leaving the exchanger can be found for any forcing function, $T(0,t)$ or $T_v(t)$, by introducing the corresponding transforms into Eq. (25.72) or (25.73). However,

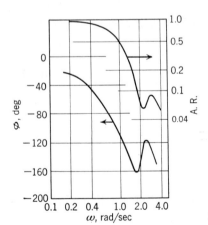

Fig. 25.13 Bode diagram of heat exchanger for variation in steam temperature. (Cohen and Johnson.)

the resulting expression is very complex and cannot be easily inverted. For the case of sinusoidal inputs, the substitution rule discussed in Chap. 18 can be used to determine the AR and phase angle of the frequency response. Cohen and Johnson give a Bode diagram corresponding to Eq. (25.73) for a specific set of heat-exchanger parameters. This diagram is shown in Fig. 25.13.

Notice that the theory predicts an interesting resonance effect at higher frequencies. The resonance effect has been observed experimentally in a steam-to-water exchanger.[1] Unfortunately, the experimental data of Cohen and Johnson do not extend to sufficiently high frequencies to exhibit resonance. The reader is referred to the original article for further details.

SUMMARY

In this chapter, several complex systems have been analyzed mathematically. The result of each analysis was a set of equations (algebraic and/or differential) which presumably describe the dynamic response of the system to one or more disturbances. The process of obtaining the set of equations is often called *modeling*, and the set of equations is referred to as the *mathematical model* of the system. In general, the model is based upon the physics and chemistry of the system. For example, in the analysis of a heat exchanger, one may write that the heat flux through a wall is equal to a convective transfer coefficient times a temperature driving force.

For a process which is not well understood, there is little chance that an accurate model can be obtained from the theoretical approach used here. For such systems, a direct dynamic test can be made. To do this, a known disturbance such as a pulse, step, or sinusoidal input is applied and the response recorded. This approach was discussed in Chap. 24. On the other hand a model which is based on a theoretical analysis is extremely valuable, for it means that the system is well understood and that the effect of changes in system design and operation can be predicted.

The analysis of a steam-jacketed kettle provided an example of a nonlinear system containing nonlinear functions of several variables. The problem was handled by linearizing these functions about an operating point and ultimately obtaining a block diagram of the system from which the transfer function of the control system could be obtained. Although this approach is relatively straightforward, the resulting linear model can only be used over a narrow range of variables.

The analysis of the gas absorber gave some insight into the dynamic character of a typical multistage process which is widely used in the chemical process industries. A linear analysis of an n-plate column leads to n ordinary differential equations which combine to give an overdamped nth-order response. Nonlinearities may be present in this system in such

[1] S. Lees and J. O. Hougen, *Ind. Eng. Chem.*, **48**:1064 (1956).

forms as a product of flow and concentration or a nonlinear equilibrium relationship. When changes in inlet flow occur, a set of differential equations describing the dynamics of the liquid flow must be added to those describing mass transfer. When the change of plate efficiency with flow is considered, the model of a gas absorber becomes even more complex. Most of the design techniques developed in the past for multistage operations (gas absorption, distillation, etc.) have applied to steady-state operation. The dynamic analysis of such processes calls for dynamic parameters which are usually unavailable. For example, the liquid-flow dynamics of trays used in distillation towers are relatively unknown. As more research is directed toward evaluation of the dynamic parameters, more accurate models can be constructed.

The discussion of distributed-parameter systems further illustrated the complexities which can arise in physical systems and emphasized the need for lumping of parameters whenever this is justified. The distributed-parameter systems lead to partial differential equations, which may be very difficult to solve for most of the forcing functions of practical interest. However, we saw that the response of distributed-parameter systems to sinusoidal forcing functions can be obtained directly by application of the substitution rule, in which s is replaced by $j\omega$. A distributed-parameter system features nonminimum phase lag characteristics. This is in sharp contrast to the lumped-parameter systems for which the phase angle approaches a limit at infinite frequency.

As systems are analyzed in more detail and with fewer assumptions, the models which describe them become more complex, although more accurate. To predict the response of the system from the model requires that equations of the model be solved for some specific input disturbance. The only practical way to solve a complex model is to use an analog or digital computer. This method of solving the mathematical model is often called computer simulation. The computer response will resemble that of the physical system if the model is accurate. Considerable attention has been given to computer simulation in the past several years. In the last section of this text, the analog computer and its use to simulate control systems will be discussed in considerable detail.

part **VII** Nonlinear Methods

26 *Phase Space*

In the previous chapters, we have confined our attention to the behavior of linear systems or to the analysis of linearized equations representative of nonlinear systems in the vicinity of the steady-state condition. While much useful information can be obtained from such analysis, it frequently is desirable or necessary to consider nonlinearities in control system design.

No real physical system is truly linear, particularly over a wide range of operating variables. Hence, to be complete, a control system design should allow for the possibility of a large deviation from steady-state behavior and resulting nonlinear behavior. The purpose of the next four chapters is to introduce some of the tools which can be used for this purpose and to indicate some of the complications which arise when nonlinear systems are considered.

Definition of a Nonlinear System   *A nonlinear system is one for which the principle of superposition does not apply.* Thus, by superposition, the response of a linear system to the sum of two inputs is the same as the sum of the responses to the individual inputs. This behavior, which allows

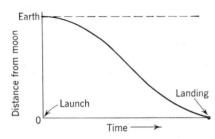

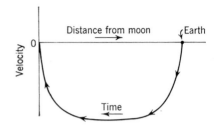

Fig. 26.1 Distance-time plot for moon rocket.

Fig. 26.2 Velocity-distance plot for moon rocket.

us to characterize completely a linear system by a transfer function, is not true of nonlinear systems.

As an example, consider a liquid-level system. If the outflow is proportional to the square root of the tank level, superposition does not hold and the system is nonlinear. If the tank will always operate near the steady-state condition, the square-root behavior may be adequately represented by a straight line and superposition applied, as we have done before. On the other hand, if the tank level should fall to half the steady-state value, we would no longer expect the transfer function derived on the linearized basis to apply. The analysis becomes more complicated, as we shall see in our introduction to the study of nonlinear systems.

The Phase Plane The analysis of nonlinear dynamic systems may often be conceptually simplified by changing to a coordinate system known as *phase space*. In this coordinate system, time no longer appears explicitly, it being replaced by some other property of the system. For example, consider the flight of a rocket to the moon. In a grossly oversimplified manner, we may describe this motion by a plot of the distance of the rocket from the moon versus time. If all goes well, we would like such a plot to resemble Fig. 26.1. Note the initial acceleration during launch and the final deceleration at landing. We may, however, also represent this motion by a plot of rocket velocity versus distance from earth. This plot is shown in Fig. 26.2, where velocity is defined as d(distance from moon)/dt. Figure 26.2 is called a phase diagram of the rocket motion. Time now appears merely as a parameter along the curve of the rocket motion. It has been replaced as a coordinate by the rocket velocity. Although in the present example Fig. 26.2 may not be of significant advantage over Fig. 26.1, we shall find phase diagrams very helpful in the analysis of certain nonlinear control systems.

To begin our study of phase diagrams, we convert a linear motion studied previously to the phase plane. The linear motion will be that of the spring-mass-damper system.

Phase-plane Analysis of Damped Oscillator The differential equation describing the motion of the system of Fig. 8.1 in response to a unit-step function is

$$\tau^2 \frac{d^2Y}{dt^2} + 2\zeta\tau \frac{dY}{dt} + Y = 1 \tag{26.1}$$

Equation (26.1) has previously been solved to yield the motion in the form of $Y(t)$ versus t as shown in Fig. 8.2. For phase analysis, however, we want the motion in terms of velocity versus position, $\dot{Y}$ versus Y, where the dot notation is used to indicate differentiation with respect to t. Hence, we rewrite Eq. (26.1) as

$$\frac{dY}{dt} = \dot{Y}$$
$$\frac{d\dot{Y}}{dt} = \frac{-Y - 2\zeta\tau\dot{Y} + 1}{\tau^2} \tag{26.2}$$

It is usually convenient in phase-plane analysis to write the variables in terms of deviation about the *final condition*. In this case, the spring will ultimately come to rest at $Y = 1$. Hence we define

$$X = Y - 1$$
$$\dot{X} = \dot{Y}$$

Then, Eq. (26.2) becomes

$$\frac{dX}{dt} = \dot{X}$$
$$\frac{d\dot{X}}{dt} = \frac{-X - 2\zeta\tau\dot{X}}{\tau^2} \tag{26.3}$$

These are now viewed as two simultaneous, first-order differential equations in the variables X and $\dot{X}$.

To solve Eqs. (26.3), we assume a solution of the form

$$X = Ce^{st} \tag{26.4}$$

where C and s are constants. Then

$$\frac{dX}{dt} = sX \qquad \frac{d\dot{X}}{dt} = s\dot{X} \tag{26.5}$$

Substituting Eq. (26.5) into (26.3) yields

$$sX - \dot{X} = 0$$
$$\frac{X}{\tau^2} + \left(s + 2\frac{\zeta}{\tau}\right)\dot{X} = 0 \tag{26.6}$$

An obvious solution to Eq. (26.6) is

$$X = \dot{X} = 0 \tag{26.7}$$

which, of course, satisfies the original differential equation for the initial conditions $X(0) = \dot{X}(0) = 0$. A well-known theorem of algebra states that

Eqs. (26.6) cannot have a solution other than Eq. (26.7) unless the determinant of the system of equations is zero, i.e., unless

$$\begin{vmatrix} s & -1 \\ \dfrac{1}{\tau^2} & \left(s + \dfrac{2\zeta}{\tau}\right) \end{vmatrix} = 0 \tag{26.8}$$

Since we are looking for other solutions, we must have Eq. (26.8) hold. Expanding this yields

$$\tau^2 s^2 + 2\zeta\tau s + 1 = 0$$

which, of course, is just the characteristic equation of the original differential equation. This quadratic equation has two roots

$$s_{1,2} = \frac{-\zeta \pm \sqrt{\zeta^2 - 1}}{\tau} \tag{26.9}$$

Referring to Eq. (26.4) we see that, since there are two values of s and the differential equations are linear, the solution is of the form

$$\begin{aligned} X &= C_1 e^{s_1 t} + C_2 e^{s_2 t} \\ \dot{X} &= s_1 C_1 e^{s_1 t} + s_2 C_2 e^{s_2 t} \end{aligned} \tag{26.10}$$

The linearity of Eqs. (26.3) guarantees that these linear combinations of solutions will also be solutions.

The constants C_1 and C_2 are, as usual, determined by the initial conditions, which are assumed to be given as

$$X(0) = X_0, \ \dot{X}(0) = \dot{X}_0$$

Thus, Eq. (26.10) yields

$$\begin{aligned} X_0 &= C_1 + C_2 \\ \dot{X}_0 &= s_1 C_1 + s_2 C_2 \end{aligned}$$

which has the solution

$$C_1 = \frac{s_2 X_0 - \dot{X}_0}{s_2 - s_1} \qquad C_2 = \frac{\dot{X}_0 - s_1 X_0}{s_2 - s_1}$$

If we take s_2 as the root with the positive sign

$$s_2 = \frac{-\zeta + \sqrt{\zeta^2 - 1}}{\tau}$$

the constants take the form

$$\begin{aligned} C_1 &= \frac{\tau}{2\sqrt{\zeta^2 - 1}} (s_2 X_0 - \dot{X}_0) \\ C_2 &= \frac{\tau}{2\sqrt{\zeta^2 - 1}} (\dot{X}_0 - s_1 X_0) \end{aligned} \tag{26.11}$$

Equations (26.10) and (26.11) together give $X(t)$ and $\dot{X}(t)$ for all possible initial conditions X_0 and $\dot{X}_0$. For a given set of initial conditions, we compute C_1 and C_2 from (26.11), and then each value of t in Eq. (26.10) yields a pair of values for X and $\dot{X}$. These may be plotted as a point on an $\dot{X}X$ diagram (i.e., a phase plane). The locus of these points as t varies from zero to infinity will be a curve in the $\dot{X}X$ plane. As an example, consider the case $X_0 = -1$, $\dot{X}_0 = 0$, $\zeta < 1$. The solution is already known to us in the form of X versus t (Chap. 8) and is replotted in Fig. 26.3 for convenience, together with a plot of $\dot{X}$ versus t. If these curves are replotted as $\dot{X}$ versus X, with t as a parameter, the result is as shown in Fig. 26.4. The reader should carefully compare Figs. 26.3 and 26.4 to satisfy himself that they are indeed equivalent. The relationship between the two may be expressed by the statement that Fig. 26.3 is a parametric representation of Fig. 26.4. Having only the curve X versus t of Fig. 26.3, one can construct Fig. 26.4.

To explore the phase-diagram concept further, note that division of the second of Eqs. (26.3) by the first yields

$$\frac{d\dot{X}}{dX} = \frac{-X - 2\zeta\tau\dot{X}}{\tau^2\dot{X}} \qquad (26.12)$$

in which the variable t has been eliminated. Equation (26.12) may be recognized as a homogeneous first-order differential equation. Hence, the substitution $\dot{X} = VX$ yields

$$\frac{X\,dV}{dX} = \frac{-1 - 2\zeta\tau V}{\tau^2 V} - V = \frac{-(1 + 2\zeta\tau V + \tau^2 V^2)}{\tau^2 V}$$

an equation which is separable in X and V. This can then be easily solved for V in terms of X. Finally, replacing $V = \dot{X}/X$ gives the solution for

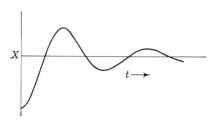

Fig. 26.3 Typical motion of second-order system.

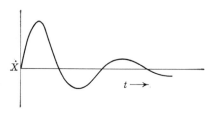

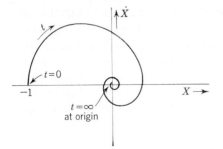

Fig. 26.4   **Phase plane corresponding to motion of Fig. 26.3.**

$\dot{X}$ versus X, or the equation for the curve of Fig. 26.4. The algebraic details of this rather tedious process are omitted.[1] The point of the discussion is to emphasize further the equivalence between the description of the motion as X versus t or $\dot{X}$ versus X.

A convenient feature of the phase diagram is that several motions, corresponding to different initial conditions, can be readily plotted on the same diagram. Thus, if we add to Fig. 26.4 a curve for the motion under the initial condition $X_0 = 1$, $\dot{X}_0 = 0$, we obtain Fig. 26.5. This new trajectory represents the motion of the system after it is stretched 2 units and released from rest. (This follows from the definition $X = Y - 1$.) Furthermore, we have also interpolated in Fig. 26.5 to obtain the motion corresponding to $X_0 = 0$, $\dot{X}_0 = 1$. As we shall see later, this interpolation is justified. Hence, it is evident that the phase diagram gives us the "big picture" of the motion of the underdamped spring-mass-damper system. No matter where the system starts, it spirals to the condition $X_0 = \dot{X}_0 = 0$, the steady-state position. This spiral motion in the phase plane corresponds to the oscillatory nature of the X versus t curve of Fig. 26.3.

Before beginning a more detailed study of the mechanics of phase analysis, it may be worthwhile to see how situations amenable to such analysis arise naturally in the physical world.

[1] See, for example, D. Graham and D. McRuer, "Analysis of Nonlinear Control Systems," pp. 287–289, John Wiley & Sons, Inc., New York, 1961.

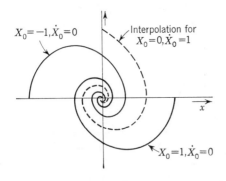

Fig. 26.5   **Interpolation on the phase plane.**

Motion of a Pendulum   Consider the pendulum of Fig. 26.6.   As
the pendulum is moving in the direction shown, there are two forces acting
to oppose its motion.   These forces, which act tangentially to the circle
of motion, are (1) the gravitational force $mg \sin \theta$ and (2) the friction in the
pivot which we suppose to be proportional to the tangential velocity of the
mass, $BR(d\theta/dt)$.   We shall assume the air resistance to be negligible and
the rod to be of negligible mass.   Application of Newton's second law gives

$$-mR \frac{d^2\theta}{dt^2} = mg \sin \theta + BR \frac{d\theta}{dt}$$

Rearrangement leads to

$$\frac{d^2\theta}{dt^2} + D \frac{d\theta}{dt} + \omega_n{}^2 \sin \theta = 0 \qquad (26.13)$$

where

$$D = \frac{B}{m}$$

$$\omega_n{}^2 = \frac{g}{R}$$

This equation resembles the equation for the motion of the spring-mass-
damper system.   However, the presence of the term involving $\sin \theta$ makes
the equation nonlinear.

Equation (26.13) has the following form in phase coordinates:

$$\frac{d\theta}{dt} = \dot{\theta}$$

$$\frac{d\dot{\theta}}{dt} = -\omega_n{}^2 \sin \theta - D\dot{\theta} \qquad (26.14)$$

and a phase diagram would be a plot of angular velocity $\dot{\theta}$ versus position θ.
At this point, we can gain some insight by simple analysis of Eq. (26.14)
without actually obtaining a solution.

Referring for the moment back to the spring-mass-damper system,

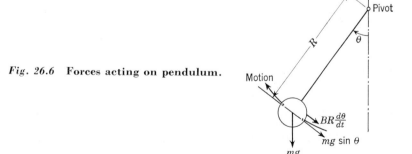

Fig. 26.6   Forces acting on pendulum.

we saw that the system ceased to oscillate when the point $X = \dot{X} = 0$ was reached. That is, all curves stopped at the origin of Fig. 26.5. The reason for this is quite clear; when $X = \dot{X} = 0$ is substituted into Eqs. (26.3), there is obtained

$$\frac{dX}{dt} = \frac{d\dot{X}}{dt} = 0$$

Since neither X nor $\dot{X}$ is changing with time, the motion ceases. Further examination of Eqs. (26.3) shows that $X = \dot{X} = 0$ is the *only* point at which both dX/dt and $d\dot{X}/dt$ are zero. Thus, we see that the mass will come to rest *only* when the situation of zero displacement and zero velocity is reached.

Now we perform a similar analysis on Eqs. (26.14). We are asking the following question: At what point or points in the phase plane ($\dot{\theta}$ versus θ diagram) do both $d\theta/dt$ and $d\dot{\theta}/dt$ become zero? From the first of these equations, we see that this can happen[1] only when $\dot{\theta} = 0$. Using this result in the second equation, it can be seen that it is also necessary that

$$\sin \theta = 0 \tag{26.15}$$

Equation (26.15) is satisfied at any of the points

$$\theta = n\pi$$

where n is a positive or negative integer or zero. However, from a physical standpoint, we can really distinguish between only two of these points, which we take as $\theta = 0$ and $\theta = \pi$. Thus, the positions $\theta = 0, 2\pi, 4\pi, -2\pi$, etc., all look the same to us; i.e., the pendulum is hanging straight down. Similarly, the points $\theta = \pi, 3\pi$, etc., all correspond to the pendulum standing straight up.

Thus, the analysis leads to the conclusion that the motion will cease when the pendulum comes to rest in either of the positions shown in Fig. 26.7. In addition, it is clear from Eqs. (26.14) that, if the pendulum stops at any other point, the motion continues. Of course, this analysis agrees with our physical intuition. However, we expect to find a distinction between the stability characteristics of the two equilibrium points, since the position at π is likely to be hard to attain and maintain. This distinction will be explored in more detail in Chap. 27.

---

[1] The reader should not be lulled into a false sense of security at this point. It would be wise to disregard the fact that $d\theta/dt$ and $\dot{\theta}$ are, in fact, the same quantity. θ should be thought of as a coordinate in the phase plane, and $d\theta/dt$ as the rate of change with time of the other coordinate. The virtue of making this distinction will become clear in the next example, a chemical reactor.

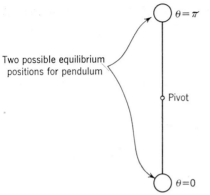

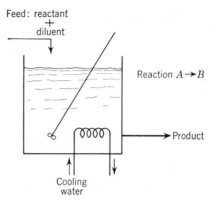

Fig. 26.7 Equilibrium positions for pendulum.

Fig. 26.8 Schematic of exothermic chemical reactor.

A Chemical Reactor[1] Consider the stirred-tank chemical reactor of Fig. 26.8. The contents of the reactor are assumed to be perfectly mixed, and the reaction taking place is

$$A \rightarrow B \tag{26.16}$$

which occurs at a rate

$$R_A = kC_A e^{-E/RT} \tag{26.17}$$

where R_A = moles of A decomposing per hour per cubic foot of reacting mixture
 k = reaction velocity constant, hr^{-1}
 C_A = concentration of A in reacting mixture, moles/ft^3
 E = activation energy, a constant, Btu/mole
 R = universal gas law constant
 T = absolute temperature of reacting mixture
  The reaction is exothermic; ΔH Btu of heat are generated for each mole of A which reacts. Hence, in order to control the reactor, cooling water is supplied to a cooling coil. The actual reactor temperature is compared with a set point, and the rate of cooling-water flow adjusted accordingly. To indicate this control mathematically, we write that $Q(T)$ Btu/hr of heat are removed through the cooling coil. In Chap. 28 we shall make a more detailed analysis of the dynamic behavior of the reactor. For the present preliminary analysis, it is not necessary to look

[1] This example is based upon the work of R. Aris and N. R. Amundson, *Chem. Eng. Sci.*, **7**:121–155 (1958).

carefully at $Q(T)$, and hence it is merely assumed that, as T rises, more heat is removed in the coil.

Let x_{A_0} = mole fraction of A in feed stream

x_{B_0} = mole fraction of B in feed stream

Then $(1 - x_{A_0} - x_{B_0})$ is the fraction of inerts in the feed stream. A mass balance on A,

$(A$ in feed$) - (A$ in product$) - (A$ reacting$)$
$$= (A \text{ accumulating in reactor})$$

takes the form

$$F\rho x_{A_0} - F\rho x_A - k\rho V e^{-E/RT} x_A = \rho V \frac{dx_A}{dt} \qquad (26.18)$$

where F = feed rate, ft³/hr

x_A = mole fraction of A in reactor

ρ = density of reacting mixture, moles/ft³

V = volume of reacting mixture, ft³

To arrive at Eq. (26.18) we have used Eq. (26.17) and made the following assumptions:

1. The density of the reacting mixture is constant, unaffected by the conversion of A to B.

2. The feed and product rates F are equal and constant.

3. Together, 1 and 2 imply that V, the volume of reacting mixture, is constant.

4. Perfect mixing occurs, so that x_A is the same in the reactor and product stream.

A similar mass balance may be derived for substance B. However, Eq. (26.16) shows that one mole of B appears for every mole of A destroyed. Hence

$$x_B - x_{B_0} = x_{A_0} - x_A \qquad (26.19)$$

Equation (26.19) permits us to circumvent the mass balance for x_B, since knowing x_A we can calculate x_B directly.

The energy balance on the reactor

(Sensible heat in feed) $-$ (sensible heat in product)
$+$ (heat generated by reaction) $-$ (heat removed in cooling coil)
$$= (\text{energy accumulating in reactor})$$

can be written as

$$F\rho C_p(T_0 - T) + k\rho V(\Delta H)e^{-E/RT} x_A - Q(T) = \rho V C_p \frac{dT}{dt} \qquad (26.20)$$

where T_0 = temperature of feed stream

T = temperature in reactor

C_p = specific heat of reacting mixture

In writing Eq. (26.20), it is assumed that

1. The specific heat of the reacting mixture is constant, unaffected by the conversion of A to B.

2. The perfect mixing means that the temperatures of the reacting mixture and product stream are the same.

3. The heat of reaction ΔH is constant, independent of temperature and composition.

We remark here that these assumptions, as well as those made in Eq. (26.18), may be relaxed without affecting the conceptual aspects of the phase analysis. They are made only to keep the example as uncluttered as possible, without being trivial.

Equations (26.18) and (26.20) may be rearranged to the system

$$\frac{dx_A}{dt} = \frac{F}{V} (x_{A_0} - x_A) - ke^{-E/RT} x_A$$

$$\frac{dT}{dt} = \frac{F}{V} (T_0 - T) + \frac{k(\Delta H)}{C_J} e^{-E/RT} x_A - \frac{Q(T)}{\rho V C_p} \tag{26.21}$$

As a typical application of this system of equations, we might consider starting up the reactor, initially filled with a mixture at composition $x_A(0)$ and temperature $T(0)$. Suppose the feed rate, feed composition, feed temperature, and flow rate of cooling water are held constant and the reactor is operated in this manner until steady state is reached. To describe the transient behavior of the chemical reactor, one can solve Eqs. (26.21) by integrating them numerically, using a typical stepwise procedure such as the Euler or Runge-Kutta method. This will result in functions $x_A(t)$ and $T(t)$ for values of t from zero to some value (if one exists) at which, for practical purposes, $x_A(t)$ and $T(t)$ cease to change with t.

Alternatively, we may consider a phase-plane analysis of Eqs. (26.21) and seek solutions in the form of x_A versus T curves. Note that division of the first of Eqs. (26.21) by the second gives

$$\frac{dx_A}{dT} = \frac{(F/V)(x_{A_0} - x_A) - ke^{-E/RT} x_A}{(F/V)(T_0 - T) + \dfrac{k(\Delta H)}{C_p} e^{-E/RT} x_A - \dfrac{Q(T)}{\rho V C_p}} \tag{26.22}$$

The parameter t has been eliminated in Eq. (26.22), which is simply a differential equation relating x_A and T. As we shall see in Chap. 28, this phase-plane analysis of the chemical reactor offers significant advantages over the ordinary analysis.

In the chemical reactor, we no longer have the special relationship among the phase variables that we had in both previous cases. For both the spring and pendulum problems, we more or less artificially changed a second-order differential equation to two first-order equations by introducing the phase variable $\dot{X}$ (or $\dot{\theta}$). This phase variable was directly related to the other phase variable X (or θ) by the equation

$$\dot{X} = \frac{dX}{dt}$$

For the chemical reactor, there is no such simple relation between x_A and T.

We can study the steady-state solutions to Eqs. (26.21) without solving the equations, much as was done in the case of the damped pendulum of the previous example. As before, we note that steady state requires that x_A and T simultaneously cease to change with time,

$$\frac{dx_A}{dt} = \frac{dT}{dt} = 0$$

From Eqs. (26.21), this implies that

$$\frac{F}{V}(x_{A_0} - x_{A_s}) - ke^{-E/RT_s}x_{A_s} = 0$$

$$\frac{F}{V}(T_0 - T_s) + \frac{k(\Delta H)}{C_p}e^{-E/RT_s}x_{A_s} - \frac{Q(T_s)}{\rho V C_p} = 0 \tag{26.23}$$

where x_{A_s} and T_s are the steady-state values of x_A and T.

The first of Eqs. (26.23) can be solved for x_{A_s}, yielding

$$x_{A_s} = x_{A_0}\frac{1}{1 + (kV/F)e^{-E/RT_s}} \tag{26.24}$$

Substitution of (26.24) into the second of Eqs. (26.23) yields

$$\frac{k(\Delta H)x_{A_0}/C_p}{e^{E/RT_s} + kV/F} = \frac{Q(T_s)}{\rho V C_p} + \frac{F}{V}(T_s - T_0) \tag{26.25}$$

Equation (26.25) is implicit in T_s, the steady-state temperature. In physical terms, it expresses an equality between the heat generated by the reaction and the heat removed in the cooling coil and product stream. To emphasize this, we have arranged it so that the left side is the heat generation and the right side is the heat removal.

Solution of Eq. (26.25) for T_s requires numerical values for the various parameters. Without going into this much detail at present, we may obtain some qualitative information. To do this, we sketch the right and left sides of this equation as functions of T_s. A typical shape for the left side[1] is given by the sigmoidal curve of Fig. 26.9. The unusual curvature, of course, is caused by the e^{E/RT_s} term in the denominator. To plot the right side, we must know $Q(T)$. While we have avoided specifying the form of $Q(T)$, we know it increases with T. If there were no control action, i.e., if the flow rate of cooling water were maintained constant regardless of T, then $Q(T)$ would increase almost linearly with T. This is because at

[1] *Ibid.*, p. 121.

constant water rate, the heat transfer in the coil is approximately proportional to the difference between T and the mean temperature of the cooling water. This latter temperature would not vary so rapidly as T at practical flow rates. However, since we expect to have control action, we know that the cooling-water flow rate will be increased with increasing T. Therefore, $Q(T)$ may be expected to increase faster than linearly with T, which means that the right side of (26.25) increases faster than linearly. Several typical curves of this right side are shown in Fig. 26.9.

A solution of Eq. (26.25) requires that the graphs of the right and left sides intersect. As shown in Fig. 26.9 there may be one, two, or three such intersections, depending on the relative locations of the heat generation (left side) and heat removal (right side). This means that there may be one, two, or three possible steady states for the reactor.

As we shall see in Chap. 28, the steady state actually attained by the reactor depends on initial conditions $x_A(0)$ and $T(0)$. The steady-state temperature T_s is then the temperature at the pertinent intersection, and the steady-state composition can be determined from Eq. (26.24). We shall also see that some of the steady states are unstable. In fact, the low-temperature steady state for curve (c) of Fig. 26.9, occurring as a point of tangency, is to be regarded with suspicion. Practically speaking, a perfect tangency would not occur. Minor variations in operating conditions (i.e., noise), which occur continually in actual process operation, may shift the curve (c) slightly to the left or right, resulting in two or zero low-temperature intersections, respectively.

Summary   In this chapter, we have introduced the concept of a phase analysis and some of its basic elements. We have seen how physical situations give rise naturally to phase solutions. Furthermore, we have had our first look at true nonlinear behavior. In so doing, we have come to at least one interesting conclusion: A nonlinear motion or control system

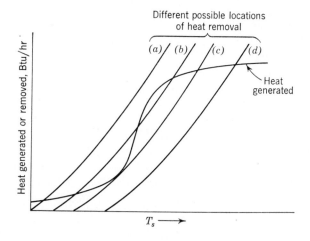

Fig. 26.9 Steady-state generation and removal functions for exothermic chemical reactor.

response may have more than one steady-state solution. This was true for the chemical reactor and for the pendulum. In contrast, the linear motions and control system responses we studied in the previous chapters had only one steady-state solution. In the next chapter, we shall discover still more differences which render nonlinear analysis more difficult than linear analysis.

27 Methods of Phase-plane Analysis

The advantages of the phase analysis introduced in Chap. 26 can be more fully appreciated after some acquaintance with the tools available for such analysis. We do not intend to give here a detailed exposition of all, or even most, of the aspects of this subject. Instead, we hope to indicate its flavor and to stimulate further study.

Phase Space Consider the general nth-order ordinary differential equation, written in a form which is solved for the highest derivative

$$\frac{d^n x}{dt^n} = f\left(x, \frac{dx}{dt}, \frac{d^2 x}{dt^2}, \cdot \cdot \cdot, \frac{d^{n-1} x}{dt^{n-1}}, t\right) \tag{27.1}$$

If we define the new variables $(x_1, x_2, \ldots, x_n)$ by the relations

$$x_1 = x$$

$$x_2 = \frac{dx}{dt}$$

$$x_3 = \frac{d^2 x}{dt^2} \tag{27.2}$$

$$\cdot \ \cdot \ \cdot \ \cdot \ \cdot$$

$$x_n = \frac{d^{n-1} x}{dt^{n-1}}$$

then Eq. (27.2) may be rewritten as a system of n simultaneous first-order equations:

$$\frac{dx_1}{dt} = x_2$$

$$\frac{dx_2}{dt} = x_3$$

. . . . .   (27.3)

$$\frac{dx_{n-1}}{dt} = x_n$$

$$\frac{dx_n}{dt} = f(x_1, x_2, \ . \ . \ . \ , x_n, t)$$

The reader will note that this is precisely what we did in the previous chapter to obtain Eqs. (26.2) from Eq. (26.1).

Example 27.1   Consider the equation

$$x^3 \frac{d^3x}{dt^3} + te^x \frac{dx}{dt} - x^2 \sin t = t$$

To put this in the form of (27.1) we solve for d^3x/dt^3:

$$\frac{d^3x}{dt^3} = \frac{t + x^2 \sin t - te^x \, dx/dt}{x^3}$$

We now define

$$x_1 = x$$

$$x_2 = \frac{dx}{dt}$$

$$x_3 = \frac{d^2x}{dt^2}$$

Then

$$\frac{dx_1}{dt} = x_2$$

$$\frac{dx_2}{dt} = x_3$$

$$\frac{dx_3}{dt} = \frac{t + x_1^2 \sin t - te^{x_1}x_2}{x_1^3}$$

is a system of equations in x_1, x_2, and x_3, equivalent to the original third-order equation.

The system (27.3) is called the *phase-space* version of Eq. (27.1). The variables $(x_1, x_2, \ . \ . \ . \ , x_n)$ are coordinates of the phase space. A specification of a value for each coordinate at some instant of time t_1 is referred to as the state of the system at t_1. Solutions of Eq. (27.3) are curves in the phase space, with t a parameter on the curves. Each value of t, say t_1, defines a point in the space, $x_1(t_1)$, $x_2(t_1)$, $\ . \ . \ . \ , x_n(t_1)$. The solution curve is a locus of these points for all values of t. It is called a *trajectory* and connects successive states of the system.

Example 27.2 Equations (26.3) are the *phase-space* version of (26.1). Their solution is given by (26.10):

$$X = C_1 e^{s_1 t} + C_2 e^{s_2 t}$$
$$\dot{X} = s_1 C_1 e^{s_1 t} + s_2 C_2 e^{s_2 t} \qquad (26.10)$$

A set of constants (C_1, C_2) can be determined from the initial conditions by Eq. (26.11). Equations (26.10) then define a solution in the *phase space*. In this case, the phase space is the two-dimensional $\dot{X}X$ plane. For each value of t, we calculate a point $(X, \dot{X})$ from Eq. (26.10). The result of connecting the points for all values of t is a *trajectory*, such as the one of Fig. 26.4. Another set of initial conditions will determine another set (C_1, C_2) and hence another trajectory. Several trajectories are shown in Fig. 26.5. At time t_1, the *state* of the system is given by the point

$$X(t_1) = C_1 e^{s_1 t_1} + C_2 e^{s_2 t_1}$$
$$\dot{X}(t_1) = s_1 C_1 e^{s_1 t_1} + s_2 C_2 e^{s_2 t_1}$$

We can now see one of the advantages of the phase space. It serves as a unifying concept for the treatment of ordinary differential equations. An equation of any order may be treated in terms of first-order equations. Solutions may be generalized to trajectories in phase space. Initial conditions merely select a particular trajectory. Of course, equations of fourth- or higher-order require treatment in a space which is of too many dimensions to be visualized. The graphic aspects of phase representation are advantageous primarily in the case of two dimensions (the *phase plane*) and to a limited extent for three dimensions.

Problems suited for phase analysis are by no means confined to those generated by a single differential equation. Indeed, as we saw in the case of the stirred-tank chemical reactor of Chap. 26, systems of first-order equations may arise quite naturally from the physical problem. The most general system of equations for phase analysis has the form

$$\frac{dx_1}{dt} = f_1(x_1, x_2, \ldots, x_n, t)$$

$$\frac{dx_2}{dt} = f_2(x_1, x_2, \ldots, x_n, t)$$

$$\cdots \cdots \cdots \cdots \cdots \cdots \qquad (27.4)$$

$$\frac{dx_n}{dt} = f_n(x_1, x_2, \ldots, x_n, t)$$

Example 27.3 To write Eqs. (26.21) describing the stirred-tank chemical reactor in the form (27.4), we treat x_A as x_1 and T as x_2. Then (26.21) becomes

$$\frac{dx_A}{dt} = f_1(x_A, T, t)$$
$$\frac{dT}{dt} = f_2(x_A, T, t)$$

If the inlet temperature and composition T_0 and x_{A_0} do not vary with time, then t does not appear in f_1 or f_2, and we can write

$$\frac{dx_A}{dt} = f_1(x_A, T)$$

$$\frac{dT}{dt} = f_2(x_A, T)$$

$$(27.5)$$

where

$$f_1(x_A, T) = \frac{F}{V}(x_{A_0} - x_A) - ke^{-E/RT} x_A$$

$$f_2(x_A, T) = \frac{F}{V}(T_0 - T) + \frac{k(\Delta H)e^{-E/RT}}{C_p} x_A - \frac{Q(T)}{\rho V C_p}$$

Equations (27.5), in which the time variable does not appear explicitly, are referred to as *autonomous*. The most general system of autonomous equations is given by

$$\frac{dx_1}{dt} = f_1(x_1, x_2, \ldots , x_n)$$

$$\frac{dx_2}{dt} = f_2(x_1, x_2, \ldots , x_n)$$

$$\ldots \ldots \ldots \ldots \ldots \ldots$$

$$\frac{dx_n}{dt} = f_n(x_1, x_2, \ldots , x_n)$$

$$(27.6)$$

The bulk of practical use of phase-space analysis has been made in the two-dimensional autonomous case:

$$\frac{dx_1}{dt} = f_1(x_1, x_2)$$

$$\frac{dx_2}{dt} = f_2(x_1, x_2)$$

$$(27.7)$$

For this reason, we largely confine our attention in the remainder of this study to systems which may be written in the form of Eqs. (27.7). As we have seen, there is no loss in *conceptual generality*, but we cannot expect the *graphical aspects* of the material we shall develop to generalize to higher-dimensional phase space. The solution of the system (27.7) may be presented as a family of trajectories in the $x_2 x_1$ plane. If we are given the initial conditions

$$x_1(t_0) = x_{10}$$

$$x_2(t_0) = x_{20}$$

the initial state of the system is the point (x_{10}, x_{20}) in the $x_2 x_1$ plane and the trajectory may, in principle, be traced from this point.

By dividing the second of Eqs. (27.7) by the first, we obtain

$$\frac{dx_2}{dx_1} = \frac{f_2(x_1, x_2)}{f_1(x_1, x_2)}$$

$$(27.8)$$

Now dx_2/dx_1 is merely the slope of a trajectory, since a trajectory is a plot of x_2 versus x_1 for the system. Hence, at each point in the phase plane (x_1,x_2), Eq. (27.8) yields a unique value for the slope of a trajectory through the point, namely, $f_2(x_1,x_2)/f_1(x_1,x_2)$. This last statement should be amended to exclude any point (x_1,x_2) at which $f_1(x_1,x_2)$ and $f_2(x_1,x_2)$ are *both* zero. These important points are called *critical points* and will be examined in more detail below. Since the slope of the trajectory at a point, say (x_1,x_2), is by Eq. (27.8) unique, it is clear that trajectories cannot intersect except at a critical point, where the slope is indeterminate.

  The Method of Isoclines   Let us now utilize this information about the trajectory slope to approximate the trajectory. We shall illustrate the technique with an example.

  Example 27.4   Find the trajectory of the system

$$\frac{dx_1}{dt} = x_2$$

$$\frac{dx_2}{dt} = -5x_1 - 2x_2 \qquad\qquad (27.9)$$

which passes through the point

$$x_1 = 1 \qquad x_2 = 0$$

The slope of any trajectory is given by

$$\frac{dx_2}{dx_1} = -\frac{5x_1 + 2x_2}{x_2}$$

We search for all points through which the trajectories must have the same slope. If this slope is called S, then

$$-\frac{5x_1 + 2x_2}{x_2} = S$$

is the equation which must be satisfied by all points at which the slope is to be S. This may be rearranged to

$$x_2 = \frac{-5x_1}{S + 2}$$

which is the equation of a line through the origin in the x_2x_1 plane. Thus, for example,

$$x_2 = -x_1$$

is the locus of all points at which the trajectories have slope 3. Similarly, the x_1 axis is the locus of points at which the slope of the trajectory is infinite. Such loci, which in this special case are straight lines, are called *isoclines*. Several isoclines, with the slopes indicated, are plotted on Fig. 27.1.

  To sketch the desired trajectory, we first note that it starts at the point $(1,0)$. At this point, according to Eqs. (27.9),

$$\frac{dx_1}{dt} = 0$$

$$\frac{dx_2}{dt} = -5$$

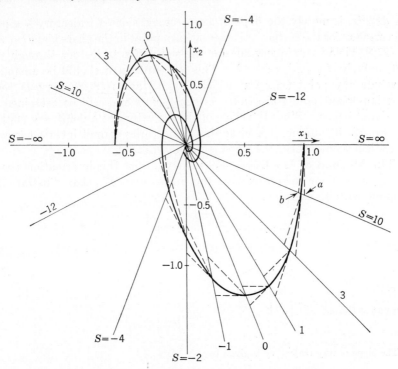

Fig. 27.1 Isocline construction of phase plane for Eqs. (27.9).

Hence, the trajectory starts out vertically downward. Between the isoclines $S = \infty$ and $S = 10$, the slope of the trajectory must vary between infinity and 10. The points on the $S = 10$ isocline, which would be reached if the trajectory had a constant slope of infinity or 10, are labeled a and b, respectively, in Fig. 27.1. The actual point at which the trajectory reaches the $S = 10$ isocline is taken as midway between a and b, which is equivalent to an averaging of the slopes. The construction is continued in this manner, and the trajectory sketched so as to connect the indicated points and to have the correct slope as it passes through each isocline. The short dashed marks between isoclines, indicating the correct slope, are also helpful in satisfying this latter condition.

Another trajectory, starting from the point $(-0.6,0)$, is shown on Fig. 27.1. This serves to emphasize that, once the isoclines have been located, interpolation is possible on the phase plane, and many trajectories representing various initial conditions are easily visualized or sketched.

There are other graphical techniques for construction of phase portraits. These are discussed in, for example, Thaler and Pastel.[1] The method of isoclines is usually suitable when the isocline equation

$$\frac{f_2(x_1,x_2)}{f_1(x_1,x_2)} = S$$

[1] G. J. Thaler and M. P. Pastel, "Analysis and Design of Nonlinear Feedback Control Systems," McGraw-Hill Book Company, New York, 1962.

is not overly complicated and where a good overall knowledge of the phase plane is required. In practice, for more complex systems such as the chemical reactor of Chap. 26, the phase plane is often obtained by use of an analog computer. This technique will be illustrated in Chap. 32.

Analysis of Critical Points In the situations of most interest to us, Eq. (27.8) will represent the behavior of a (nonlinear) control system, as in Eq. (26.22). Therefore, we shall be interested in maintaining the system at or near a steady state. Since, from Eq. (27.7), a steady-state point is defined by

$$f_1(x_1,x_2) = f_2(x_1,x_2) = 0$$

it is clear that the steady states are critical points. At the critical points, the slope of the trajectory is undefined; hence, many trajectories may intersect at these points. In Fig. 27.1 the origin is a critical point. It can be seen from the isoclines that, in this case, all trajectories spiral into the origin. Hence, this particular system is such that, no matter what the initial state (i.e., for any disturbance which is applied), the system returns to steady state at the critical point.

The critical point of Fig. 27.1 is called a *focus*, because the trajectories spiral into it. This spiral motion of the trajectories corresponds to the oscillatory approach of the system to steady state. The oscillatory motion occurs because the system of Eqs. (27.9) is underdamped, as indicated by the characteristic equation [see Eq. (26.8)]

$$\begin{vmatrix} -s & 1 \\ -5 & -2-s \end{vmatrix} = s^2 + 2s + 5 = 0$$

When put into standard form, this characteristic equation has parameters

$$\tau = \frac{1}{\sqrt{5}} \qquad \zeta = \frac{1}{\sqrt{5}}$$

Since $\zeta < 1$, the system is underdamped.

An overdamped system, such as that generated by the system

$$\frac{dx_1}{dt} = x_2$$

$$\frac{dx_2}{dt} = -5x_1 - 6x_2$$

having characteristic equation

$$s^2 + 6s + 5 = 0$$

so that

$$\tau = \frac{1}{\sqrt{5}} \qquad \zeta = \frac{3}{\sqrt{5}}$$

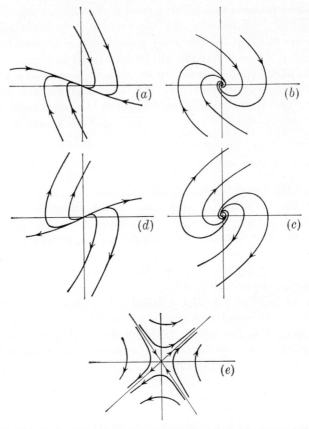

Fig. 27.2 Second-order critical points. (*a*) **Stable node;** (*b*) **stable focus;** (*c*) **unstable focus;** (*d*) **unstable node;** (*e*) **saddle point.**

has a critical point such as that of Fig. 27.2*a*. Here the trajectories enter the critical point directly, without oscillation. This type of critical point is called a node. For comparison, a typical focus is sketched in Fig. 27.2*b*. In fact, other types of behavior may be exhibited by critical points of a second-order system, depending upon the nature of the roots of the characteristic equation. These are summarized for *linear* systems in Table 27.1 and sketched in Fig. 27.2. The distinction between stable and unstable nodes or foci is made to indicate that the trajectories move toward the stable type of critical point and away from the unstable point. The saddle point arises when the roots of the characteristic equation are real and have opposite sign. In this case there are only two trajectories which enter the critical point, and after entering, the trajectories may leave the critical point (permanently) on either of two other trajectories. No other trajectory can enter the critical point, although some approach it very closely. The qualitative verification of the behavior sketched in Fig. 27.2 is left to the problems.

Table 27.1   **Classification of Critical Points**

Type of critical point	Characteristic equation	Pertinent values of ζ	Nature of roots	Sign of roots
Stable node	$\tau^2 s^2 + 2\zeta\tau s + 1 = 0$	$\zeta > 1$	Real	Both $-$
Stable focus	$\tau^2 s^2 + 2\zeta\tau s + 1 = 0$	$0 < \zeta < 1$	Complex	Real parts both $-$
Unstable focus	$\tau^2 s^2 + 2\zeta\tau s + 1 = 0$	$-1 < \zeta < 0$	Complex	Real parts both $+$
Unstable node	$\tau^2 s^2 + 2\zeta\tau s + 1 = 0$	$\zeta < -1$	Real	Both $+$
Saddle point	$\tau^2 s^2 + 2\zeta\tau s - 1 = 0$	All	Real	One $+$, one $-$

This categorization of critical points according to the particular linear system is often of value in the analysis of nonlinear systems. The reason for this is that, in a sufficiently small vicinity of a critical point, a nonlinear system behaves approximately linearly. Thus, the system of Eq. (26.14) for the pendulum is nonlinear. It has two physically distinguishable steady states, corresponding to the pendulum pointing up or down. The nonlinear term $\sin\theta$ may be linearized around each steady state. Near the steady state at $\theta = 0$,

$$\sin\theta \approx \theta$$

and near the steady state at $\theta = \pi$, a Taylor series yields

$$\sin\theta = -(\theta - \pi)$$

Therefore, near $\theta = 0$, Eqs. (26.14) are closely approximated by the linear equations

$$\frac{d\theta}{dt} = \dot\theta$$

$$\frac{d\dot\theta}{dt} = -\omega_n^2\theta - D\dot\theta \qquad (27.10)$$

and near $\theta = \pi$, by

$$\frac{dx}{dt} = \dot x$$

$$\frac{d\dot x}{dt} = \omega_n^2 x - D\dot x \qquad (27.11)$$

where $x = \theta - \pi$. These linearized versions of Eqs. (26.14) can be easily solved to determine the nature of the *linear approximations* to the critical points. Thus, the characteristic equation for Eqs. (27.10) is

$$s^2 + Ds + \omega_n^2 = 0 \qquad (27.12)$$

while that for Eqs. (27.11) is

$$s^2 + Ds - \omega_n^2 = 0 \qquad (27.13)$$

As shown in Table 27.1, Eq. (27.12) yields a stable critical point, which

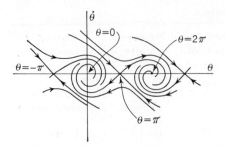

Fig. 27.3 Phase portrait of lightly damped pendulum.

may be a node or focus depending on the degree of damping. (Note that, as the damping is increased, the behavior changes from focus to node, or from oscillatory to nonoscillatory.) On the other hand, Eq. (27.13) indicates a saddle point for the motion near $\theta = \pi$.

These conclusions apply strictly only to the linearized phase equations, Eqs. (27.10) and (27.11). To compare them with the behavior of the true system of Eqs. (26.14), the actual phase diagram is sketched for a lightly damped case in Fig. 27.3. For simplicity, this diagram is extended beyond the range $0 \leq \theta \leq 2\pi$ even though this is the only region of physical significance. Actually, the section for $0 \leq \theta \leq 2\pi$ should be cut out and rolled into a cylinder so that the lines corresponding to $\theta = 0$ and $\theta = 2\pi$ coincide. This phase cylinder would more realistically represent the motion of the pendulum. As seen from Fig. 27.3, the point at $\theta = \pi$ is, indeed, a saddle point and the point $\theta = 0$ (or 2π) is a stable focus. If the system were more heavily damped, this latter point would be a stable node.

A greater understanding of the saddle point may now be obtained by analyzing the $\theta = \pi$ point in terms of what we know to be the physical behavior of the pendulum at this point. That is, the point may be approached from either of two directions. When the pendulum is at the point, an infinitesimal disturbance will cause it to fall in either of two directions. Other trajectories narrowly miss this point, indicating that just the right initial velocity must be imparted to the pendulum at a given initial point to cause it to stop in the $\theta = \pi$ position.

In summary, it can be concluded that in this case the linearized equations give valuable, accurate information about the behavior of the nonlinear system in the vicinity of the critical points. Because the linearized equations are more easily solved, it is always desirable to be able to relate the behavior of the actual system to the behavior of the linearized solutions in the vicinity of the operating point. In fact, in our previous work on control systems, we have assumed for nonlinear systems that design of a stable control system based on the linearized equations was adequate to ensure stable operation of the actual system. The basis for this assumption is given by the following theorem of Liapunov:[1]

[1] A. M. Letov, "Stability in Nonlinear Control Systems," Princeton University Press, Princeton, N. J., 1961.

Let the nonlinear equations of a motion be linearized by expansion in deviation variables around a particular critical point. If the linearized solution for the deviation variables is stable, the actual motion will be stable in some vicinity of the critical point. If the linearized solution is neutrally stable (i.e., its characteristic equation has roots on the imaginary axis), no statement can be made about the actual motion. If the linearized solution is unstable, then the actual motion will be unstable.

It is necessary to define what is meant by stability and instability of the actual nonlinear motion in the vicinity of the critical point. Although stability in nonlinear systems is a complex subject, for our purposes it will suffice to state that a stable nonlinear motion in the vicinity of a critical point is one for which all phase-plane trajectories in this vicinity travel toward and end at the critical point. An unstable motion is one for which trajectories move away from the critical point. This would mean that, while theoretically the state of the system may remain at the critical point indefinitely, any slight disturbance causes the unstable system to move away from the critical point. These conclusions agree with our physical understanding of the pendulum motion, since the steady condition at $\theta = \pi$ is easily destroyed.

It is because of Liapunov's theorem that linear control theory is so successful in control system design. One really hopes to control the system so that it remains permanently in the vicinity of a particular point (i.e., a steady state). However, when serious upsets occur in an automatically controlled plant, moving it far from steady state, it is often necessary to return the plant to manual control until conditions are again close to steady state. This is because the controllers are designed for satisfactory operation in the linear range only. One of the great drawbacks of linear control theory is the fact that stability of the linearized equations guarantees stability of the nonlinear system only in *some* vicinity of the particular critical point. No information about the size of this vicinity or about the behavior outside this vicinity is obtained. If the linear vicinity is extremely small, then unknown to the designer who has used linear methods, almost any plant disturbance of practical size may result in control system failure. An example of this behavior will be given in the next chapter.

A great deal of effort in present research on nonlinear control theory is an attempt to find the extent of the vicinity in which the linear conclusions apply. The theory of Liapunov[1] is useful in this respect but is mathematically beyond the scope of this text.

Limit Cycles The first major difference between linear and nonlinear motions is the possible existence of more than one critical point in the latter type. The second is the possible existence of limit cycles.

A *limit cycle* is defined as a periodic oscillation whose amplitude and frequency depend only on the properties of the system and not on the

[1] *Ibid.*

initial state of the system (provided the initial state lies in a certain non-trivial region of the phase space). In the phase plane, *stable* limit cycles are recognized as closed curves which are approached asymptotically by all nearby trajectories. *Unstable* limit cycles are closed curves from which all nearby trajectories diverge. An example of a stable limit cycle is the "steady-state" behavior of a home heating system when controlled by a thermostat. A periodic oscillation in house temperature is always reached, and the amplitude and frequency of the oscillation are independent of the temperature which existed in the house at the time that the furnace was started. Unstable limit cycles can never be realized physically for any system by definition. However, as will be seen in the next chapter, they divide the phase plane into regions of totally different dynamic behavior and hence are of considerable importance.

It is important to distinguish between limit cycles and other closed curves which may occur. The linear system

$$\tau^2 \frac{d^2x}{dt^2} + x = 0$$

has phase-space solution

$$x^2 + \tau^2(\dot{x})^2 = C^2 \tag{27.14}$$

where $\dot{x} = dx/dt$ and the constant C depends upon initial conditions. Equation (27.14) defines a family of concentric ellipses in the phase plane. However, *these are not limit cycles*, because the closed curve which is followed by the system depends upon the initial state of the system through the constant C. In Chap. 28, we shall study some limit cycles occurring in typical control systems.

Other Aspects We have presented only those aspects of phase-plane analysis which will be of use in the examples of the next chapter. This can be considered only as a brief introduction to the subject, and the interested reader is referred to the references already cited for more information. Among the important subjects which we have omitted are graphical methods for determination of time along a trajectory, various aspects of phase-plane topology, and the mathematical aspects of stability.

PROBLEMS

27.1 Using the method of isoclines, sketch the phase plane, showing several trajectories, for a system with transfer function

$$\frac{Y(s)}{X(s)} = \frac{1}{\tau^2 s^2 + 2\zeta\tau s + 1}$$

Do this by rewriting the transfer function into the form of two first-order differential equations. Recommended values of ζ for different phase planes are: *a.* 0.1; *b.* 0.2; *c.* 0.5; *d.* 1.0; *e.* 2.0; *f.* 5.0.

28 Examples of Phase-plane Analysis

In this chapter, we shall consider two different examples of the use of the phase plane to analyze nonlinear control systems. The first is a simple on-off control system for a stirred-tank heater. The second is the chemical reactor of Chap. 26. In both cases, the systems are second-order and autonomous, so that they are ideal situations for use of the phase plane.

On-Off Control of Stirred-tank Heater   The use of on-off control offers significant economic advantages over proportional control or other more sophisticated modes of control. The control mechanism is simply a relay which turns on or off depending on the value of the measured variable. The disadvantage is usually that the quality of control is inferior to that realized with proportional control.

Consider the stirred-tank heater of Fig. 28.1. Water is being heated to a controlled temperature by mixing with steam. It is assumed for the analysis that the cold-water input rate is constant. Heated water overflows into an outlet pipe at the top of the tank, so that no accumulation of mass occurs in the tank. Most of the steam is added, at a fixed flow rate, from the main steam supply. However, this amount of steam is set at a value

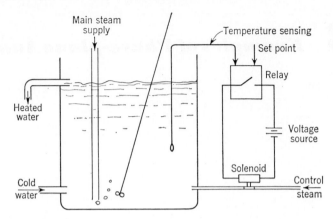

Fig. 28.1   On-off control of stirred-tank heater.

somewhat less than the amount required to heat the cold water to the desired temperature.   An additional amount of steam may be added whenever the solenoid valve is opened.   When this additional steam is admitted, the sum of the two steam inputs is enough to heat the water to a temperature somewhat in excess of the desired temperature.   A temperature-measuring device such as a thermocouple or vapor-pressure bulb transmits the tank temperature to the relay.   When this temperature is below the set point, the relay closes, which opens the solenoid valve, thus admitting more steam. Eventually, the additional steam will result in the temperature exceeding the set point, the relay will open, the valve will close cutting off the additional steam, and the temperature will fall again.

It is apparent that an oscillating control will be achieved.   In fact, from the discussion of the previous chapter, we recognize that a limit cycle will occur.   We consider now a numerical example of this type of control system.

Water at 40°F, at a rate of 100 lb/min, is to be heated to 150°F.   The main steam supply is to be set so that it will heat this much water to 125°F, while additional steam, through the controlled solenoid valve, is available to heat the water another 50°F.   This means that the steady-state temperatures with the solenoid closed and open, 125 to 175°F, are equally spaced about the set point.   Heat losses to the surroundings are negligible.   The volume of the tank is 1.6 ft³.   The relay control system has a vapor-pressure bulb for measurement of temperature.   This measuring system has a time constant of 30 sec.   The solenoid valve is very rapid in response.

We first analyze this system considering the relay to behave ideally. This means that it opens precisely at the instant the temperature exceeds the set point and closes similarly.   Later, we shall correct this to conform more closely to the behavior of actual relays.

If the tank is perfectly stirred, it is a first-order system with a time constant of

$$\tau = \frac{\rho V}{w} = \frac{(62)(1.6)}{100} = 1.0 \text{ min}$$

and its transfer function relating changes in the steam input rate to temperature is

$$G_p(s) = \frac{10}{s+1}$$

where 10 (°F)(min)/(lb) is the change in steady-state temperature per unit change in steady-state steam flow. The necessary fixed and controlled steam rates are (using 1,000 Btu/lb for latent heat)

$$Q_{\text{fixed}} = \frac{(125 - 40)(100)}{1,000} = 8.5 \text{ lb/min}$$

$$Q_{\text{controlled}} = \frac{(175 - 125)(100)}{1,000} = 5.0 \text{ lb/min}$$

The amount of steam which would be necessary to maintain the water at a steady-state temperature of 150°F is

$$Q_s = \frac{(150 - 40)(100)}{1,000} = 11.0 \text{ lb/min}$$

Hence, in terms of deviation variables, the controller output may be taken as ± 2.5 lb/min of steam.

A block diagram may now be constructed for this system, as shown in Fig. 28.2. This diagram uses deviations from 150°F as temperature variables, so the set point is taken as zero. The action of the relay is symbolized by the input-output relations, indicating that $+2.5$ lb/min of steam are admitted when the error is positive and -2.5 lb/min when the error is negative, again in deviation variables. The transduction from the vapor-pressure bulb to a temperature reading is included implicitly in Fig. 28.2 in the comparator. The comparator is physically a device which balances the pressure generated by the bulb against a mechanical tension caused by positioning the set point. It need not be explicitly shown because its dynamics are very fast.

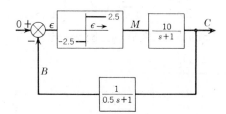

Fig. 28.2 Block diagram for system of Fig. 28.1.

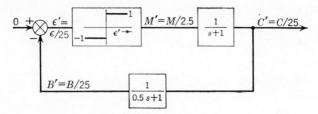

Fig. 28.3 **Dimensionless block diagram for system of Fig. 28.1.**

It is convenient to use a dimensionless version of Fig. 28.2. This is provided in Fig. 28.3, where the changes

$$M' = \frac{M}{2.5}$$

$$\epsilon' = \frac{\epsilon}{25}$$

$$C' = \frac{C}{25}$$

$$B' = \frac{B}{25}$$

have been made.

The usual methods of linear control theory are not applicable to the block diagram of Fig. 28.3. The relay does not obey the principle of superposition in its input-output relation. It is necessary to revert to the differential equations describing the control loop. These are

$$M' = \frac{dC'}{dt} + C' \tag{28.1}$$

$$C' = \frac{1}{2}\frac{dB'}{dt} + B' \tag{28.2}$$

$$\epsilon' = -B' \tag{28.3}$$

In addition we have

$$M' = \begin{cases} 1 & \epsilon' > 0 \\ -1 & \epsilon' < 0 \end{cases} \tag{28.4}$$

Combination of Eqs. (28.1) to (28.4) yields

$$\frac{1}{2}\frac{d^2\epsilon'}{dt^2} + \frac{3}{2}\frac{d\epsilon'}{dt} + \epsilon' = \begin{cases} -1 & \epsilon' > 0 \\ 1 & \epsilon' < 0 \end{cases} \tag{28.5}$$

Equation (28.5) can be rewritten in phase notation as

$$\frac{d\epsilon'}{dt} = \dot{\epsilon}'$$

$$\frac{d\dot{\epsilon}'}{dt} = \begin{cases} -(3\dot{\epsilon}' + 2\epsilon' + 2) & \epsilon' > 0 \\ -(3\dot{\epsilon}' + 2\epsilon' - 2) & \epsilon' < 0 \end{cases} \tag{28.6}$$

Equation (28.6) breaks up into two regions, the region for which $\epsilon' > 0$ will be referred to as R, and that for which $\epsilon' < 0$ as L. The critical point for R occurs at

$$\epsilon' = -1 \qquad \dot{\epsilon}' = 0$$

and that for L at

$$\epsilon' = 1 \qquad \dot{\epsilon}' = 0$$

Note that each critical point is outside the region to which it pertains. In region R, the isocline equation is

$$-\frac{2 + 2\epsilon' + 3\dot{\epsilon}'}{\dot{\epsilon}'} = S_R$$

or

$$\dot{\epsilon}' = \frac{-2(\epsilon' + 1)}{S_R + 3} \tag{28.7}$$

The corresponding isocline equation in L is

$$\dot{\epsilon}' = \frac{-2(\epsilon' - 1)}{S_L + 3} \tag{28.8}$$

The isoclines in R, which is the right half of the $\dot{\epsilon}'\epsilon'$ plane, radiate from the R critical point $(-1,0)$ and have slopes $-2/(S_R + 3)$. The isoclines in L radiate from the critical point $(1,0)$ and have slopes $-2/(S_L + 3)$. These isoclines are indicated in Fig. 28.4. Note that, in this figure, the ϵ' scale has been expanded by a factor of 10 to magnify the behavior near the origin.

A typical trajectory has been constructed, using the method of isoclines. When the trajectory crosses from one region to the other on the $\dot{\epsilon}'$ axis, the applicable isoclines also change. It can be seen from Fig. 28.4 that the trajectory approaches the origin. Since the trajectories must be vertical as they cross the ϵ' axis, the final state is a limit cycle of zero amplitude and infinite frequency about the origin. In other words, the relay alternately opens and closes at very high frequency, a condition known as *chattering*.

Physically, this condition will never be realized because the dynamics of the solenoid valve and the relay itself would become important. Instead, the final condition will be a limit cycle of high, rather than infinite, frequency and low, rather than zero, amplitude.

However, the basic idealization which has led us to this suspect conclusion is in the behavior of the relay. True relays have input-output characteristics more similar to that shown in Fig. 28.5. There is a dead band around the set point, of width $2\epsilon_0$, over which the relay is insensitive to changes in the error signal. Anyone who has made fine adjustments in the setting of a home thermostat has observed this behavior.

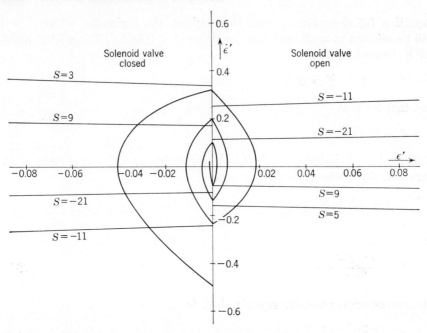

Fig. 28.4 Phase-plane trajectory for on-off control of system of Fig. 28.1.

Consider as an example the case for $\epsilon'_o = 0.01$. The effect of this dead zone is to change the dividing line between R and L to that shown in Fig. 28.6. The new dividing line has the equation:

$$\epsilon' = \begin{cases} 0.01 & \dot{\epsilon}' > 0 \\ -0.01 & \dot{\epsilon}' < 0 \end{cases}$$

Now, as shown in Fig. 28.6, all trajectories approach a limit cycle, for which the error amplitude is approximately 0.03. The frequency is finite and is

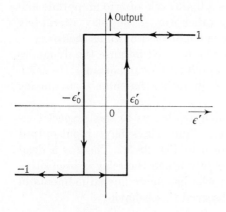

Fig. 28.5 Characteristics of true relay with dead zone.

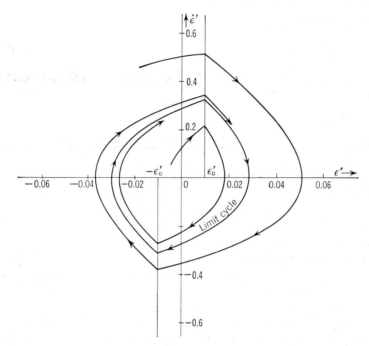

Fig. 28.6 **Phase plane for system of Fig. 28.1 using relay with characteristics of Fig. 28.5.**

obtained by computing the time around the limit cycle. Although we have not presented here the graphical methods for determining this time, it can always be calculated by noting from the first of Eqs. (28.6) that

$$t = \int dt = \int \frac{d\epsilon'}{\dot{\epsilon}'} \tag{28.9}$$

Thus, time around the limit cycle can be computed by graphical evaluation of the integral in Eq. (28.9). The only difficulty is near the ϵ' axis, where $\dot{\epsilon}'$ goes to zero. To circumvent this, we may use the second of Eqs. (28.6)

$$t = -\int \frac{d\dot{\epsilon}'}{3\dot{\epsilon}' + 2\epsilon' \pm 2}$$

over a small segment of the trajectory as it crosses the ϵ' axis. The result of this graphical calculation is $\omega = 9.2$ rad/min.

The frequency thus computed for the error signal is, for obvious physical reasons, the same as the frequency of the controlled signal, C'. However, the amplitude of C', which is of more direct interest, is not the same as the amplitude of ϵ'. It may be found in this case by noting from

Eqs. (28.2) and (28.3) that

$$C' = -\tfrac{1}{2}\dot{\epsilon}' - \epsilon' \tag{28.10}$$

It is therefore clear from Fig. 28.6 that C' attains a maximum value near the switching points where

$$C' \approx \pm 0.17$$

Reverting to the original variables, it follows that the water temperature will oscillate with an amplitude of

$$(0.17)(25) = 4.25°F$$

The effect of a small dead zone, $2\epsilon_0 = 2(.01)(25) = 0.5°F$, is thus quite significant.

In practice, the width of this dead zone is usually an adjustable design parameter. This width is always chosen as a compromise. The wider it is made, the lower will be the limit-cycle frequency, thus saving excessive switching or chatter. However, the limit-cycle amplitude increases with dead-zone width, decreasing the quality of control.

The Exothermic Chemical Reactor   We now wish to consider the phase-plane behavior of the chemical reactor of Chap. 26. This study is based upon the paper by Aris and Amundson.[1] For convenience, the dynamic equations are reproduced below:

$$\frac{dx_A}{dt} = \frac{F}{V}(x_{A_0} - x_A) - ke^{-E/RT}x_A$$
$$\frac{dT}{dt} = \frac{F}{V}(T_0 - T) + \frac{k(\Delta H)e^{-E/RT}}{C_p}x_A - \frac{Q(T)}{\rho V C_p} \tag{26.21}$$

Defining the dimensionless variables

$$\tau = \frac{Ft}{V} \qquad y = \frac{x_A}{x_{A_0}} \qquad \theta = \frac{C_p T}{x_{A_0}(\Delta H)} \qquad \theta_0 = \frac{C_p T_0}{x_{A_0}(\Delta H)}$$

these equations become

$$\frac{dy}{d\tau} = 1 - y - r(y,\theta)$$
$$\frac{d\theta}{d\tau} = \theta_0 - \theta + r(y,\theta) - q(\theta) \tag{28.11}$$

where

$$r(y,\theta) = \frac{kVy}{F}e^{-EC_p/Rx_{A_0}(\Delta H)\theta}$$
$$q(\theta) = \frac{Q(T)}{F\rho x_{A_0}(\Delta H)}$$

[1] R. Aris and N. R. Amundson, *Chem. Eng. Sci.*, **7**:121–155 (1958).

As a control heat-removal function $q(\theta)$, Aris and Amundson chose the form

$$q(\theta) = U(\theta - \theta_c)[1 + K_c(\theta - \theta_s)] \tag{28.12}$$

where θ_c is the dimensionless mean temperature of water in the cooling coil. This indicates that the heat removal is always proportional to the difference between the reactor temperature and mean cooling-water temperature. In addition, the term in brackets indicates that proportional control on the cooling-water flow rate is present. The flow rate is increased by an amount proportional to the difference between the actual reactor temperature θ and the desired steady-state temperature θ_s. This increase in cooling-water flow rate is assumed for convenience to cause an approximately proportional increase in heat removal. The constant U is a dimensionless analog of U_0A, the overall heat-transfer rate.

As a specific numerical example, Aris and Amundson selected the following values for constants:

$$\frac{kV}{F} = e^{25}$$

$$\frac{EC_p}{Rx_{A_0}(\Delta H)} = 50$$

$$\theta_s = 2$$

$$\theta_0 = \theta_c = 1.75$$

$$U = 1$$

Under these conditions, Eqs. (28.11) become

$$\frac{dy}{d\tau} = 1 - y - ye^{50(\frac{1}{2}-1/\theta)}$$

$$\frac{d\theta}{d\tau} = 1.75 - \theta + ye^{50(\frac{1}{2}-1/\theta)} - (\theta - 1.75)[1 + K_c(\theta - 2)] \tag{28.13}$$

It can be seen that there is a critical point of Eqs. (28.13) at

$$y = \tfrac{1}{2} = y_s$$
$$\theta = 2 = \theta_s$$

and this is the location at which control is desired. This point has the correct steady-state temperature and a 50 percent conversion of reactant. In addition, there may be two more critical points of Eq. (28.13) depending on the proportional control constant K_c, as will be discussed below.

Since we are primarily interested in control about θ_s, we make use of Liapunov's theorem on local stability, presented in Chap. 27. Linearizing Eq. (28.13) in deviation variables $\theta - \theta_s$ and $y - y_s$ by using Taylor's series yields

$$\frac{d(y - y_s)}{d\tau} = -2(y - y_s) - 6.25(\theta - \theta_s)$$

$$\frac{d(\theta - \theta_s)}{d\tau} = (y - y_s) + \left(4.25 - \frac{K_c}{4}\right)(\theta - \theta_s) \tag{28.14}$$

where $y_s = \frac{1}{2}$. As we have seen before, the solution to this linear system is

$$y - y_s = C_1 e^{s_1 t} + C_2 e^{s_2 t}$$
$$\theta - \theta_s = C_3 e^{s_1 t} + C_4 e^{s_2 t}$$

where, in this case, s_1 and s_2 are the roots of [see Eq. (26.8) *et seq.*]

$$s^2 + \frac{K_c - 9}{4} s + \frac{2K_c - 9}{4} = 0 \tag{28.15}$$

According to the Routh criteria, all coefficients in this characteristic equation must be positive in order that the real parts of the roots s_1 and s_2 be negative. Hence, we can see immediately from Eq. (28.15) that, in order to achieve a stable node or focus, it is necessary that $K_c > 9$.

However, Aris and Amundson obtained the phase plane for (among other values) a value of K_c slightly greater than 9. This was accomplished by numerical solution of Eqs. (28.13). It was found that, in the vicinity of the steady-state point, the situation is as depicted in Fig. 28.7. There are two limit cycles surrounding the stable focus critical point. The inner limit cycle is unstable, and the outer limit cycle is stable, according to the definitions given in Chap. 27. It may be seen that any disturbance (or initial condition) which moves the system no further from the critical point than the unstable limit cycle can be controlled. That is, the control system will eventually bring the system back to steady state. However, once the system is forced outside this limit cycle, it will eventually spiral out to the stable limit cycle. Control cannot be restored, and the reactor temperature and concentration oscillate continuously. This example illustrates very well the limitations of linear control theory. All that the linear investigation could reveal is that, for $K_c > 9$, the system will be stable in some vicinity of the control point. The phase-plane analysis shows that, for K_c slightly greater than 9, this vicinity is inside the unstable limit cycle of Fig. 28.7. If K_c is increased further, the two limit cycles disappear and

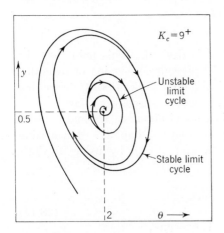

Fig. 28.7 Stable and unstable limit cycles in exothermic chemical reactor.

good control can be achieved. This example points out the importance of unstable limit cycles. Although a physical system can never follow an unstable limit cycle, the limit cycle divides the phase plane into distinct dynamic regions for the physical system.

Other values of K_c were analyzed by Aris and Amundson. For low values of K_c, there are two other critical points besides the control point. For example, for $K_c = 0.8$, there are critical points at

$$y = 0.95 \qquad \theta = 1.77$$

and

$$y = 0.15 \qquad \theta = 2.15$$

Linear analysis shows that both these are stable, but for $K_c < 9$, the control point ($y = 0.5$, $\theta = 2$) is not. Phase-plane analysis shows that, if the reactor is started at high temperatures, it will come to steady state at the high-temperature critical point and vice versa. Starting the reactor at the desired control point will be of no avail, as it will leave and go to one of the other steady-state points, depending on the direction of the initial disturbance. For high values of K_c, there is only one critical point, which is at the control point. Phase-plane analysis shows that K_c must exceed approximately 30 before rapid return to steady state at the desired control point, following all disturbances, is achieved. Some phase-plane portraits for this system that were obtained by means of an analog computer are shown in Figs. 32.26–32.28.

This discussion is only a rather brief introduction to the extensive work by Aris and Amundson. The reader is strongly urged to consult the original paper for a more comprehensive treatment of the problem.

Summary We have seen that phase-plane analysis can be used for two typical nonlinear control problems. The results of this analysis give extensive information about the control system behavior. The responses to various disturbances can be visualized by sketching only a few trajectories.

On the other hand, the method is effectively limited to second-order systems. Furthermore, analysis is considerably more laborious than the linear analysis, and a decision regarding the value of the additional information must be made. It is to be expected that increasing use of the phase concept will be made in future work in process control.

PROBLEMS

28.1 Suppose that no fixed steam supply is used in the stirred-tank heater. Instead, the solenoid will admit, when opened, all the steam necessary to heat the tank to 175°F, or 25°F above the set point. Using the relay with a 0.5°F dead space, what will be the new limit cycle amplitude? Compare this with the symmetric case discussed in the chapter. Discuss advantages and disadvantages of each case.

29 *The Describing Function Technique*

In Chap. 28, an on-off temperature-control system was studied in the phase plane. This work led to information about the limit cycle of the system as well as about the manner in which trajectories approached the limit cycle. Very often, this latter information about the transient approach to the limit cycle is unnecessary. Of primary interest to the designer are the amplitude and frequency of the limit cycle. The describing function method facilitates rapid, accurate estimates of these quantities without construction of the phase plane.

In this chapter we shall study application of the describing function method to the analysis of the on-off controller for the temperature bath of Chap. 28. The treatment will be introductory only and largely confined to a single example. The purpose is to indicate the existence of the method and to show how it complements the phase-plane technique. The reader desirous of a more extensive treatment is referred to the text by Graham and McRuer.[1]

---

[1] D. Graham and D. McRuer, "Analysis of Nonlinear Control Systems," John Wiley & Sons, Inc., New York, 1961.

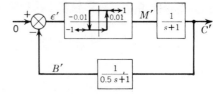

Fig. 29.1 **Block diagram for control of stirred-tank heater using relay with dead zone.**

Harmonic Analysis   Consider the block diagram for the on-off control of the stirred-tank heater of Chap. 28, shown in dimensionless form in Fig. 29.1.   In the following analysis, we omit the primes from the variables of Fig. 29.1.   Our objective is to find the amplitude and frequency of the limit cycle which occurs in the control loop.   The describing function method assumes that the error signal, in the limit cycle condition, is sinusoidal:

$$\epsilon = A \sin \omega t \tag{29.1}$$

A glance at Fig. 28.6 shows that the error signal is not actually sinusoidal, since a sinusoidal signal appears as an ellipse in the phase plane.   However, the difference between the actual limit cycle and an ellipse is not great, particularly if only the amplitude and frequency are of interest.

If the error signal is sinusoidal, the relay output M can be derived from Fig. 29.2, where it can be seen from the input-output relations that $M(t)$ is a square wave which lags $\epsilon(t)$ by a time $(1/\omega) \sin^{-1}(\epsilon_0/A)$.   The

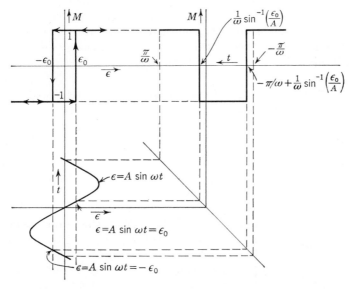

Fig. 29.2 **Result of application of sinusoidal error signal to relay with dead zone.**

time lag is due to the dead zone in the relay.   Thus,

$$M(t) = S\left(t - \frac{1}{\omega}\sin^{-1}\frac{\epsilon_0}{A}, \frac{2\pi}{\omega}\right) \tag{29.2}$$

where

$$S(t,P) = \begin{cases} 1 & 0 < t < P/2 \\ -1 & P/2 < t < P \\ S(t + P, P) & \text{all } t \end{cases} \tag{29.3}$$

is the undelayed unit square wave of period P shown in Fig. 29.3.

As is well known from Fourier series analysis,[1] $S(t, 2\pi/\omega)$ may be expanded in a series of sine waves to give

$$S\left(t, \frac{2\pi}{\omega}\right) = \frac{4}{\pi}\left(\sin \omega t + \frac{1}{3}\sin 3\omega t + \frac{1}{5}\sin 5\omega t + \cdots\right) \tag{29.4}$$

Hence, by Eq. (29.2)

$$M(t) = \frac{4}{\pi}\left[\sin\left(\omega t - \sin^{-1}\frac{\epsilon_0}{A}\right) + \frac{1}{3}\sin\left(3\omega t - 3\sin^{-1}\frac{\epsilon_0}{A}\right) \right.$$
$$\left. + \frac{1}{5}\sin\left(5\omega t - 5\sin^{-1}\frac{\epsilon_0}{A}\right) + \cdots\right] \tag{29.5}$$

According to Eq. (29.5), $M(t)$ contains a fundamental and odd harmonics.   Let us consider what happens to these components of M as they pass around the control loop.   Assuming that ω is sufficiently large, the harmonics are much more heavily attenuated by the two first-order elements than is the fundamental, because the harmonic frequencies are higher.   For example, if ω is 9 rad/min, the relative attenuation of the fundamental and

---

[1] R. V. Churchill, "Fourier Series and Boundary Value Problems," Chap. 4, McGraw-Hill Book Company, New York, 1941.

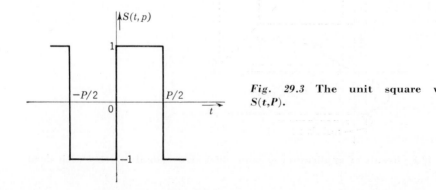

Fig. 29.3 The unit square wave $S(t,P)$.

third harmonic between M and B is expressed by the quotient

$$\frac{\dfrac{1}{\sqrt{1 + (27)^2}\,\sqrt{1 + (27/2)^2}}}{\dfrac{1}{\sqrt{1 + (9)^2}\,\sqrt{1 + (9/2)^2}}} = 0.11$$

Since the initial amplitude of the third harmonic in $M(t)$ is one-third of the fundamental, it is clear that the amplitude of the third harmonic will be less than 4 percent of the amplitude of the fundamental in $B(t)$. The amplitudes of the higher harmonics will be even less. To all intents and purposes, $B(t)$ is sinusoidal and, hence, so is $\epsilon(t)$. Furthermore, the presence of harmonics in $M(t)$ may be ignored, and the approximation

$$M(t) \cong \frac{4}{\pi} \sin\left(\omega t - \sin^{-1}\frac{\epsilon_0}{A}\right) \tag{29.6}$$

is acceptable because the higher harmonics are filtered out by the rest of the loop.

In order for a limit cycle to be maintained, it is necessary that

$$B(t) = -\epsilon(t) = -A \sin \omega t \tag{29.7}$$

However, if $M(t)$ is given by Eq. (29.6), $B(t)$ can be calculated by frequency response. The AR between M and B is

$$\left|\frac{B}{M}\right| = \frac{1}{\sqrt{1 + (\omega)^2}\,\sqrt{1 + (\omega/2)^2}} \tag{29.8}$$

and the phase difference between B and M is

$$\angle B - \angle M = -\tan^{-1}\omega - \tan^{-1}\frac{\omega}{2} \tag{29.9}$$

According to Eq. (29.7), the overall amplitude ratio between B and ϵ must be unity and the overall phase lag 180°. Also, according to Eq. (29.6), the AR between ϵ and M is $4/\pi A$ and the phase lag is $\sin^{-1}(0.01/A)$. Combining these facts results in

$$\frac{4}{\pi A}\,\frac{1}{\sqrt{1 + (\omega)^2}\,\sqrt{1 + (\omega/2)^2}} = 1$$
$$-\sin^{-1}\frac{0.01}{A} - \tan^{-1}\omega - \tan^{-1}\frac{\omega}{2} = -180° \tag{29.10}$$

Equations (29.10) are a system of two equations in the unknowns A and ω. Trial-and-error solution yields

$$A = 0.03$$
$$\omega = 9 \text{ rad/min}$$

in excellent agreement with the results of the phase-plane method presented in Chap. 28. The reason for the accuracy of these results is the high attenuation of harmonics provided by the linear elements in the loop. The labor saving of this method over the phase-plane method is apparent.

Of more direct interest is the amplitude of the signal C in the limit cycle. This may now be estimated by frequency response to be

$$|C| = \sqrt{1 + \left(\frac{9}{2}\right)^2}\,(0.03) = 0.14$$

The true result derived by the phase-plane method is 0.17, so that an error of 18 per cent is attributed to the neglect of harmonics in C. The reason for the decreased accuracy in the amplitude of C over that in the amplitude of ϵ is that only one of the linear elements has acted on the square-wave output of the relay before it reaches C. Hence, the harmonics are not fully attenuated in C and the signal C will be less sinusoidal than ϵ. However, for engineering purposes the error in C is probably not excessive.

The Describing Function Because the basic technique of harmonic analysis often yields accurate results with modest effort, it is profitable to systematize the procedure. To do this, a describing function is defined for the nonlinear loop element. This function assumes a sinusoidal input to the nonlinearity and gives the AR and phase lag of the *fundamental* in the output. Thus, for the relay considered in the last section, the describing function is defined by

$$N = \frac{4}{\pi A}\ \angle - \sin^{-1}\frac{\epsilon_0}{A}$$

where N is used as the symbol for the describing function.

In general, the loop diagram for a relay control system appears as in Fig. 29.4. As shown previously, the necessary condition for a limit cycle, ignoring harmonics, is

$$|N|\,|G_pH| = 1$$
$$\angle N + \angle G_pH = -180° \tag{29.11}$$

As in the case of the relay, the magnitude and angle of N in general depend upon the amplitude A of the input to N. The magnitude and angle of

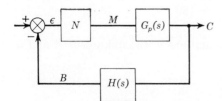

Fig. 29.4 **Typical control loop containing nonlinear element.**

G_pH depend upon ω. Equations (29.11) can be rewritten

$$|G_pH(\omega)| = \frac{1}{|N(A)|}$$
$$\angle G_pH(\omega) = -180° - \angle N(A)$$

(29.12)

Equations (29.12) can be solved graphically on a gain-phase plot. This is a plot of the log of AR versus phase, as shown in Fig. 29.5 for the case treated in the previous section. The linear elements are plotted as $|G_pH|$ versus $\angle G_pH$, with ω plotted as a parameter on the curve. The relay is plotted as $1/|N(A)|$ versus $-180° - \angle N(A)$, with A plotted as a parameter on the curve. According to Eqs. (29.12), a limit cycle occurs at the intersection of the two curves, where the amplitude and frequency can be read from the parametric labeling of A and ω.

The advantages of the gain-phase plot are (1) elimination of trial-and-error solution of equations such as Eqs. (29.10) and (2) ease of treatment of complex linear systems G_pH. In addition, the gain-phase plot can be used to estimate the occurrence or nonoccurrence of a limit cycle, according to whether or not an intersection occurs.

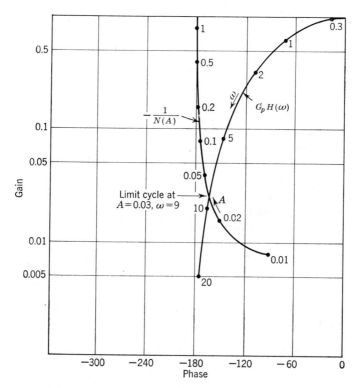

Fig. 29.5 **Gain-phase plot for system of Fig. 29.1.**

Summary   The describing function can be used to good advantage for estimation of amplitude and frequency of limit cycles in systems similar to the one studied here.   The success of the method depends upon the presence of a sufficient number of linear elements in the loop to filter out the harmonics generated by the nonlinear element.   No information about the transient response is obtained.   However, the method requires considerably less labor than does the phase-plane method, and the limit cycle amplitude and frequency are often the quantities of primary interest.

It should also be noted that the describing function method is not limited to second-order systems, as is the phase-plane method.   In fact, the higher the order of G_pH in Fig. 29.4, the more accurate will be the describing function results.

part VIII Analog Computer

30 *Linear Computer Operations*

The analog computer is now recognized as an important tool in the study of transients and control. The main purpose of this chapter is to present the basic concepts of analog computation with sufficient detail to help the reader apply the computer to the solution of differential equations. In Chap. 32, the application of the computer to control systems will be discussed. It is recommended that a computer be available during the study of this material.

A large number of chemical companies are now adding computer facilities. A recent survey article[1] describes this advance of analog computers into chemical industry and lists the principal manufacturers of general-purpose analog computers.

INTRODUCTION

Amplifier The basic component of an analog computer is a high-gain d-c operational amplifier. For the purpose of this discussion, the amplifier

[1] Analog Computers Penetrate the Chemical Industry, *Chem. Eng. News*, Feb. 6, 1961, pp. 118–126.

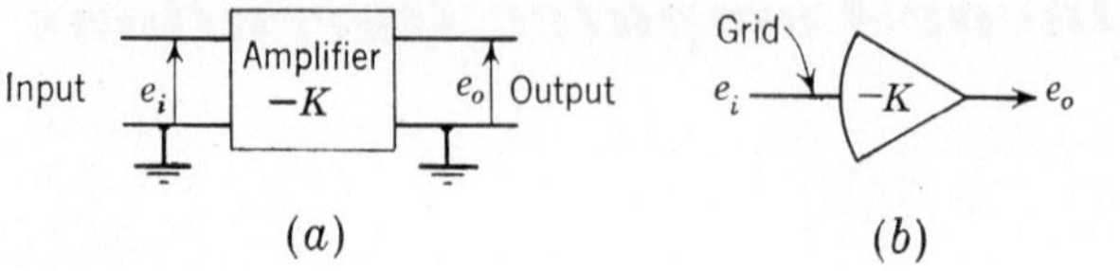

Fig.   30.1   Operational amplifier and symbol.

(a)   (b)

may be considered as a box with two pairs of terminals as shown in Fig. 30.1a. If a voltage is applied to the input terminals, the voltage at the output terminals is some multiple K of the input but reversed in sign. For simplicity, the high-gain amplifier is represented by the symbol in Fig. 30.1b in which the ground leads are omitted. The voltages e_i and e_o shown in Fig. 30.1b are with respect to ground.

The main linear operations in analog computation are summation, integration, and multiplication by a constant. The first two operations are obtained by adding appropriate circuit elements (resistors and capacitors) to an operational amplifier, as described in the following paragraphs. A description of nonlinear components (function generator, multiplier, diode) will be considered in the next chapter.

Summation   The circuit shown in Fig. 30.2a performs "summation" according to the relationship

$$e_o = - \left(\frac{R_f}{R_1} e_1 + \frac{R_f}{R_2} e_2 \right)$$ (30.1)

This equation states that voltages e_1 and e_2 are multiplied by constants R_f/R_1 and R_f/R_2, respectively, and then summed. (Notice that the sign is reversed in the operation of summation.) The circuit of Fig. 30.2a is too awkward to write when many summers are present, and the more compact symbol shown in Fig. 30.2b is generally used. The numbers next to the input lines of Fig. 30.2b (R_f/R_1 and R_f/R_2) are referred to as *gains*. The circuit shown in Fig. 30.2 will be called a summer even though this term is appropriate only when the gains are all unity. If more than two voltages are to be summed, one simply adds an additional input resistor for each additional input voltage.

To show that the circuit in Fig. 30.2 performs summation, we can write

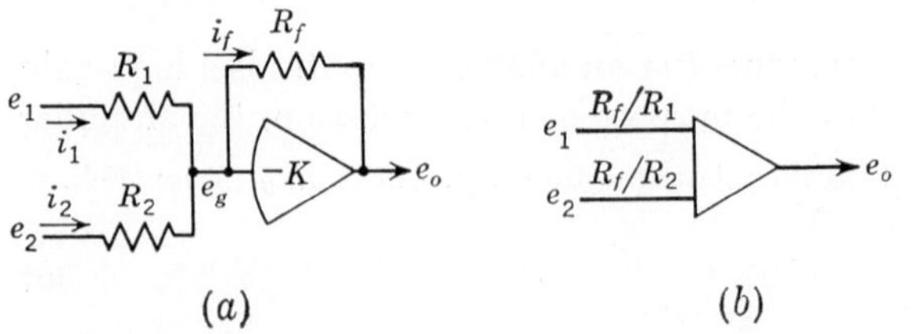

Fig.   30.2   Circuit for summation.

(a)   (b)

the following equations based on Ohm's law:

$$e_1 - e_g = i_1 R_1 \tag{30.2}$$
$$e_2 - e_g = i_2 R_2 \tag{30.3}$$
$$e_g - e_o = i_f R_f \tag{30.4}$$

where e_g is the grid voltage and i_1, i_2, and i_f are the currents through R_1, R_2, and R_f as shown in Fig. 30.2a. A current balance at the grid gives

$$i_f = i_1 + i_2 \tag{30.5}$$

Notice that, in writing the current balance at the grid, we have taken the grid current to be zero. This is an electrical characteristic of the amplifier which will be used without further justification throughout this discussion. Solving Eqs. (30.2) to (30.4) for i_1, i_2, and i_f and substituting the resulting expressions into Eq. (30.5) give

$$\frac{e_g - e_o}{R_f} = \frac{e_1 - e_g}{R_1} + \frac{e_2 - e_g}{R_2} \tag{30.6}$$

The amplifier provides the relationship

$$e_g = -\frac{e_o}{K} \tag{30.7}$$

Replacing e_g in Eq. (30.6) by the right side of Eq. (30.7) gives, after slight rearrangement,

$$-e_o\left(1 + \frac{1}{K}\right) = \left(\frac{e_1}{R_1} + \frac{e_o}{KR_1} + \frac{e_2}{R_2} + \frac{e_o}{KR_2}\right) R_f \tag{30.8}$$

The amplifier gain K is very high. For an inexpensive operational amplifier, K may vary from 30,000 to 50,000. For amplifiers used in more expensive computers K may be as large as 10^8. Since K is so large, the terms in Eq. (30.8) involving K become very small and can be neglected, with the result

$$-e_o = \frac{R_f}{R_1} e_1 + \frac{R_f}{R_2} e_2$$

which is the same as Eq. (30.1). The reader who desires to know more about the errors in analog computing caused by the finite gain of the amplifier should consult the textbook by Howe.[1]

Example 30.1 Determine the output voltage for the summer shown in Fig. 30.3a, and draw the compact symbol for this operation. All the resistors are in megohms.

[1] R. M. Howe, "Design Fundamentals of Analog Computer Components," D. Van Nostrand Company, Inc., Princeton, N.J., 1961.

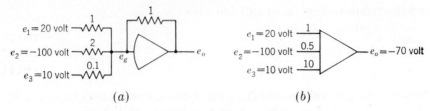

(a)   (b)

Fig. 30.3   Computer circuit for Example 30.1.

According to Fig. 30.2, we write

$$e_o = -(1e_1 + 0.5e_2 + 10e_3)$$

or

$$e_o = -(20 + (0.5)(-100) + 10(10)) = -70 \text{ volts}$$

The compact symbol is shown in Fig. 30.3b.

Integration   An integrator can be formed by adding a capacitor and a resistor to an amplifier as shown in Fig. 30.4. This circuit performs integration according to the expression

$$e_o = -\frac{1}{RC}\int_0^t e_1(t)\,dt + e_o(0) \qquad (30.9)$$

where $e_o(0)$ is the initial voltage on the capacitor before the signal e_1 is applied. A compact symbol for the integrator is shown in Fig. 30.4b. The term $1/RC$ is called the gain of the integrator. The method used to impose initial conditions will be considered later.

The integration performed by the circuit of Fig. 30.4 is with respect to time measured in seconds if R is in ohms and C is in farads. This follows from the basic electrical laws involved and the definitions for resistance and capacitance. The electrical design of computers is such that the resistors actually used often range from 0.1 to 10 megohms and the capacitors range from 0.01 to 1 microfarad (μf). With R in megohms and C in microfarads, the product RC again has the units of seconds. Throughout the remainder of this discussion the sizes of resistors and capacitors will be measured in megohms and microfarads.

To show that the circuit of Fig. 30.4 performs integration, we may

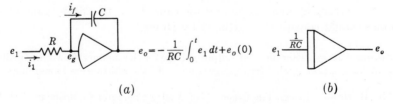

(a)   (b)

Fig. 30.4   Circuit for integration.

write the following circuit equations which follow from the elementary rules of circuit analysis and the characteristics of the amplifier.

$$e_1 - e_g = i_1 R \tag{30.10}$$

$$e_g - e_o = \frac{1}{C} \int i_f \, dt \tag{30.11}$$

$$e_g = -\frac{e_o}{K} \tag{30.12}$$

$$i_f = i_1 \tag{30.13}$$

Combining these equations gives

$$-e_o \left(1 + \frac{1}{K}\right) = \frac{1}{C} \int \left(\frac{e_1}{R} + \frac{e_o}{RK}\right) dt$$

Since K is very large as described in the previous section, the terms involving K are negligible, and the previous equation can be written

$$e_o = -\frac{1}{RC} \int e_1 \, dt \tag{30.14}$$

If initially $e_o = e_o(0)$, then Eq. (30.14) is written in terms of a definite integral as in Eq. (30.9).

Example 30.2 A step change of magnitude 10 volts is applied to the integrator shown in Fig. 30.5a. If the initial condition voltage $e_o(0)$ is zero, plot e_o as a function of time for (1) $R = 1$ and (2) $R = 0.5$.

From Fig. 30.4, we can write

1. $$e_o = -\frac{1}{(1)(1)} \int_0^t 10 \, dt + 0$$

or

$$-e_o = 10t$$

The output response is shown in Fig. 30.5*b*.

2. $$e_o = -\frac{1}{(0.5)(1)} \int_0^t 10 \, dt + 0$$

or

$$-e_o = 20t$$

This response is shown in Fig. 30.5b. For this example, the rate of change of e_o is proportional to the gain $(1/RC)$ of the integrator.

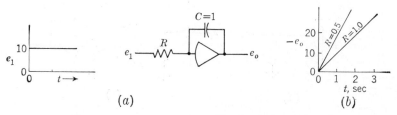

(a)   (b)

Fig. 30.5 Computer circuit for Example 30.2.

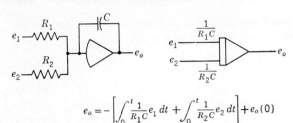

Fig. 30.6 **Combining summation and integration.**

$$e_o = -\left[\int_0^t \frac{1}{R_1 C} e_1 \, dt + \int_0^t \frac{1}{R_2 C} e_2 \, dt\right] + e_o(0)$$

When several input resistors are added to the integrator circuit of Fig. 30.4, summation and integration may be combined as shown in Fig. 30.6.

Coefficient Potentiometer The coefficient potentiometer is used to multiply a voltage by a constant k, which can be varied continuously from 0 to 1. The potentiometer is a voltage divider with a movable wiper or contact as shown in Fig. 30.7a. The input-output relation is $e_o = ke_1$. The symbol for the potentiometer is shown in Fig. 30.7b.

When the potentiometer is used in a computer circuit, the wiper of the potentiometer is generally connected to the input resistor of an amplifier as shown in Fig. 30.7c. This input resistor places a load on the potentiometer, and since the grid voltage is very close to ground potential ($e_g \cong 0$), the computer circuit is equivalent to that shown in Fig. 30.7d. Let kR_p be the potentiometer resistance between the wiper and the ground, where R_p is the total resistance of the potentiometer. If R_1 is not present (no loading), $e_o = ke_1$. However, if R_1 is present, e_o is *not* simply ke_1 but a current balance at the wiper-arm contact point yields

$$\frac{e_1 - e_o}{(1 - k) R_p} = \frac{e_o}{k R_p} + \frac{e_o}{R_1}$$

(a) $\qquad\qquad\qquad\qquad$ (b) $\qquad\qquad\qquad\qquad$ (c)

(d) $\qquad\qquad\qquad\qquad\qquad\qquad$ (e)

Fig. 30.7 **Coefficient potentiometer.**

which can be simplified to give

$$\frac{e_o}{e_1} = \frac{k}{1 + k(R_p/R_1) - k^2(R_p/R_1)}$$

We see that loading causes e_o/e_1 to vary with the position of the wiper k and the size of the loading resistance R_1. Notice that, as $R_p/R_1 \to 0$, $e_o/e_1 \to k$. This means that, as the loading resistance R_1 becomes very large relative to the potentiometer resistance R_p, the effect of loading becomes negligible.

In most computers, there is no error caused by loading because the desired voltage ratio (e_o/e_1) is obtained by positioning the wiper with its loading resistor attached. To understand how loading error is avoided, we must discuss the "pot set" mode, which refers to the computer circuit configuration when the potentiometers are set. When the computer is placed in this mode, switches A and B (Fig. 30.7e) move to position 2, so that all input resistors are grounded and all potentiometers are connected to a reference voltage of $+100$ volts. A potentiometer *with its loading resistor attached* is then set by means of a null-balance circuit in which the voltage at the wiper is balanced against an adjustable known voltage of $100k$. The wiper is moved to a position where the wiper voltage just balances the voltage of $100k$, this condition being indicated by a null reading on a meter. This procedure sets the potentiometer to its desired effective value of k. For computing, the switches are moved to position 1.

The computer operator should be aware of the loading problem in setting coefficient potentiometers in order to avoid errors in computation. To illustrate such an error, suppose that in a particular problem a potentiometer is set to 0.1 when it feeds an integrator having an input resistance of 1 megohm. If the time scaling[1] is changed to speed up the problem by a factor of 10, the input resistor would have to be changed from 1 to 0.1 megohm and the potentiometer would have to be reset to 0.1 to correct for the new loading condition. It is easy to commit an error by failing to realize that a potentiometer must be reset after changing the input resistor to which it is connected.

Analog Computer A general-purpose electronic analog computer is an orderly collection of amplifiers, potentiometers, reference power supplies, and nonlinear components that are to be described later. These components are housed in a cabinet which includes a power supply for driving the amplifiers. The connections from the various components are brought to a panel called a *patch board*, where the components can be connected together by wires called *patch cords*. In many computers, the patch board

---

[1] Time scaling will be discussed later in this chapter, at which time the example described above will be more meaningful.

Fig. 30.8 (a) A large-size (100-amplifier) computer. (Courtesy of Electronic Associates, Inc.)

is removable; this feature offers a great advantage, for a wired computer circuit can be stored and used later. Also, problems can be wired on a spare patch board while the computer is operating on another problem.

In high-precision computers, the resistors and capacitors used to construct summers and integrators are not accessible to the user. These components are located inside the cabinet and wired so that certain ampli-

Fig. 30.8 (*b*) **A 32-amplifier desk-top computer.** (Courtesy of Applied Dynamics, Inc.)

fiers perform as summers and others as integrators. This protects the components from exposure and mistreatment.

A reference-voltage power supply is provided for establishing initial conditions on the integrators and furnishing constant voltages to the computer circuit when needed. In many computers, the reference voltage is accurately maintained at ± 100 volts. Any voltage less than 100 volts is obtained by connecting the reference voltage to a potentiometer, which is set to lower the voltage to the desired value. Computer cabinets capable of holding up to 200 amplifiers are available commercially. If a large computer facility is needed, several cabinets can be slaved (connected) together and used as a single computer. Examples of commercially available computers are shown in Fig. 30.8. Each computer is furnished with an operator's guide which gives detailed instructions for operating the computer. Since these instructions vary from one computer to another, no attempt is made here to discuss operating techniques. Only principles which are common to all computers will be presented.

Recording Equipment A computer solution to a problem is in the form of a time-varying voltage signal, and instruments are needed to indicate and record this voltage-time relationship. A very simple indicating device is a voltmeter. If a record is wanted, a strip-chart recorder or xy recorder is used. A strip-chart recorder is a device in which the signal voltage drives a pen across a strip of chart paper, which is moving at a known speed. This known chart speed provides the time axis. An xy recorder is a two-axis graphic voltage recorder which provides a flat-bed recording surface that holds a sheet of ordinary graph paper. Recording is accomplished by a pen supported in a dual-axis carriage. Each axis of the pen carriage is controlled by separate self-balancing servomechanisms, which displace the pen in proportion to the signal voltage received. By connecting two time-varying voltages to this device, one obtains a recording. A photograph of a typical xy recorder is shown in Fig. 30.9. To obtain a recording of a voltage as a function of time on an xy recorder, it is necessary to connect a voltage which varies linearly with time (ramp function) to the x drive; in this way a second of computer time is made equivalent to a known distance along the x axis of the recording. The ramp function for providing the time base can be obtained from the computer by integrating a step function. Because of the inertia of the mechanical parts of an xy recorder, the ability of the device to respond to rapidly changing signals is somewhat limited. Some of the best recorders can record accurately at speeds as high as 10 to 20 in./sec.

Another popular recorder in analog computation is the common two-channel oscilloscope. The principle of recording is the same as that for the

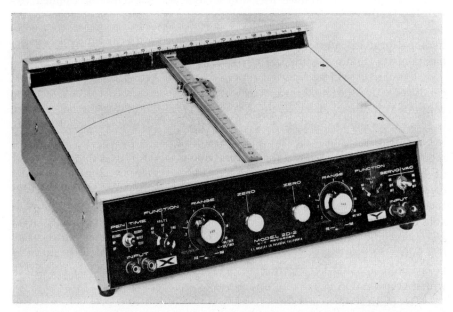

Fig. 30.9 **A typical *xy* plotter. (Courtesy of F. L. Moseley Company.)**

xy recorder; in this case the signal is traced by a spot of light generated by an electron beam moving across the face of the cathode-ray tube. In contrast to an *xy* recorder, the oscilloscope has no practical limitation on speed of response.

Because of its fast response, the oscilloscope is frequently used to record signals by a method called *repetitive operation* in which the solution is repeated over and over at a rapid rate, with the result that the spot of light following the signal leaves a continuous visible trace on the screen of the oscilloscope. The solution is repeated at a rate which may vary from about 10 to 100 times per second. The computer contains a special time-sequencing device which will automatically perform the operation of applying initial condition voltages, starting the solution, applying initial condition voltages, etc. Because of the rapid rate at which the solution is repeated, one can observe immediately the change in the shape of the solution curve as problem parameters are changed. For this reason repetitive operation is extremely useful in the study of the transient behavior of a control system as controller settings are varied. It is also useful in solving curve-fitting problems and boundary-value problems, for which initial conditions must be found by trial and error. An example of such a boundary problem is given in Chap. 31.

SOLUTION OF DIFFERENTIAL EQUATIONS

Computer Circuit To understand how the components described in the previous section can be used to solve a differential equation, consider the example

$$\frac{dx}{dt} = ax + f(t) \tag{30.15}$$

where a is a positive constant and $f(t)$ is an arbitrary function of time. Assume that x and dx/dt may be represented by voltages from appropriate amplifiers and that t is equivalent to time (seconds). Suppose that the voltage $-dx/dt$, which varies with t, is available from some amplifier. If this voltage is connected to an integrator as shown in Fig. 30.10a, x is obtained at the output.

According to Eq. (30.15), dx/dt is equal to the sum of two quantities: ax and $f(t)$. By means of a summer, x from amplifier 2 and $f(t)$ can be

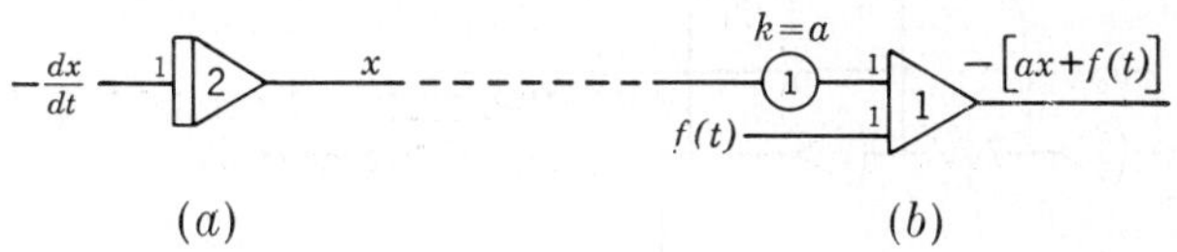

Fig. 30.10 Connecting computer components to solve a differential equation.

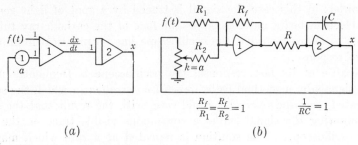

$$\frac{R_f}{R_1} = \frac{R_f}{R_2} = 1 \qquad\qquad \frac{1}{RC} = 1$$

(a) (b)

Fig. 30.11   Computer circuit for solving $dx/dt = ax + f(t)$.

combined to form $ax + f(t)$ as shown.   By attaching a patch cord from the output of the summer, amplifier 1, to the input of the integrator, amplifier 2, we establish electrically the equality sign of Eq. (30.15) and obtain the computer circuit shown in Fig. 30.11a.   An equivalent diagram showing all circuit elements is given in Fig. 30.11b.

There is no fixed procedure for constructing a computer circuit to solve a differential equation.   There are usually several alternate circuits which will be satisfactory.   For a higher-order linear differential equation, one usually solves for the highest derivative, which is equal to a linear combination of the dependent variable and its successive derivatives.   The derivatives are obtained on the computer by repeated integration and then combined by means of potentiometers and summers to satisfy the differential equation.

Initial Conditions   The circuits shown in Fig. 30.11 represent the electrical system at some instant during the solution of the differential equation.   To solve a differential equation on the computer, boundary conditions must be specified.   For the first-order equation under consideration [Eq. (30.15)] assume that the initial condition of x, written as $x(0)$, is specified; this requires that the voltage from the integrator be held at $x(0)$ until the problem is started.

A simple method for establishing the initial condition (IC) voltage is to connect a battery of the proper voltage between the grid and the amplifier output as shown in Fig. 30.12.   Since the grid voltage is very close to ground potential, the output of the integrator will be $x(0)$ at the instant the problem starts.   A switch operated by a relay is placed in the circuit as shown in Fig. 30.12.   When this switch is closed, the battery charges the capacitor

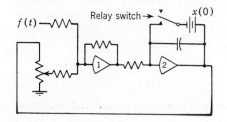

Fig. 30.12   Establishing the IC voltage on an integrator.

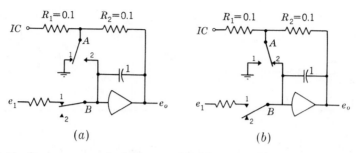

Fig. 30.13 **Common method for establishing the IC voltage.** **(a) Operate mode; (b) IC mode.**

to the IC voltage. The computer solution is started by opening the relay switch. A computer contains many relay switches, all of which are opened and closed simultaneously by means of a master switch, so that all IC voltages can be removed at the same instant. In an actual computer the battery is replaced by a variable d-c power supply.

An alternate method for obtaining the IC voltage, which is more commonly used, is shown in Fig. 30.13. When the computer is placed in the IC mode, relays A and B are in position 2. A voltage of desired magnitude and polarity is attached to the IC connection, and the amplifier output voltage e_o changes to minus the IC voltage according to a first-order response[1] with a time constant of RC, which is 0.1 sec for the values of the components used in Fig. 30.13. This means that after 1 sec (10 time constants), e_o is essentially equal to the negative of the IC voltage. When the computer is placed in the operate mode, switches A and B go to position 1 and the voltage e_1 is integrated in the usual manner. When the solution is being computed, resistor R_1 provides a load for the IC voltage, since one side of the resistor is grounded through switch A.

When the compact symbol for an integrator is used, the IC voltage will be indicated by the conventions shown in Fig. 30.14. When the convention of Fig. 30.14a is used, the method of applying the IC voltage is not suggested. However, the convention of Fig. 30.14b shows that the IC voltage is obtained by reducing the reference voltage of -100 volts to the desired value by means of a potentiometer; this method would be appropri-

---

[1] In Chap. 32, it will be shown that the circuit of Fig. 30.13, when in the IC mode, is equivalent to a first-order process.

Fig. 30.14 **Conventions used to indicate IC voltage.**

$IC = 10$

$e_o(0) = 10$

(a)

-100

$k = 0.1$

$e_o(0) = 10$

(b)

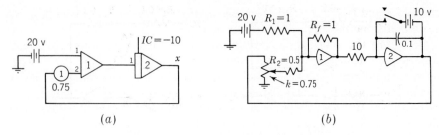

(a)   (b)

Fig. 30.15   **Computer circuit for Example 30.3.**

ate when the voltage is applied by the method shown in Fig. 30.13.   Notice that the reference voltage is opposite in sign to the initial voltage wanted at the amplifier output because of the sign reversal in the amplifier.

Example 30.3   Draw a complete circuit diagram for solving the differential equation

$$\frac{dx}{dt} = 1.5x + 20$$

$$x(0) = -10$$

This equation is simply a particular form of Eq. (30.15), in which $a = 1.5$ and $f(t) = 20$.   We can therefore use Fig. 30.11 with slight modification as shown in Fig. 30.15.

Since a potentiometer setting must be less than 1, it is necessary to set potentiometer 1 at $a/2$ and the corresponding gain of amplifier 1 at 2.   It should be clear that the resistors and capacitor selected in Fig. 30.15b represent only one particular set of values. For example, the gains of 1 and 2 for amplifier 1 may be obtained by another combination such as $R_f = 2$, $R_1 = 2$, $R_2 = 1$.

Magnitude Scaling   In the previous section, the problem of scaling was ignored.   To obtain satisfactory results from an analog computer, scaling is of utmost importance.   The problem of scaling a differential equation for solution on the analog computer can be very confusing unless a systematic approach is used.   The following method of scaling is recommended because of its simplicity and increasing popularity among computer operators.

Consider a second-order differential equation

$$\frac{d^2x}{dt^2} + a\frac{dx}{dt} + bx = f(t) \tag{30.16}$$

where x is the dependent problem variable and t is the independent problem variable which may represent time, distance, etc.   The coefficients a and b are positive constants.   The dependent variable and its derivatives $(x, dx/dt, d^2x/dt^2, \text{etc.})$ will be available as voltages from the outputs of amplifiers on the computer.   Magnitude scale factors are used to relate the voltages from amplifiers to the dependent variable and its derivatives. For example, if x represents feet of displacement in the original problem, then we could label the output voltage of an amplifier as $k_x x$, where k_x is the scale factor having units of volts per foot.

During the solution of a problem on an analog computer, the output voltage from any amplifier should not exceed the operating range of the amplifier. For many computers, this nominal range is -100 to $+100$ volts, and for convenience, we shall use this range throughout this discussion. If the amplifier output exceeds 100 volts, the amplifier response may not be linear, with the result that the summer or integrator will not perform correctly. If we know in advance, through rough calculation or practical experience, that the maximum value of x which occurs during the solution is x_m, we have a basis for determining k_x as

$$k_x = \frac{100}{x_m}$$

For example, if $x_m = 0.5$ ft, then

$$k_x = \frac{100 \text{ volts}}{0.5 \text{ ft}} = 200 \text{ volts/ft}$$

and the amplifier output voltage corresponding to x is labeled $[200x]$. With this technique, we make the best use of the voltage range available and ensure that the amplifier will never overload (exceed 100 volts at output) during the computer solution. In practice, we may round off k_x to a value somewhat less than $100/x_m$ if this gives a scale factor which is more convenient.

To select scale factors for the derivatives of the dependent variable ($\dot{x}$, $\ddot{x}$, etc.), we also must have estimates of the maximum values of these variables, namely, $\dot{x}_m$, $\ddot{x}_m$, etc. (Here the dot notation has been used for the derivatives; thus $\dot{x} = dx/dt$, $\ddot{x} = d^2x/dt^2$, etc.). The scale factors are

$$k_{\dot{x}} = \frac{100}{\dot{x}_m}$$

$$k_{\ddot{x}} = \frac{100}{\ddot{x}_m}$$

etc.

Equation (30.16) may be written in terms of $[k_x x]$, $[k_{\dot{x}} \dot{x}]$, $[k_{\ddot{x}} \ddot{x}]$, etc. The brackets are used to indicate that the enclosed quantity is a voltage at the output of some amplifier. Rewriting Eq. (30.16) in terms of the dot notation gives

$$\ddot{x} = -a\dot{x} - bx + f(t) \tag{30.17}$$

Equation (30.17) may be written to include the scale factors; thus

$$\frac{[k_{\ddot{x}} \ddot{x}]}{k_{\ddot{x}}} = -a \frac{[k_{\dot{x}} \dot{x}]}{k_{\dot{x}}} - b \frac{[k_x x]}{k_x} + f(t) \tag{30.18}$$

or

$$[k_{\ddot{x}}\ddot{x}] = -a\frac{k_{\ddot{x}}}{k_{\dot{x}}}[k_{\dot{x}}\dot{x}] - b\frac{k_{\ddot{x}}}{k_x}[k_x x] + k_{\ddot{x}}f(t) \tag{30.19}$$

A circuit for solving Eq. (30.19) is shown in Fig. 30.16. Notice that a potentiometer is placed between amplifiers that generate x and its derivatives to provide for the different scale factors for the successive derivatives. For example, the potentiometer between the amplifiers generating $[k_x x]$ and $[k_{\dot{x}}\dot{x}]$ is set to

$$P_5 = \frac{k_x}{k_{\dot{x}}}$$

or in general

$$\text{Potentiometer setting} = \frac{\text{scale factor for } dx^{n-1}/dt^{n-1}}{\text{scale factor for } dx^n/dt^n}$$

The justification for this can be seen from the following argument. In general an integrator follows the expression

$$x = \int \dot{x}\,dt$$

Multiplying both sides of this expression by k_x gives

$$[k_x x] = \int k_x \dot{x}\,dt$$

Multiplying the term on the right by $k_{\dot{x}}/k_{\dot{x}}$ and **bracket**ing the term $k_{\dot{x}}\dot{x}$ give

$$[k_x x] = \frac{k_x}{k_{\dot{x}}}\int [k_{\dot{x}}\dot{x}]\,dt$$

The coefficient $k_x/k_{\dot{x}}$ is the gain of the integrator, **which can be** adjusted conveniently by a potentiometer placed before it.

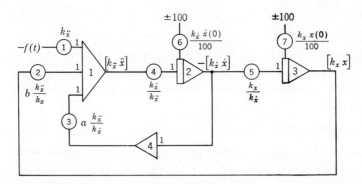

Fig. 30.16   Computer circuit for solving $\ddot{x} + a\dot{x} + bx = f(t)$**.**

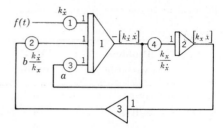

Fig. 30.17 Alternate computer circuit for solving $\ddot{x} + a\dot{x} + bx = f(t)$.

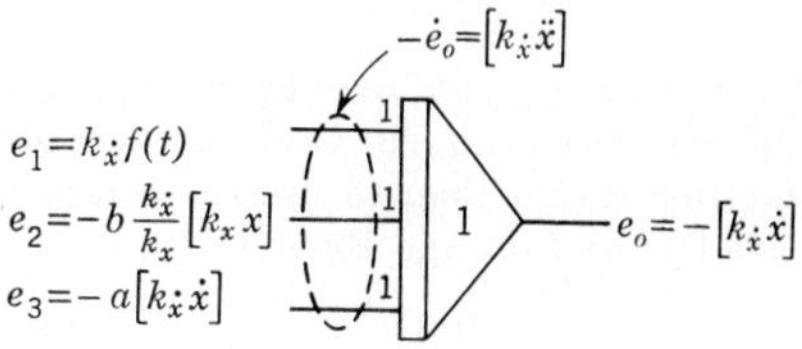

Fig. 30.18

$$e_1 = k_{\dot{x}} f(t)$$
$$e_2 = -b\frac{k_{\dot{x}}}{k_x}[k_x x]$$
$$e_3 = -a[k_{\dot{x}}\dot{x}]$$

The circuit shown in Fig. 30.16 is a generalized computer circuit for solving the second-order differential equation of Eq. (30.17). If $[k_{\ddot{x}}\ddot{x}]$ is not needed as part of the computer solution, we can simplify Fig. 30.16 and thereby save one amplifier as shown in Fig. 30.17. In this change, one amplifier has been used to combine summation and integration. (In Fig. 30.16, summation and integration were performed separately by amplifiers 1 and 2, respectively.) The equation used to draw Fig. 30.17 is obtained by multiplying both sides of Eq. (30.19) by $k_{\dot{x}}/k_{\ddot{x}}$ to give

$$[k_{\dot{x}}\ddot{x}] = -a[k_{\dot{x}}\dot{x}] - b\frac{k_{\dot{x}}}{k_x}[k_x x] + k_{\dot{x}}f(t) \tag{30.20}$$

The settings for potentiometers 1, 2, and 3 are equal to the coefficient terms on the right side of Eq. (30.20). We may show that Fig. 30.17 is correct by consideration of Fig. 30.18, in which it can be seen that the total input to integrator 1 is equivalent to $[k_{\dot{x}}\ddot{x}]$, according to Eq. (30.20).

To illustrate this procedure of magnitude scaling, consider the following example.

Example 30.4 Draw a complete circuit diagram for solving the differential equation

$$\ddot{x} + 0.5\dot{x} + 4x = 0 \tag{30.21}$$
$$x(0) = 2$$
$$\dot{x}(0) = 0$$

Estimates of the maximum values are

$$x_m = 2 \text{ ft}$$
$$\dot{x}_m = 4 \text{ ft/sec}$$
$$\ddot{x}_m = 8 \text{ ft/sec}^2$$

First calculate the scale factors as shown in the following table:

Problem variable	Max value	$\dfrac{100}{\text{Max value}}$	k
x, ft	2	50	50
$\dot{x}$, ft/sec	4	25	20
$\ddot{x}$, ft/sec^2	8	12.5	10

Notice that we have rounded off $k_{\dot{x}}$ and $k_{\ddot{x}}$ to 20 and 10 for convenience. In general, it will be more convenient to obtain the potentiometer settings directly from the differential equation rather than to use the generalized circuits of Fig. 30.16 or 30.17. Using the scale factors from the above table, Eq. (30.21) can be written in the form

$$[10\ddot{x}] = -\frac{(10)(0.5)}{20}[20\dot{x}] - \frac{(10)(4)}{50}[50x] \tag{30.22}$$

or

$$[10\ddot{x}] = -0.25[20\dot{x}] - 0.8[50x] \tag{30.23}$$

A systematic procedure used to obtain Eq. (30.23) consists of the following steps:

1. Having solved for the highest derivative, multiply both sides of the resulting equation by 10, since $[10\ddot{x}]$ is wanted as a separate quantity.

2. Multiply the first term on the right side by 20/20 and bracket $20\dot{x}$, and multiply the second term by 50/50 and bracket $50\dot{x}$.

From Eq. (30.23), the computer circuit with correct potentiometer settings can be drawn immediately as shown in Fig. 30.19. Notice that the scale factor ratio $k_x/k_{\dot{x}}$ is 50/20 or 2.5. Since a potentiometer setting cannot exceed 1, potentiometer 3 is set at 0.25 and the gain of amplifier 3 is made 10. (Of course, any other combination of potentiometer setting and amplifier gain may be used for which the product is 2.5.) A similar argument holds for the setting of potentiometer 2.

Time Scaling The solution of many problems by means of the analog computer will require time scaling, which makes it possible to speed up or slow down the problem on the computer.

It is important to remember that the analog computer integrates with respect to time measured at the computer. This time is simply the actual number of seconds that have elapsed since the computer began to integrate.

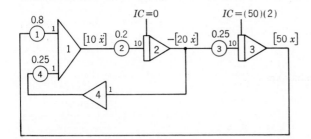

Fig. 30.19 Computer circuit for Example 30.4.

To distinguish this computer time from the independent problem variable t, the symbol τ is defined as

$\tau =$ time measured at computer, sec

In the previous discussion in which time scaling was ignored, 1 sec at the computer was equivalent to one unit of independent variable. For example, if the differential equation for a damped vibrator were written with time t in seconds, then 1 sec of computer time would equal 1 sec of problem time. Similarly if the differential equation were written with t in units of minutes, 1 sec of computer time would now equal 1 min of problem time. To take another example, if a differential equation for a heat exchanger were written with the independent variable in feet of length, then 1 sec of computer time would equal 1 ft of exchanger length.

The need for time scaling arises from several considerations. For example, if the time needed for the solution of a particular problem requires 1,200 sec, then 1,200 sec (20 min) of computer time would be needed if no time scaling were used. It should be obvious that to wait so long for the solution is wasteful of operator's time and machine utilization. In this example, we see that it is desirable to speed up the solution.

The need for time scaling is often governed by the type of instrument which is used to record the computer solution. For example, if an x-y recorder is used to record a computer signal which is varying rapidly, the recording pen may lag behind the computer signal because of friction and mechanical inertia of the recorder. In this case, it is desirable to slow down the solution in order to obtain a faithful record of the computer signal. These examples should be sufficient to suggest the importance of time scaling.

The time-scale factor β is defined as the ratio of time at the computer τ to the independent problem variable t; thus

$$\beta = \frac{\tau}{t} = \frac{\text{computer time, sec}}{\text{independent problem variable}} \qquad (30.24)$$

The method of time scaling to be described here has become widely accepted. The rule for scaling will be presented first and then applied to examples. A justification of the rule will be given later.

Rule for time scaling:

1. First scale the problem for magnitude only as described earlier. In this case $\beta = 1$ (i.e., the time scaling is such that 1 second of computer time equals one unit of independent variable).

2. To time-scale, multiply the gain of each input to an integrator by $1/\beta$. If $1/\beta > 1$, the problem is speeded up; if $1/\beta < 1$, the problem is slowed down.

3. Change any forcing function of time $f(t)$ to $f(\tau/\beta)$.

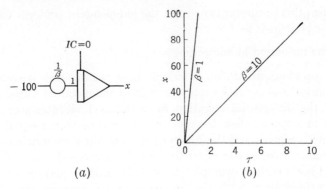

(a)   (b)

Fig. 30.20 Effect of time scaling on computer response.

To show how this rule works for a simple example, consider the differential equation

$$\frac{dx}{dt} = 100 \qquad x(0) = 0$$

for which the solution is to be found up to $t = 1.0$. The solution is simply $x = 100t$, and the maximum value of x is 100. A computer circuit consists of an integrator to which is attached a 100-volt signal as shown in Fig. 30.20a.

If we want to obtain the computer solution in 10 sec ($\tau = 10$), $\beta = 10$. According to the rule, the gain of the amplifier is multiplied by $1/\beta = 0.1$. The speed of solution is now one-tenth the speed obtained before, and the output of the amplifier reaches 100 volts in 10 sec as shown in Fig. 30.20b.

As another simple example consider the differential equation

$$\frac{dx}{dt} = f(t)$$
$$f(t) = 0.02t$$
$$x(0) = 0$$

for which a solution is desired up to $t = 100$. The analytical solution is

$$x = 0.01t^2$$

and the maximum value of x is 100. The computer circuit consists of a

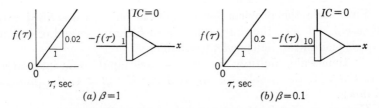

(a) $\beta = 1$ (b) $\beta = 0.1$

Fig. 30.21 Time-scaling a forcing function.

single integrator to which is attached a ramp function input signal as shown in Fig. 30.21a.

If we want to obtain the computer solution in a shorter time, say 10 sec, then we select $\beta = 0.1$, which requires that the gain of the integrator be $1/\beta$ or 10; furthermore, the rate of change of the ramp function must be increased by a factor of 10 as indicated in Fig. 30.21b.

The method of time scaling presented here has the following advantages:

1. Except for multiplication of gains to integrators by $1/\beta$, none of the potentiometer settings need to be changed as β is varied.
2. None of the IC voltages need be changed as β varies.
3. The shape of the response curve from the computer will remain the same as β varies. This means that the only effect of time scaling is to compress or expand the time scale. For example, if a computer solution at $\beta = 1$ gives a maximum value of x of 90 at 1 sec (computer time), then the solution for $\beta = 0.5$ will also have a maximum value of x of 90 but it will occur at 0.5 sec (computer time).

These features of time scaling make it possible to scale at the computer without having to make any changes in the computer circuit, except for multiplying the gains of integrators by $1/\beta$. The time-scale factor can be selected to match the response characteristics of the recording equipment and to accommodate the wishes of the user. For example, if many recordings are to be made, it may be expedient to speed up the solution. However, if the computer is being used in a demonstration, it may be necessary to slow down the solution for ease of observation.

Justification of Time-scaling Rule (The reader may omit this section on first reading and go directly to Example 30.5.)

The rule for time scaling will be justified from the mathematical viewpoint for the second-order differential equation used previously [Eq. (30.16)]:

$$\frac{d^2x}{dt^2} + a \frac{dx}{dt} + bx = f(t)$$

From the calculus,[1] one can show that a function of t and its derivatives $[f(t), df/dt, d^2f/dt^2, \text{etc.}]$ can be written in terms of a new independent variable τ if the relationship between t and τ is given; thus, if we have $t = ftn(\tau)$, the relationships are

$$\frac{df}{d\tau} = \frac{df}{dt} \frac{dt}{d\tau} \tag{30.25}$$

$$\frac{d^2f}{d\tau^2} = \frac{d^2f}{dt^2} \left(\frac{dt}{d\tau}\right)^2 + \frac{df}{dt} \frac{d^2t}{d\tau^2} \tag{30.26}$$

---

[1] The reader will find these rules for differentiation in any book on advanced calculus. For example, see L. A. Pipes, "Applied Mathematics for Engineers and Physicists," p. 308, McGraw-Hill Book Company, 2d ed., 1958.

For this particular problem, t is a linear function of τ, as given in Eq. (30.24); thus,

$$t = \frac{1}{\beta}\tau \qquad (30.27)$$

Introducing this into Eqs. (30.25) and (30.26) gives

$$\frac{df}{d\tau} = \frac{1}{\beta}\frac{df}{dt} \qquad (30.28)$$

$$\frac{d^2f}{d\tau^2} = \frac{1}{\beta^2}\frac{d^2f}{dt^2} \qquad (30.29)$$

and in general, one can show that

$$\frac{d^nf}{d\tau^n} = \frac{1}{\beta^n}\frac{d^nf}{dt^n} \qquad (30.30)$$

We can now apply Eq. (30.30) to the magnitude-scaled equation [Eq. (30.19)] and replace each derivative with respect to t by a derivative with respect to τ. [For example, $\ddot{x}\ (= d^2x/dt^2)$ will be replaced by $\beta^2 d^2x/d\tau^2$.] The result is

$$\left[\beta^2 k_{\ddot{x}}\frac{d^2x}{d\tau^2}\right] = -a\frac{k_{\ddot{x}}}{k_{\dot{x}}}\left[\beta k_{\dot{x}}\frac{dx}{d\tau}\right] - b\frac{k_{\ddot{x}}}{k_x}[k_x x] + k_{\ddot{x}}f(\tau/\beta) \qquad (30.31)$$

The computer circuit which solves Eq. (30.31) is shown in Fig. 30.22. (For convenience, the following dot notation has been used in Fig. 30.22: $\dot{x}_\tau = dx/d\tau$, $\ddot{x}_\tau = d^2x/d\tau^2$, etc.) Notice that, in Eq. (30.31), β and β^2 have been enclosed by the brackets and are thereby included as part of the scale factor. (For example, $k_{\ddot{x}}$ becomes $\beta^2 k_{\ddot{x}}$.) This grouping of the terms of Eq. (30.31) has left the form identical with that of Eq. (30.19). The only change in the circuit of Fig. 30.16 which is needed in order that it solve Eq. (30.31) is to multiply each of the settings of potentiometers 4 and 5 by

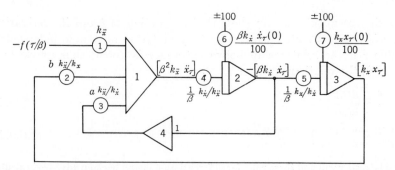

Fig. 30.22   **A magnitude- and time-scaled computer circuit for solving a second-order differential equation.**

$1/\beta$. This is required because the new scale factors of Eq. (30.31) include the β terms.

We have now shown why the gain of each integrator is multiplied by $1/\beta$; however, there are other conclusions to be drawn as a result of this time-scaling procedure. We shall first show why the IC voltages remain the same after time scaling. For example, before time scaling potentiometer 6 (Fig. 30.22) is to be set to

$$P'_6 = k_{\dot{x}} \frac{\dot{x}(0)}{100} \tag{30.32}$$

After scaling, potentiometer 6 should be set to

$$P''_6 = \beta k_{\dot{x}} \frac{\dot{x}_\tau(0)}{100} \tag{30.33}$$

Using $\dot{x}_\tau = \dot{x}/\beta$ from Eq. (30.28) in Eq. (30.33) gives

$$P''_6 = \beta k_{\dot{x}} \frac{\dot{x}(0)}{100\beta} = k_{\dot{x}} \frac{\dot{x}(0)}{100} \tag{30.34}$$

Therefore, the potentiometer setting remains the same.

We shall next show that the new scale factors of Eq. (30.31) and Fig. 30.22 will produce a solution curve having the same shape after time scaling as it did before. By the same shape is meant that, for corresponding times, the voltage from a particular amplifier is the same. To understand this, consider the voltage from amplifier 2 of Fig. 30.22. Before scaling, the voltage at some arbitrary time t_1 is

$$e_1 = -k_{\dot{x}}\dot{x}(t_1) \tag{30.35}$$

After scaling, the voltage at the corresponding time, which is $\tau_1 = \beta t_1$, is

$$e_2 = -\beta k_{\dot{x}}\dot{x}_\tau(\beta t_1) \tag{30.36}$$

Using the expression $\dot{x}_\tau = \dot{x}/\beta$ from Eq. (30.28) in Eq. (30.36) gives

$$e_2 = -\beta k_{\dot{x}} \frac{\dot{x}(t_1)}{\beta} = -k_{\dot{x}}\dot{x}(t_1) \tag{30.37}$$

which shows that e_1 and e_2 are equal at corresponding times t_1 and τ_1.

To illustrate further the rules of scaling and to show a different type of problem which can be solved by the analog computer, we conclude this chapter with the following example:

Example 30.5 Consider the following isothermal, gaseous chemical reactions occurring at constant volume:

$$\begin{array}{cc} k_1 & k_3 \\ A \rightleftharpoons B \rightleftharpoons C \\ k_2 & k_4 \end{array} \tag{30.38}$$

Let x, y, and z be the moles of A, B, and C, respectively, at any instant, and let k_1, k_2, k_3, and k_4 be the velocity constants for the various reactions as shown in Eq. (30.38). The time rates of change of A and B are given by the equations

$$\frac{dx}{dt} = -k_1x + k_2y \tag{30.39}$$

$$\frac{dy}{dt} = k_1x - k_2y - k_3y + k_4z \tag{30.40}$$

We also have the relationship expressing the fact that the total number of moles of gases remains constant: thus

$$n = x + y + z \tag{30.41}$$

where n is total moles, a constant.

Eliminating z from Eqs. (30.39) to (30.41) gives the pair of simultaneous differential equations

$$\frac{dx}{dt} = -k_1x + k_2y \tag{30.42}$$

$$\frac{dy}{dt} = (k_1 - k_4)x - (k_2 + k_3 + k_4)y + k_4n \tag{30.43}$$

In order to solve a specific problem, let $k_1 = 4$, $k_2 = 1$, $k_3 = 4$, $k_4 = 1$, and $n = 100$; also let $x(0) = 100$ (i.e., only pure A is present initially). Using these constants in Eqs. (30.42) and (30.43) gives

$$\frac{dx}{dt} = -4x + y \tag{30.44}$$

$$\frac{dy}{dt} = 3x - 6y + 100 \tag{30.45}$$

$$x(0) = 100$$
$$y(0) = 0$$

We now wish to draw a computer circuit to solve these simultaneous differential equations.

Suppose that we are given estimates of the maximum values to be $x_m = 100$, $y_m = 100$, and $\dot{x}_m = \dot{y}_m = 400$. Then the magnitude scale factors can be determined from the following table:

Problem variable	Max value	$\dfrac{100}{\text{max value}} = k$
x	100	1.0
y	100	1.0
$\dot{x}$	400	0.25
$\dot{y}$	400	0.25

Using the scale factors from the table, Eqs. (30.44) and (30.45) can be written as follows:

$$[0.25\dot{x}] = -4(0.25)[x] + (0.25)[y] \tag{30.46}$$
$$[0.25\dot{y}] = (3)(0.25)[x] - (6)(0.25)[y] + (0.25)(100) \tag{30.47}$$

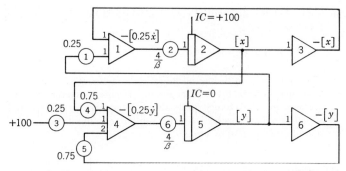

Fig. 30.23 Computer circuit for solving a problem in chemical kinetics, Example 30.5.

or

$$[0.25\dot{x}] = -[x] + 0.25[y] \tag{30.48}$$
$$[0.25\dot{y}] = 0.75[x] - 1.5[y] + 25 \tag{30.49}$$

The circuit for solving Eqs. (30.48) and (30.49) is shown in Fig. 30.23. In this circuit $1/\beta$ has been included in the settings of potentiometers 2 and 6. If β is 10, we set potentiometers 2 and 6 to a value of 4/10.

PROBLEMS

30.1 The circuit shown in Fig. P30.1 will solve the differential equation

$$\frac{dx}{dt} = 2t - 1.5x \qquad x(0) = 0$$

    a. A solution is wanted for t up to 2 units. For this condition the maximum values of x and dx/dt are estimated to be 4 and 2, respectively. If the time-scale factor β is 2, determine the potentiometer settings k_1 and k_2 and the gains a and b.

    b. What is the rate at which the voltage from the function generator increases with time as measured by an observer at the computer?

    c. Draw an analog computer circuit which will produce the required function of time.

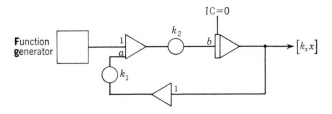

Fig. P30.1

30.2 *a.* Show that the differential equation which is solved by the circuit in Fig. P30.2 is

$$\frac{dx}{dt} = x + 20 \qquad x(0) = 0$$

b. The solution is to be slowed down by a factor of 10 (that is, $\beta = 10$) by changing R_1 from 0.1 to 1 megohm. If potentiometer $P1$ is not reset to account for the change in loading, determine the differential equation which is actually solved by the computer. The total resistance of the potentiometer is $100K$ ohms.

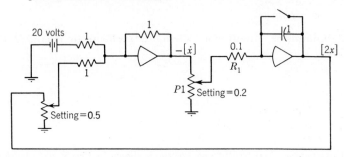

Fig. P30.2

30.3 Show that the circuit of Fig. P30.3 will produce the function

$$e_1 = 10 \cos 0.4\tau$$

Determine the potentiometer settings and the IC voltages.

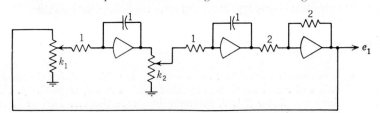

Fig. P30.3

30.4 Show that the differential equations which describe the steady-state behavior of a countercurrent, double-pipe heat exchanger are

$$u_1 \frac{dT_1}{dz} = \frac{UA}{C_{1}\rho_1 V} (T_2 - T_1)$$

$$u_1 \frac{dT_2}{dz} = \frac{UA}{C_{1}\rho_1 V} \frac{F_1 C_1}{F_2 C_2} (T_2 - T_1)$$

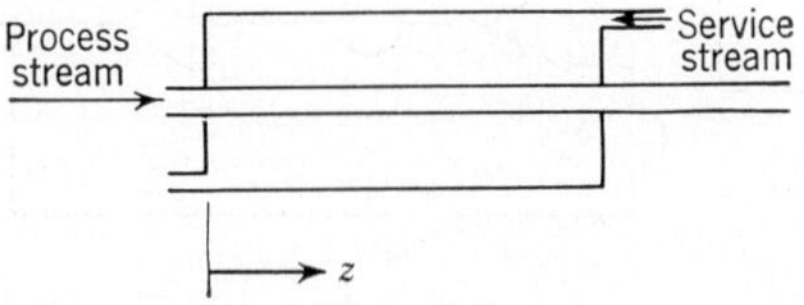

Process stream → Service stream

→ z

Fig. P30.4

where T = temperature of stream, °F

$\quad z$ = length of exchanger, ft

$\quad U$ = overall heat-transfer coefficient, Btu/(hr)(ft²)(°F)

$\quad u$ = velocity of stream, ft/sec

$\quad F$ = flow rate of stream, lb/sec

$\quad C$ = heat capacity of stream, Btu/(lb)(°F)

$\quad \rho$ = density of stream, lb/ft³

A/V = ratio of heat transfer area per unit length to volume per unit length of process side of heat exchanger

1 refers to process stream
2 refers to service stream

Devise a magnitude-scaled computer circuit to solve these equations when the following conditions hold:

$\quad U$ = 190 Btu/(hr)(ft²)(°F)

$\quad \rho_1 = \rho_2$ = 62.3 lb/ft³

A/V = 50 ft⁻¹

$\quad T_{1_0}$ = entering process stream temperature = 60°F

$\quad T_{2_0}$ = entering service stream temperature = 180°F

$\quad T_{2_f}$ = leaving service stream temperature, 135°F

F_1/F_2 = 0.5

$\quad C_1 = C_2$ = 1.0 Btu/(lb)(°F)

$\quad u_1$ = 1.0 ft/sec

Describe how the computer results are to be interpreted to obtain the length of exchanger required.

30.5 An interesting problem from mechanics is the determination of the motion of the coupled oscillators shown in Fig. P30.5. The two masses, which are attached to

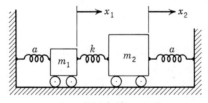

Fig. P30.5

massless springs, are free to move horizontally on a frictionless table. Applying Newton's law to each mass gives the following equations of motion.[1]

$$m_1\ddot{x}_1 = -ax_1 - k(x_1 - x_2) \qquad (30.50)$$
$$m_2\ddot{x}_2 = -ax_2 + k(x_1 - x_2) \qquad (30.51)$$

where x_1 and x_2 are the displacements of the masses from their rest positions. To solve a specific problem, let $m_1 = m_2 = 1$ slug (32.2 lb$_m$), a = 900 lb$_f$/ft, k = 100 lb$_f$/ft, $x_1(0) = 2$, $x_2(0) = \dot{x}_1(0) = \dot{x}_2(0) = 0$.

To estimate the maximum values of x and $\dot{x}$, assume that there is no coupling between the masses (i.e., the center spring is absent or $k = 0$). In this case each mass would oscillate as a single linear oscillator. Draw a magnitude-scaled computer circuit to solve this problem.

[1] A discussion of this mechanical system may be found in K. R. Symon, "Mechanics," pp. 165–173, Addison-Wesley Publishing Company, Inc., Reading, Mass., 1953.

31 *Nonlinear Computer Operations*

The nonlinear components used in analog computation consist of multipliers, function generators, diodes, etc. These devices perform the functions implied by their names. The nonlinear elements make it possible to solve nonlinear differential equations as well as linear differential equations with time-dependent coefficients. These nonlinear elements extend the usefulness of a computer tremendously, for solving a nonlinear problem on a computer is not significantly more difficult than solving a linear one. In this chapter we shall describe briefly the multiplier, the function generator, and diodes. There are many details about these components which can be found in several textbooks devoted solely to analog computation.[1]

---

[1] A. S. Jackson, "Analog Computation," McGraw-Hill Book Company, New York, 1960; C. L. Johnson, "Analog Computer Techniques," McGraw-Hill Book Company, New York, 1956; A. E. Rogers and T. W. Connolly, "Analog Computation in Engineering Design," McGraw-Hill Book Company, New York, 1960.

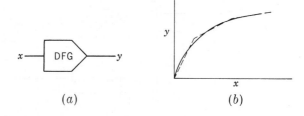

Fig. 31.1 **Function generator.**

(a) (b)

FUNCTION GENERATOR

In many problems a variable y may be a function of another variable x; for example, heat capacity C is a function of temperature T. Furthermore, the functional relationship may be known analytically or graphically. The function generator is a device which relates two voltages according to a particular functional relationship. A convenient symbol for a function generator is shown in Fig. 31.1a.

The most common function generator now used is the diode function generator. This device actually approximates a function $y = f(x)$ by a number of straight-line segments which are joined together as illustrated in Fig. 31.1b. It should be clear that, for a simple curve such as the one shown, only a few line segments are needed to fit the curve quite closely. In setting up the function generator, one can adjust the slope of each line segment and the break point (point at which adjacent lines meet) by simply adjusting potentiometers built into the device. The electrical components which are used in a diode function generator consist of potentiometers, resistors, and diodes. The actual circuits and principle of operation, which require considerable explanation, will not be presented here; such information can be found in textbooks on analog computers.[1] Actually, the setting up of a function generator does not require the detailed knowledge of the circuits.

MULTIPLIER

As the name suggests, a multiplier is a device which gives the instantaneous product of two time-varying voltages. A convenient symbol representing multiplication is shown in Fig. 31.2. Notice that the product z is $xy/100$. In this discussion, we shall assume that the maximum voltage from any amplifier is 100 volts. In order that z always be less than

[1] See Johnson, *op. cit.*, pp. 153–157.

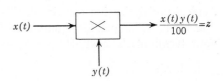

Fig. 31.2 **Symbol for multiplication.**

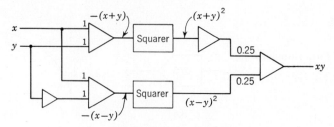

Fig. 31.3   Quarter-square multiplication.

100 volts, it is necessary that the product xy be divided by 100. The electrical device which performs the multiplication automatically includes the 100 in the divisor.

There are three types of multipliers used in computers; each one is distinctly different in operation from the others. These are called (1) the quarter-square multiplier, (2) the servomechanical multiplier, and (3) the time-division multiplier. The first two are widely used and will be described here. The time-division multiplier is now less common and will not be considered further; the interested reader will find a description of it in Jackson.[1]

Quarter-square Multiplier   The quarter-square multiplier is based on the following identity which the reader can easily verify:

$$xy = \frac{1}{4}[(x + y)^2 - (x - y)^2] \tag{31.1}$$

If one has a function generator which will produce the square of an input voltage, then Eq. (31.1) can be mechanized by the circuit shown in Fig. 31.3. The only new components required are the two squaring devices, which can be formed from diode function generators as described in the previous section. When sufficient segments are provided, the approximation of the square function by straight-line segments can be very close to the true function. A high-quality quarter-square multiplier will contain a squaring circuit which uses from 10 to 20 segments.

In practice, the quarter-square multiplier now available from most computer manufacturers consists of a device which produces a current that is proportional to the product of two voltages. This current device is used in conjunction with a high-gain amplifier to produce the multiplication. The current device is called a quarter-square card because the diodes and resistors which produce the squaring functions are mounted on a printed circuit card which can be bought separately for a computer. A typical multiplication circuit which uses a quarter-square card is shown in Fig. 31.4. The quarter-square card, which is represented by the block, has four terminals labeled $+X$, $-X$, $+Y$, and $-Y$. Each factor (x and y)

---

[1] Jackson, *op. cit.*

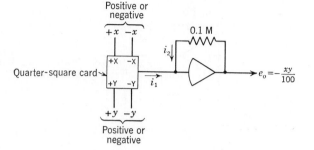

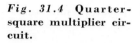

Fig. 31.4 **Quarter-square multiplier circuit.**

of the product must be available as a positive and negative signal. The card accepts $+x$, $-x$, $+y$, and $-y$ and produces a current which enters the grid of a high-gain amplifier. A typical quarter-square card[1] produces a current which is

$$i = \frac{xy}{10} \qquad \mu a \tag{31.2}$$

If xy is positive, the current flows out of the card into the grid as shown by the arrow in Fig. 31.4. If xy is negative, the current flows into the card from the grid (opposite to the direction shown by the arrow in the figure).

With this knowledge of the current-voltage relationship for the card, one can show that the circuit of Fig. 31.4 is correct by writing a current balance at the grid; thus

$$i_1 + i_2 = 0$$

Replacing i_1 by $xy/10$ from Eq. (31.2) and replacing i_2 by $e_o/0.1$, since the grid voltage is zero, give

$$\frac{xy}{10} + \frac{e_o}{0.1} = 0$$

or

$$e_o = - \frac{xy}{100} \tag{31.3}$$

When the quarter-square card is used as a multiplier, the voltage entering the input terminals can be of either polarity.

One can also use the multiplication circuit to add other voltages to the product as shown by Fig. 31.5. The reader may check this circuit by writing a current balance at the grid.

Servomultiplier The servomultiplier is a mechanized coefficient potentiometer. The coefficient potentiometer multiplies a time-varying

---

[1] The relationship given by Eq. (31.2) applies to the quarter-square card (No. 164) manufactured by Applied Dynamics, Inc., Ann Arbor, Mich.

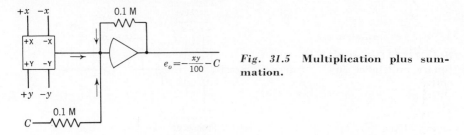

$$e_o = -\frac{xy}{100} - C$$

Fig. 31.5 Multiplication plus summation.

voltage $x(t)$ by a constant k which depends on the wiper position (Fig. 31.6a). If the wiper arm is automatically adjusted so that its position is proportional to a voltage $y(t)$, one has the basis for a servomultiplier. Assume that x and y can vary from 0 to $+100$ volts and that the wiper positioner (Fig. 31.6b) moves the wiper arm linearly from the lower end to the upper end of the potentiometer as y varies from 0 to 100 volts. In this case,

$$k = \frac{y(t)}{100}$$

and the voltage at the wiper arm is

$$e_0 = \frac{x(t)y(t)}{100}$$

Although the idea of the servomultiplier presented in Fig. 31.6 is quite simple, the mechanization of it entails a complicated design of an electromechanical servomechanism. The mechanization of Fig. 31.6 is shown by the schematic diagram of Fig. 31.7, which indicates two identical potentiometers A and P, the wipers of which are ganged so that they move together. The feedback system, which adjusts the wiper of P so that the mechanical position of P is $y/100$, will first be explained.

To the upper end of P is attached 100 volts, and the lower end of P is attached to ground. Assume that y is 50 volts, for which case the wiper of P will be positioned halfway between the ends of potentiometer P, and the voltage tapped off by the wiper is 50 volts. The error ϵ will be zero, and the servomotor will be stationary. Now consider what happens

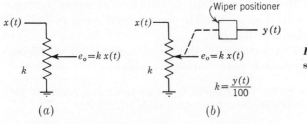

Fig. 31.6 Principle of a servomultiplier.

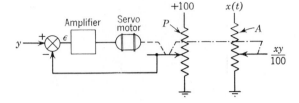

Fig. 31.7 Servomechani-
cal multiplier.

when y changes to 60 volts. The error ϵ will become positive and thereby cause the servomotor to drive the wiper of potentiometer P upward toward the positive end of P. Eventually, the wiper will find a new position where the tapped voltage is 60 volts, for which case ϵ will again be zero and the servomotor will remain stationary. It should be clear that the inertia of the mechanical parts (motor, gears, etc.) of the feedback system will limit the rate of response of a servomultiplier. A high-quality servo-multiplier can follow sinusoidal voltages having a frequency up to about 10 cycles/sec. (The frequency limitation is placed only on the signal y which enters the feedback mechanism.)

The principal cost in a servomultiplier is the feedback mechanism which drives potentiometer P of Fig. 31.7. For a slight additional cost, one can gang additional potentiometers to the servomotor and obtain several multiplications, each one of which has the common factor y. It is quite common to have as many as five potentiometers in addition to the feedback potentiometer P.

In Fig. 31.7, x may be positive or negative, but y can be only positive. Note that the sign of the output voltage will be correct. If both factors of the product xy are to take on positive or negative values, the wiring in Fig. 31.7 must be slightly modified by connecting $+100$ volts to the top of P and -100 volts to the lower end of P. Furthermore, $+x$ must be connected to the upper end of A and $-x$ to the lower end of A.

Division A circuit used to perform division consists of the combination of a multiplier and a high-gain amplifier. The division circuit using the quarter-square card described by Eq. (31.2) is shown in Fig. 31.8.

To show that this circuit is correct, we write a current balance at the grid of amplifier 1; thus

$$i_1 + i_2 = 0 \tag{31.4}$$

Replacing i_1 by $-xe_0/10$ from Eq. (31.2)[1] and recognizing that i_2 is $y/0.1$

---

[1] To see why i_1 is replaced by $-xe_0/10$, note that the polarities of the signals to the $-Y$ and $+Y$ terminals are fixed as indicated in Fig. 31.8. This means that the direction of current flow produced by the card depends only on the polarity of e_0. When e_0 is positive, the wiring to the $-X$ and $+X$ terminals is such that current flows *into* the card with the result that i_1 is negative. Similarly, when e_0 is negative, current flows *out* of the card, and i_1 is positive. These observations are in agreement with writing $i_1 = -xe_0/10$.

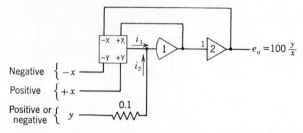

Fig. 31.8   **Division circuit using quarter-square card.**

(since the grid voltage is zero) give

$$-\frac{xe_0}{10} + \frac{y}{0.1} = 0$$

or

$$e_0 = 100\,\frac{y}{x} \qquad\qquad (31.5)$$

Notice that the voltage entering the $-Y$ terminal of the card must be negative and the voltage entering the $+Y$ terminal must be positive. The voltage representing y can be of either polarity.

To illustrate the use of a multiplier in the solution of a nonlinear differential equation, consider the following example.

Example 31.1   The following third-order nonlinear differential equation arises in boundary-layer theory in fluid mechanics:[1]

$$f(\eta)f''(\eta) + 2f'''(\eta) = 0 \qquad\qquad (31.6)$$
$$f(0) = f'(0) = 0$$
$$f'(\infty) = 1$$

where

$$f'(\eta) = \frac{df}{d\eta} \qquad f''(\eta) = \frac{d^2f}{d\eta^2} \qquad \text{etc.}$$

The function $f(\eta)$ is known as the *Blasius function*, and $f'(\eta)$ is the velocity profile which exists in the boundary layer on a flat plate.

Assuming that the maximum values are $f_m = 5$, $f'_m = 1$, $f''_m = 1$, we can obtain the scale factors shown in the following table:

Problem variable	Max value	$\dfrac{100}{\text{Max value}} = k$
f	5	20
f'	1	100
f''	1	100

[1] Any advanced text in fluid mechanics will give Eq. (31.6). For example, see H. Schlichting, "Boundary Layer Theory," p. 103, McGraw-Hill Book Company, 1955.

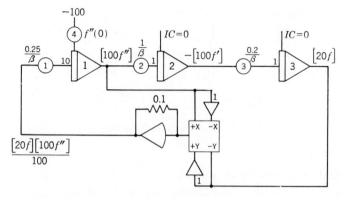

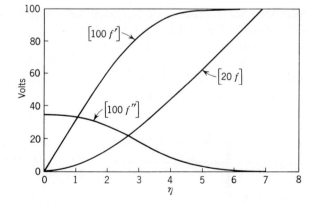

Fig. 31.9   Computer circuit for Blasius function.

Using these scale factors, Eq. (31.6) can be written

$$[100f'''] = \frac{-(0.5)(100)}{(20)} \frac{[20f][100f'']}{100}$$

or

$$[100f'''] = -2.5 \frac{[20f][100f'']}{100} \tag{31.7}$$

A circuit which solves Eq. (31.7) is shown in Fig. 31.9.

In this problem, we are faced with a new situation in that the boundary conditions are not all initial conditions; that is, $f''(0)$ is not specified, but rather $f'(\infty) = 1$. The approach becomes one of trial and error, in which solutions are run for various settings of potentiometer 4 until $f'(\infty)$ approaches 1. When we find this setting, we then have the correct solution as a plot of f, f', and f'' versus η. A computer solution based on the circuit of Fig. 31.9 is shown in Fig. 31.10.

The use of repetitive operation is very convenient for this type of problem. The trial-and-error determination of $f''(0)$ goes quickly because f' approaches 1 very quickly as shown in Fig. 31.10.

Fig. 31.10   Blasius function and its derivatives.

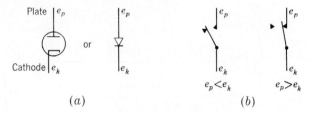

Plate e_p e_p e_p e_p

or

Cathode e_k e_k e_k e_k

$e_p < e_k$ $e_p > e_k$

(a)   (b)

Fig. 31.11   Symbols for a diode.

DIODES

A number of nonlinear phenomena can be represented by circuits using diodes. A diode is a two-element device which contains a plate and a cathode. Two symbols used to represent a diode are shown in Fig. 31.11a. For this discussion, the diode will be considered as a "switch" which conducts when the plate voltage exceeds the cathode voltage and does not conduct when the reverse is true. These operating characteristics will be called ideal, for they are only approximated by an actual diode. Only a few examples of the use of diodes are given here; additional applications will be found in Jackson.[1] In the following paragraphs, some typical nonlinear phenomena which can be simulated by diode circuits will be presented.

Coulomb (Dry) Friction Generator   Coulomb friction refers to a force existing between two bodies which slide in contact with each other. The force is constant, independent of the magnitude of the velocity and opposes the motion. The force is often associated with the coefficient of friction μ as shown in Fig. 31.12. Mathematically, the friction force F_c is related to velocity by the expressions (g_c is abbreviated to g)

$$F_c = -\mu \frac{W}{g} \qquad \dot{Y} > 0$$
$$= +\mu \frac{W}{g} \qquad \dot{Y} < 0$$

Note that the friction force has been defined so that, when the force acts to the right in Fig. 31.12, it is taken as positive. A graphical representation of these expressions is shown in Fig. 31.13.

A circuit which produces this relationship between force and velocity is shown in Fig. 31.14. An unusual feature of this circuit is that amplifier 1

[1] Jackson, *op. cit.*, pp. 194–212.

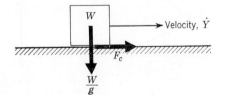

W → Velocity, $\dot{Y}$

F_c

$\dfrac{W}{g}$

Fig. 31.12   Sliding friction represented by Coulomb friction.

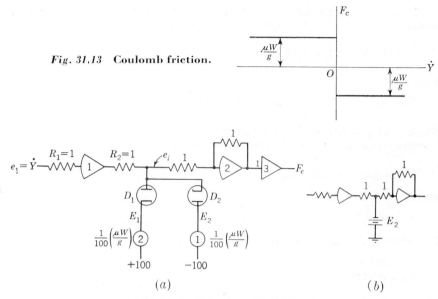

Fig. 31.13 **Coulomb friction.**

Fig. 31.14 **Simulation of Coulomb friction.**

contains no feedback element. A voltage e_1 representing the velocity $\dot{Y}$ is attached to the input resistor of amplifier 1. If e_1 becomes positive, the voltage e_i will become high and negative and cause D_2 to conduct because the plate of D_2 will be more positive than the cathode. When D_2 conducts, the circuit is equivalent to that shown in Fig. 31.14b, in which a negative voltage of magnitude E_2 is connected to an inverter. If e_1 becomes negative, e_i becomes high and positive and D_1 conducts with the result that a positive voltage of magnitude E_1 is attached to the inverter. By setting E_1 and E_2 equal to $\mu(W/g)$, we obtain the function shown in Fig. 31.13. The sizes of the resistors R_1 and R_2 in Fig. 31.14 are not important for the operation of the circuit. The 1-megohm resistors shown are satisfactory. The following example will illustrate how the Coulomb friction generator can be used in solving a problem of a mechanical oscillator.

Example 31.2 Consider the damped vibrator shown in Fig. 31.15, in which the coefficient of friction is 0.05. The symbols used are the same as those of Fig. 8.1. Initially, the mass is displaced 0.5 ft from its rest position, and we wish to draw a computer diagram which will produce the resulting Y as a function of t.

Fig. 31.15 Damped vibrator with Coulomb friction.

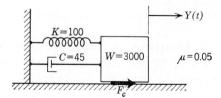

By including the friction force F_c in Eq. (8.2), the following equation of motion is obtained:

$$\frac{W}{g}\ddot{Y} + C\dot{Y} + KY = F_c(\dot{Y}) \tag{31.8}$$

where

$$F_c = -\mu\frac{W}{g} \qquad \dot{Y} > 0 \qquad \text{and} \qquad F_c = \mu\frac{W}{g} \qquad \dot{Y} < 0$$

Notice that $F(t)$ in Eq. (8.2) is zero for this problem.
Inserting the parameters into Eq. (31.8) gives the result

$$\ddot{Y} = -0.48\dot{Y} - 1.07Y + \frac{F_c}{93.2}$$

where

$$|F_c| = 4.66 \text{ lb}_f$$

If we assume the maximum values of Y and $\dot{Y}$ to be

$$Y_m = 0.5 \text{ ft}$$
$$\dot{Y}_m = 0.25 \text{ ft/sec}$$

we may obtain the scale factors shown in the following table:

Problem variable	Max value	$\dfrac{100}{\text{Max value}} = k$
Y	0.5	200
$\dot{Y}$	0.25	400

Using these scale factors, Eq. (31.8) can be written

$$[400\ddot{Y}] = -(0.48)[400\dot{Y}] - 1.07\frac{400}{200}[200Y] + \frac{400F_c}{93.2} \tag{31.9}$$

The computer circuit for solving this equation is shown in Fig. 31.16. We can check that the diode circuit produces the correct polarity at the input to amplifier 1 by the following argument: If $\dot{Y}$ is positive (motion to right), the output of amplifier 3 is highly positive and D_1 conducts with the result that the output of amplifier 4 is -20 volts. The sum of the inputs to amplifier 1 represents $[400\ddot{Y}]$, and according to Eq. (31.9), the input voltage corresponding to the friction force must be $+400F_c/93.2$. However, for $\dot{Y}$ positive, F_c is negative (friction force opposes motion) with the result that the voltage entering potentiometer 3 must be negative. In Fig. 31.17 are shown the computer responses obtained from the circuit shown in Fig. 31.16, with and without Coulomb friction. To obtain the response without Coulomb friction, one removes the input to potentiometer 3.

Limiting A limiter is a device which produces the relationship shown in Fig. 31.18. A diode circuit which will generate this relationship is shown in Fig. 31.19.

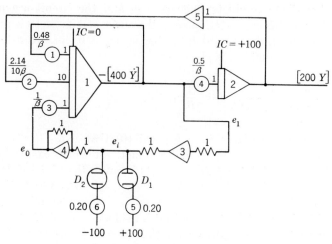

Fig. 31.16   Computer circuit for damped vibrator with Coulomb friction.

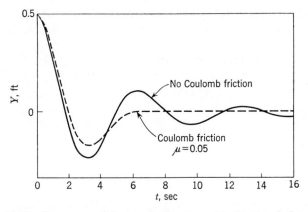

Fig. 31.17   Response of damped vibrator with Coulomb friction.

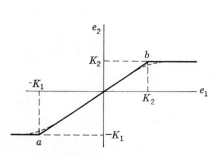

Fig. 31.18   Limiting.

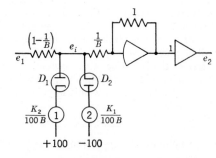

Fig. 31.19   Simulation of limiting.

If neither diode conducts, then $e_2 = e_1$ and the circuit is equivalent to an inverter; in this case, section ab of the curve of Fig. 31.18 is produced. Recalling that the grid is at ground potential ($e_g = 0$), one can show that, if neither diode conducts, e_i is given by

$$e_i = \frac{1/B}{(1 - 1/B) + 1/B}\, e_1 = \frac{1}{B}\, e_1 \tag{31.10}$$

If $-K_1 < e_1 < K_2$, we have from Eq. (31.10) that $-K_1/B < e_i < K_2/B$ and therefore neither D_1 nor D_2 will conduct, for which case $e_2 = e_1$. If $e_1 > K_2$, then $e_i > K_2/B$ and D_1 conducts because the plate of D_1 will be more positive than the cathode. As soon as D_1 can conduct, the circuit behaves as if the voltage K_2/B were applied to an amplifier having a gain of B with the result that the output is held at $e_2 = (K_2/B)B = K_2$. A similar argument can be used to show that $e_2 = -K_1$ when $e_1 < -K_1$; in this case, the operation will depend on D_2 conducting.

The diode circuit described here does not give the sharp corners shown by the ideal response of Fig. 31.18; the actual response is similar to that shown by the dotted lines. An improved circuit which generates a response which more closely matches the ideal response is described by Johnson.[1]

A common application of the limiter in process control is the representation of the saturation characteristics of a pneumatic valve. As the pressure to a linear valve increases, the flow increases linearly until the valve is wide open, after which further increase in pressure does not change the flow. These characteristics correspond to the limiting action of Fig. 31.18 if K_1 is made zero.

PROBLEMS

31.1 A tank contains 100 gal of fresh water. At time zero, brine enters the tank at a rate of 10 gal/min and solution is withdrawn from the tank at 5 gal/min. The contents of the tank are well mixed. The entering brine has a concentration of 0.2 lb salt per gallon of solution. Determine the concentration in the tank when the tank contains 200 gal of solution. Develop a scaled computer circuit to solve this problem. Assume that the solution is to be produced in 10 sec at the computer.

31.2 The control system shown in Fig. P31.2 contains a time-varying gain. If a unit-step function is applied at the input R at $t = 0$, write the differential equation that describes the behavior of the system. Draw a computer circuit to solve this equation. Prior to introducing the signal R, the value of C is zero and the gain is zero.

If the response to a unit impulse applied at $t = t_1$ [i.e., $R(t) = \delta(t - t_1)$] is desired, describe how this response can be obtained by simply applying initial conditions to the integrator without actually generating an impulse function.

[1] Johnson, *op. cit.*, p. 110.

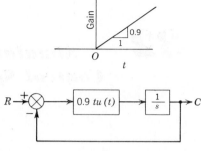

Fig. P31.2

31.3 The signal

$$e_1 = 10 \cos 4t$$

is attached to the circuit shown in Fig. P31.3. At time $t = 0$, the switch around amplifier 3 is opened. Note that amplifier 1 is a high-gain amplifier. Draw a graph showing how e_1, e_2, and e_3 vary with time.

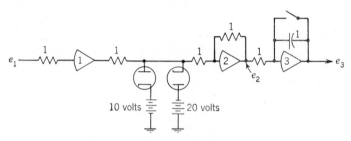

Fig. P31.3

32 *Simulation of Control Systems*

In the previous two chapters, the analog computer was introduced and applied to the solution of differential equations. In this chapter, the analog computer will be used to study the transient behavior of control systems. The computer approach to the study of control systems is often referred to as *computer simulation,* which means that the differential equations representing the dynamic characteristics of the system components are solved simultaneously to produce time-varying output voltages which resemble the transient behavior of the real system. In this sense, simulation is nothing more than the solution of a set of simultaneous differential equations.

In the previous chapters of the text, we separated the analytical treatments of linear systems from those of nonlinear systems and noted that the nonlinear system posed a considerably more difficult problem. However, from the point of view of computer simulation, both linear and nonlinear systems appear to be about the same in the degree of difficulty of programming the computer. Of course, the nonlinear problem will require nonlinear computing elements (multipliers, function generators, etc.) which are quite expensive, but the approach is essentially the same as that for a linear problem.

In the first part of this chapter, the computer simulation of linear control systems will be discussed. Since the use of block diagrams and transfer functions is convenient for linear systems, the approach will be to devise circuits for simulating the commonly used transfer functions and then to connect the appropriate circuits according to the block diagram. This approach is called *direct simulation of control systems* and offers the advantage of being able to wire a circuit directly from the block diagram. Such a procedure organizes the work and gives considerable insight into the interaction among the control system components. In the last part of the chapter, the computer simulation of nonlinear systems will be considered.

Computer simulation offers the following advantages:

1. The dynamic response of a control system for various controller settings and process parameters can be obtained with little effort once an analog circuit has been constructed. The parameters are usually changed merely by adjusting potentiometers.

2. The computer can be operated as a component in an actual physical process. For example, a pneumatic controller can be evaluated by using it to control a process which is simulated on the computer. Appropriate transducers are used to convert the pneumatic signal to an electrical signal and vice versa. A computer used in this manner is said to perform real time simulation.

3. The computer can be time-scaled so that the response from the computer is very fast compared with the response of the actual system which is being simulated. In this way many response curves can be obtained in a short time. The use of repetitive operation on the computer, as discussed in Chap. 30, is advantageous in studying the effect of controller settings on the shape of the transient response.

DIRECT SIMULATION OF TRANSFER FUNCTIONS

The usual procedure for determining the transient response of a linear control system is to draw a block diagram of the system and to place within each block the appropriate transfer function. From the block diagram, one obtains the overall transfer function which can be used to determine the response of the system for any particular change in load or set point. To obtain the transient response for a higher-order system by this approach involves very long, tedious calculations. In this section, we shall show how the analog computer can be used to simulate a control system directly from the transfer functions.

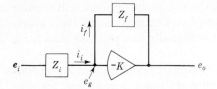

Fig. 32.1 General one-amplifier circuit.

One-amplifier Circuits We shall now develop analog computer circuits for simulating the response of some of the transfer functions which frequently occur in control systems. Consider the general one-amplifier circuit of Fig. 32.1, in which Z_i and Z_f represent impedances. For example, if Z_i is a resistor and Z_f a capacitor, the circuit is an integrator. By using various series and/or parallel combinations of resistors and capacitors for Z_i and Z_f, we can obtain a variety of transfer functions. The modern approach to circuit analysis is to use the *operational impedance* for each circuit element. For the purpose of this discussion, the operational impedance is the transfer function of a circuit element which relates output voltage to input current. In Table 32.1, the operational impedances are

Table 32.1

Circuit element	Operational impedance
Resistance R	R
Capacitance C	$\dfrac{1}{Cs}$
Inductance L	Ls

shown for the resistor, the capacitor, and the inductor. (The inductor is seldom used as a computer element.) The overall impedance of several circuit elements connected in series is found by summing the individual impedances.[1] For example, a series combination of R and C has an operational impedance of

$$Z = R + \frac{1}{Cs} = \frac{RCs + 1}{Cs}$$

The overall impedance of circuit elements connected in parallel is obtained by equating the reciprocal of the overall impedance to the sum of the reciprocals of the separate impedances. For example, the parallel combina-

---

[1] These rules for combining impedances are derived in basic texts on circuit analysis. See, for example, M. E. Van Valkenburg, "Network Analysis," Prentice-Hall, Inc., Englewood Cliffs, N.J., 1955.

tion of R and C gives

$$\frac{1}{Z} = \frac{1}{R} + \frac{1}{1/Cs} = \frac{1 + RCs}{R}$$

or

$$Z = \frac{R}{1 + RCs}$$

The following equations can be written for the circuit of Fig. 32.1:

$$e_i - e_g = i_i Z_i \tag{32.1}$$
$$e_g - e_o = i_f Z_f \tag{32.2}$$
$$i_i = i_f \tag{32.3}$$
$$e_g = -\frac{e_o}{K} \tag{32.4}$$

In writing Eq. (32.3), we have taken the grid current to be zero. Combining the above equations to eliminate e_g and neglecting e_o/K compared with e_i or e_o (because of the high gain) give the result

$$\frac{e_o}{e_i} = -\frac{Z_f}{Z_i} \tag{32.5}$$

Equation (32.5) is the basic equation for determining the transfer function of various one-amplifier circuits. A dictionary of one-amplifier circuits for simulating transfer functions can be found in several textbooks on analog computation.[1]

First-order Simulation   The first-order transfer function $A/(\tau s + 1)$ can be simulated by the circuit shown in Fig. 32.2b. For this circuit $Z_i = R_1$ and $Z_f = R/(RCs + 1)$. According to Eq. (32.5), we have

$$\frac{Y}{-X} = -\frac{1}{R_1} \frac{R}{RCs + 1} = -\frac{R/R_1}{RCs + 1}$$

Comparing this expression with the first-order transfer function shows that $A = R/R_1$ and $\tau = RC$.

An alternate computer circuit for the first-order system is shown in Fig. 32.2c. This form is useful when the gain A or time constant τ is to be varied over a range of values. If the diagram of Fig. 32.2b were used, the resistors or capacitor would have to be changed each time A or τ was changed; in the circuit of Fig. 32.2c, A and τ are changed merely by resetting potentiometers. The two circuits of Fig. 32.2 are equivalent as

[1] C. L. Johnson, "Analog Computer Techniques," pp. 59–61, McGraw-Hill Book Company, New York, 1956; A. S. Jackson, "Analog Computation," pp. 220–221, McGraw-Hill Book Company, New York, 1960.

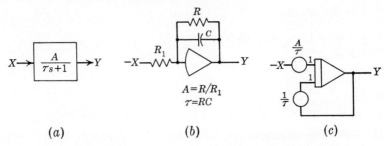

Fig. 32.2   **Simulation of first-order transfer function.**

can be seen by drawing Fig. 32.2b in the form shown in Fig. 32.3a, which can be considered as an integrator having a feedback path to resistor R.   In Fig. 32.3b, this circuit is drawn using the compact symbol for the integrator.

The gains of the integrator in Fig. 32.3b may be varied conveniently by use of potentiometers connected to the inputs of the integrator as shown in Fig. 32.3c.   In this case the input potentiometer is set equal to A/τ and the feedback potentiometer to $1/\tau$ if the gains of the integrator are each 1. Other combinations of integrator gains and potentiometer settings may be used as required.

We can also show that Fig. 32.2c simulates a first-order system by recognizing that this circuit solves the differential equation

$$-\frac{dY}{dt} = -\frac{A}{\tau} X + \frac{1}{\tau} Y$$

the transform of which is

$$-sY(s) = -\frac{A}{\tau} X(s) + \frac{1}{\tau} Y(s)$$

or

$$\frac{Y(s)}{X(s)} = \frac{A}{\tau s + 1}$$

The circuit of Fig. 32.2c will be used henceforth for the first-order system because of its increased flexibility over that of Fig. 32.2b.

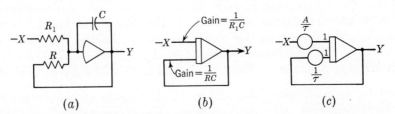

Fig. 32.3   **Equivalent circuits for simulating a first-order transfer function.**

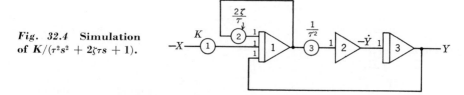

Fig. 32.4 Simulation
of $K/(\tau^2 s^2 + 2\zeta\tau s + 1)$.

Second-order Simulation   If the second-order system is of the form
$K/(\tau_1 s + 1)(\tau_2 s + 1)$, it may be simulated by connecting in series two
circuits of the type shown in Fig. 32.2.   For this case the response will
never be underdamped; that is, ζ will always be greater than 1.   However,
if we wish to simulate the second-order system

$$G(s) = \frac{Y}{X} = \frac{K}{\tau^2 s^2 + 2\zeta\tau s + 1} \tag{32.6}$$

with $\zeta < 1$, the circuit shown in Fig. 32.4 may be used.   To show that this
circuit is correct, it is convenient to use the differential-equation approach
to simulation of transfer functions.   Equation (32.6) is first rewritten as

$$\tau^2 s^2 Y + 2\zeta\tau s Y + Y = KX \tag{32.7}$$

Now Eq. (32.7) is solved for $s^2 Y$:

$$-s^2 Y = \frac{2\zeta\tau}{\tau^2} sY + \frac{Y - KX}{\tau^2} \tag{32.8}$$

Equation (32.8) is equivalent to the differential equation

$$\frac{-d^2 Y}{dt^2} = \frac{2\zeta\tau}{\tau^2} \frac{dY}{dt} + \frac{Y - KX}{\tau^2} \tag{32.9}$$

The circuit of Fig. 32.4 solves this differential equation as the reader can
readily verify.   The simulation of the transfer function given by Eq. (32.6)
requires that the IC voltages to the integrators be zero, since the transfer
function is derived for the case of zero initial conditions.

Simulation of the remaining *process* transfer function, a transportation
lag, is deferred for discussion later in the chapter.   We now proceed to
controllers.

Controller Simulation   *Proportional.*   The transfer function of a
proportional controller can be simulated by the circuit of Fig. 32.5.   The

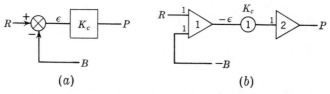

(a)   (b)

Fig. 32.5   Simulation of a proportional controller.

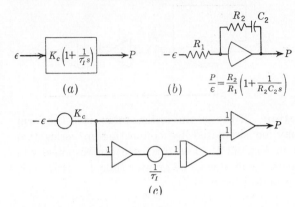

$$\frac{P}{\epsilon} = \frac{R_2}{R_1}\left(1 + \frac{1}{R_2 C_2 s}\right)$$

Fig. 32.6   Simulation of a PI controller.

error $R - B$ is obtained from amplifier 1, and the gain K_c is adjusted by means of the potentiometer. The controller output signal P is obtained at the output of amplifier 2. If $K_c > 1$, it is necessary to use a gain higher than unity on the input to amplifier 2.

Proportional-integral. A one-amplifier circuit for simulating the transfer function of the ideal PI controller is shown in Fig. 32.6b. In this figure, the amplifier which is needed for obtaining the error has been omitted for simplicity. One can show that the circuit of Fig. 32.6b simulates the transfer function

$$-\frac{P}{\epsilon} = -\frac{R_2}{R_1}\left(1 + \frac{1}{R_2 C_2 s}\right)$$

and by comparison with the ideal transfer function it can be seen that

$$K_c = \frac{R_2}{R_1} \qquad \tau_I = R_2 C_2$$

An alternate circuit is shown in Fig. 32.6c. This circuit has the advantage of providing two potentiometer adjustments which are proportional to the gain K_c and the reset rate $1/\tau_I$. We can easily verify that this last circuit is correct by recognizing that it solves the equation

$$P = -K_c\left(-\epsilon + \int \frac{-\epsilon}{\tau_I}\,dt\right)$$

the transform of which is

$$P(s) = K_c\left[\epsilon(s) + \frac{1}{\tau_I s}\,\epsilon(s)\right]$$

or

$$\frac{P(s)}{\epsilon(s)} = K_c\left(1 + \frac{1}{\tau_I s}\right)$$

While the circuit of Fig. 32.6c is more flexible, it requires two more amplifiers than does that of Fig. 32.6b.

    Proportional-derivative.   The ideal transfer function for a PD controller is

$$\frac{P(s)}{\epsilon(s)} = K_c(1 + \tau_D s) \tag{32.10}$$

As shown in Chap. 10, this is equivalent to

$$P(t) = K_c\epsilon + \tau_D K_c \frac{d\epsilon}{dt}$$

This transfer function cannot be realized in a real controller as discussed in Chap. 22; furthermore, it cannot be simulated on an analog computer. The part of the transfer function which gives the difficulty is the term representing differentiation (s or d/dt). To understand why this limitation exists, consider a step input in ϵ applied to the ideal controller. Theoretically, the output should rise instantaneously to infinity and then return instantaneously to a value corresponding to $P = K_c\epsilon$; this response is shown in Fig. 32.7 for $K_c = \tau_D = 1$. Obviously, a real controller cannot respond in this fashion. In Chap. 22, a more realistic transfer function for a PD controller was given as

$$\frac{P(s)}{\epsilon(s)} = K_c \frac{\tau_D s + 1}{(\tau_D/\gamma)s + 1} \tag{32.11}$$

where γ is a constant having a value of about 10 in many controllers. The step response to a controller having this transfer function is also shown in Fig. 32.7 for several values of γ. From this figure, it is seen that, as γ is increased, the step response approaches the ideal response.

    A computer simulation of Eq. (32.11) is feasible, and a one-amplifier circuit is shown in Fig. 32.8a, in which $K_c = R_2/R_1$, $\tau_D = R_1 C_1$, and

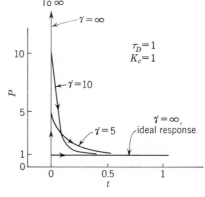

Fig. 32.7 **Unit-step response of a PD controller.**

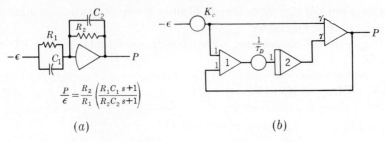

$$\frac{P}{\epsilon} = \frac{R_2}{R_1}\left(\frac{R_1 C_1 \, s+1}{R_2 C_2 \, s+1}\right)$$

(a)   (b)

Fig. 32.8   Simulation of a PD controller.

$\tau_D/\gamma = R_2 C_2$.   This circuit can be verified by application of Eq. (32.5). Another circuit is shown in Fig. 32.8b, for which K_c and τ_D can be adjusted independently.   To show that this last circuit is correct, we write the transfer function of Eq. (32.11) in the form

$$\frac{\tau_D}{\gamma} sP + P = K_c \tau_D s\epsilon + K_c\epsilon$$

Dividing this equation by $\tau_D s/\gamma$ gives

$$P = K_c\gamma\epsilon + \frac{K_c\gamma}{\tau_D}\frac{\epsilon}{s} - \frac{\gamma}{\tau_D}\frac{P}{s} \tag{32.12}$$

Recognizing that ϵ/s and P/s are equivalent to $\int\epsilon \, dt$ and $\int P \, dt$, respectively, we see that the circuit of Fig. 32.8b does solve Eq. (32.12).   In practice, $1 + \tau_D s$ can be approximated by making γ very large.

Proportional-integral-derivative.   If the circuits of Figs. 32.6c and 32.8b are combined as shown in Fig. 32.9, we have a circuit which simulates the transfer function

$$\frac{P(s)}{\epsilon(s)} = K_c\left[\frac{1 + \tau_D s}{1 + (\tau_D/\gamma)s} + \frac{1}{\tau_I s}\right]$$

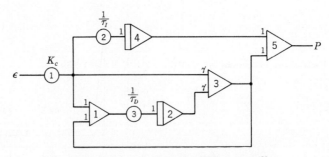

Fig. 32.9   Simulation of a PID controller.

When γ is made very large, this equation reduces to

$$\frac{P(s)}{\epsilon(s)} = K_c\left(1 + \tau_D s + \frac{1}{\tau_I s}\right)$$

which is the transfer function for an ideal three-mode controller, as presented in Chap. 10. Notice that in Fig. 32.9 potentiometers 1 and 2 are in series. To avoid loading errors, it is necessary to set 2 before setting 1.

CONSTRUCTING THE COMPUTER DIAGRAM

A simple way to describe the procedure for simulating a control system is to consider a numerical example. The liquid-level control system shown in Fig. 32.10 is used to control the level in tank 2. A load disturbance in the form of a variation in flow rate L can enter tank 1. Assume that the valve and the level-measuring element do not exhibit dynamic lag. The block diagram for the system, which is shown in Fig. 32.10*b*, is labeled in terms of the physical quantities and their units. The parameters shown in the block diagram have the following values

K_v = valve constant = 0.10 cfm/psi
R_2 = steady-state gain for tank 2 = 0.5 ft level/cfm
τ_1 = time constant of tank 1 = 2 min
τ_2 = time constant of tank 2 = 1 min
K_c = controller gain, psi/ft level
τ_I = integral time = $\frac{1}{2}$ min

Before the control system is simulated, the block diagram of Fig. 32.10 can be simplified slightly by combining the transfer functions for the valve and

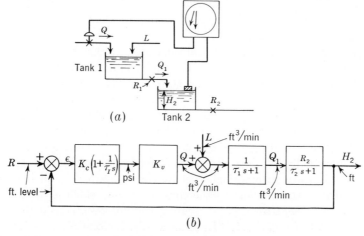

(*b*)

Fig. 32.10 **Liquid-level control system.**

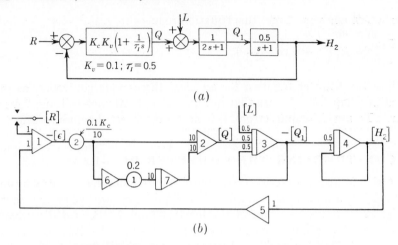

Fig. 32.11   Simulation of a liquid-level control system.

controller as shown in Fig. 32.11*a*.   To simulate this block diagram, the
blocks are replaced by the circuits of Figs. 32.2 and 32.6 and connected as
shown in Fig. 32.11*b*.   This computer circuit provides for a change in
controller gain K_c from 0 to 100.   It should be noted that, because of sign
reversal in each amplifier, an inverter (*A*5) has been placed in the feedback
path to obtain negative feedback.   Without this amplifier, the system would
have positive feedback and be unstable.

   Change in Set Point ($L = 0$)   Consider a change in set point only,
for which case no load voltage L is to be introduced into *A*3.   At time $t = 0$,
the system is at steady state.   To simulate a step change in R of 1 ft of level
we may apply a constant voltage of 1 volt to the upper input resistor of *A*1
at time zero by closing the relay switch shown in Fig. 32.11.   The subsequent variations in voltage from the other amplifiers are proportional to the
corresponding system variables.

   Notice in Fig. 32.11 that initial conditions are not specified on the
integrators.   The reason for ignoring initial conditions is that, when the
system is stable, all the variables (R, ϵ, Q, Q_1, and H_2) approach steady-state
values when the computer is placed in the operate mode.   These steady-
state values are the desired initial conditions.   After the voltages reach
steady state, the simulated system can be disturbed by making a change
in the voltage representing L or R, after which all the voltages change with
time and eventually reach a new set of steady-state values.   Hence the
initial conditions are automatically set by the computer circuit itself while
operating in the same manner as the control system which it simulates.

   Magnitude Scaling   So far, scaling has not been considered for the
computer circuit of Fig. 32.11.   In other words, the scale factors for magnitude and time are each 1.   Thus, we can label the outputs of the various

amplifiers $[Q]$, $[Q_1]$, $[H_2]$, etc., as shown in Fig. 32.11. A time-scale factor of 1 means that 1 sec of computer time is equivalent to 1 min in the real physical system, since the parameters (τ_1, τ_2, τ_I) were given in terms of minutes.

For a step change of 1 volt in R, the voltages from the amplifiers would probably be much less than the maximum value of 100 volts which the amplifiers can provide. To improve the accuracy of the computation, we should have each amplifier output near 100 volts sometime during the solution. The amplifier signals can all be increased by scaling R; for example, we may let 100 volts be equal to 1 ft in R. In this case, we have selected a scale factor of 100 for R, and the input voltage to $A1$ which represents R can be written $[100R]$. However, the other amplifiers may overload if this magnitude of input voltage is used. Hence a more comprehensive scaling analysis is required.

The method of magnitude scaling to be presented here is the same as that described in Chap. 30, where the maximum voltage available from an amplifier (100 volts in our examples) was divided by the maximum value of the variable to be represented by the output of the amplifier. From our knowledge of the transient response of a control system to a change in set point, we can *estimate* maximum values of the variables (H_2, Q, Q_1, and ϵ) as follows: For a step change in R, which we have taken to be 1, we know that H_2 will ultimately equal R for a controller having integral action. However, for an underdamped response H_2 will overshoot its ultimate value of 1. For a practical system, the maximum value of H_2 will usually be less than $2R$ (i.e., overshoot will be less than 100 per cent). Therefore, we may take the maximum value of H_2 to be 2. This value is entered in the scale factor table shown below. Immediately after the step change in R is introduced, the error ϵ will usually have a maximum which is equal in magnitude to R. Therefore, we take the maximum value of ϵ to be 1.

The maximum value of Q is estimated from the following argument: If H_2 is to change by 1 ft, the steady-state gains of the transfer functions for the tanks require that Q ultimately be 2 cfm (see Fig. 32.11a). However, if the system is underdamped and K_c is large, we can expect Q to overshoot this steady-state value of 2. We shall assume here that Q overshoots its steady-state value of 2 by 400 percent, for which case the maximum value of Q is to be 10.

To estimate the maximum value of Q_1, note that eventually Q_1 will be equal to Q. On this basis, we take the maximum value of Q_1 to be between 2 and 10, say 4.

In general, some of the estimates are often pure guesses. In particular, the estimate of the maximum value of Q is difficult because it is very sensitive to the controller gain. In practice, the circuit based on the first estimates should be tried on the computer. Any amplifier which overloads or produces a very small signal will reveal that rescaling is necessary.

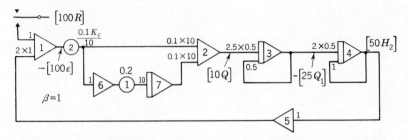

Fig. 32.12 Magnitude-scaled computer circuit for change in set point.

From the estimates of the maximum values just selected, we can obtain the scale factors as shown in the table below. To scale the computer

Variable	Max value	$\dfrac{100}{\text{Max value}} = k$
R	1	100
H_2	2	50
ϵ	1	100
Q	10	10
Q_1	4	25

diagram, the circuit of Fig. 32.11 is redrawn as shown in Fig. 32.12 and the appropriate amplifier outputs are labeled with computer variables. For example, the output of $A4$ is labeled $[50H_2]$. To provide for the different scale factors, the gains to amplifiers must be multiplied by factors which equal the ratio of scale factors. (This rule was discussed in Chap. 30.) For example, the input of $A4$ which is fed by the output of $A3$ is labeled 2×0.5, where the factor 2 is the ratio of scale factors for H_2 and Q_1. Notice that the gain on the input of $A4$ which is in the feedback loop around $A4$ is not changed. To verify this, consider the transfer function which is simulated by the circuit involving $A4$:

$$\frac{H_2}{Q_1} = \frac{0.5}{\tau_2 s + 1}$$

In terms of computer variables, this expression becomes

$$\frac{[50H_2]}{[25Q_1]} = 2\,\frac{0.5}{\tau_2 s + 1}$$

This verifies the circuit of Fig. 32.12 where only the input from $A3$ is multiplied by 2.

The comparator generates the relation

$$\epsilon = R - H_2$$

Writing this expression in terms of computer variables gives

$$[100\epsilon] = [100R] - 2[50H_2]$$

To simulate this last expression, it is necessary to use a gain of 2 for the input of $A1$ which receives the feedback signal, as shown in Fig. 32.12.

The magnitude scaling is now complete. One should note that, in scaling, we have simply distributed the steady-state gains around the loop without altering the overall gain. For example, if the unscaled diagram (Fig. 32.11b) is compared with the scaled diagram (Fig. 32.12), we see that the product of the factors by which we multiplied amplifier gains ($2 \times 0.1 \times 2.5 \times 2$) is exactly 1. The application of this rule is useful in checking the scaled computer diagram.

It should be emphasized that a unit change in set point is selected only as a matter of convenience. If any other size change were selected, the scaling procedure would remain the same. In fact, there is no need even to consider the scaling again. Since the system is linear, superposition requires that the value of any variable at a particular time be directly proportional to the magnitude of the forcing function (R or L). For example, assume that we have obtained a transient response curve for H_2 for a unit step change in R. If we double the change in R, we simply double the values of H_2.

In simulating a control system for which one has considerable operating experience, the selection of scale factors is more direct because the maximum values of the variables are known in advance. When such operating experience is lacking, one should scale the problem by estimation, as described in this example. Corrections can then be based on the operating experience gained from the estimated circuit.

In the scaled computer circuit of Figs. 32.11 and 32.12 the time-scale factor is unity. This means that 1 sec of computer time represents one unit of time in the physical system, which is 1 min for this example. In the next section, time scaling will be considered.

Time Scaling The unit of time used in this physical example has been the minute. Since the RC product which occurs in the various circuits is in seconds, we have the relationship that 1 min of duration in the physical system will take 1 sec on the computer. Using the time-scale factor β of Eq. (30.24), we see that β is 1; thus

$$\beta = \frac{\tau}{t} = \frac{\text{computer time, sec}}{\text{problem time, min}} = 1$$

If it is desired to make β different from 1, the time-scaling rule for differential equations given in Chap. 30 can generally be used. This is to be expected because the transient response of a control system is equivalent to the solution of a set of simultaneous differential equations. In fact, if the amplifiers of the computer circuit for a control system are used only as integrators and summers, the rule for time scaling is identical with that of Chap. 30; thus:

To time-scale, multiply the gain of each integrator input by $1/\beta$.

When some of the simulated components of the control system involve use of one-amplifier circuits such as those for the PI or PD controllers of Figs. 32.6b and 32.8a, the scaling for these components requires special attention, for they are not simple integrators. In such cases, each RC product which is associated with a parameter in the transfer function is multiplied by β.

Change in Load $(R = 0)$ If only load changes occur, we may wish to use scale factors which differ from those used for changes in set point. Thus, it is known that, for a step change in L, the level H_2 will at first deviate from zero but eventually it will return to zero because of the integral action in the controller.

Assume that we are interested in the response for a step change in L of 1 cfm. For convenience let 100 volts represent one unit of L; this corresponds to a scale factor of 100, and we write [100L] for the voltage representing L. If there were no control, we know from the steady-state gains of the transfer functions for the tanks that Q_1 and H_2 of Fig. 32.11 would eventually become 1 and 0.5, respectively. However, with control present, we may expect the changes in Q_1 and H_2 to be less than these values. Therefore, we shall take the maximum value of Q_1 to be 0.5 and that of H_2 to be 0.25. The maximum values are listed in the table shown below. Since $R = 0$, we know that $|\epsilon| = |H_2|$; therefore, the maximum value of ϵ is taken to be the same as that of H_2, which is 0.25. The magnitude of Q will eventually become 1 because the increase in L must be matched by the same decrease in Q if H_2 is to return to zero. Furthermore, if the system is underdamped, Q will overshoot its final value. Assume that Q may be as large as $2L$, for which case the maximum value of Q is 2. Note that this estimated change in Q is considerably less than that for the case of set-point change. This is reasonable because for a set-point change, the disturbance travels directly to the amplifier generating Q (see Fig. 32.11) whereas for load change, the disturbance must pass through the two tanks before reaching the controller. In the following table, the maximum values of the variables and the corresponding scale factors are listed. The cir-

Variable	Max value	$\dfrac{100}{\text{Max value}} = k$
L	1	100
Q_1	0.5	200
H_2	0.25	400
ϵ	0.25	400
Q	2	50

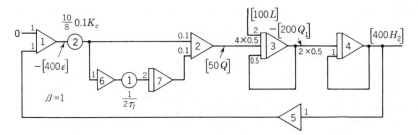

Fig. 32.13 Magnitude-scaled computer circuit for change in load.

cuit shown in Fig. 32.11*b* is redrawn in Fig. 32.13 with the amplifiers labeled according to the scale factors just found.

The response curves for the circuits of Figs. 32.12 and 32.13 are not shown here, for they will resemble those of Figs. 17.5 and 17.7. In fact, the figures of Chap. 17 were obtained from the computer circuit of Fig. 32.12.

TRANSPORTATION LAG

As shown in Chap. 17, it is quite difficult to account for a transportation lag in computing the transient response of a control system. For this reason considerable work has been done on the computer simulation of the transportation lag, which is represented by the transfer function $e^{-\tau s}$.

One approach to simulating this transfer function is to express $e^{-\tau s}$ by a Taylor series expansion:

$$e^{-\tau s} = 1 - \tau s + \frac{\tau^2 s^2}{2} - \frac{\tau^3 s^3}{3!} + \cdots \tag{32.13}$$

If only the first two terms are retained, we have as an approximation

$$e^{-\tau s} \cong 1 - \tau s \tag{32.14}$$

The most convenient way to evaluate such an approximation is to compare its frequency-response diagram with that for $e^{-\tau s}$. Recall that the frequency response of $e^{-\tau s}$ is

$$\text{Gain} = 1$$
$$\text{Phase angle } \phi = -\omega\tau \quad \text{rad}$$

On a Bode diagram, the gain curve is constant at 1 while the phase-angle curve decreases linearly with frequency. On the other hand, the frequency response of $(1 - \tau s)$ is

$$\text{Gain} = \sqrt{1 + \omega^2 \tau^2}$$
$$\phi = -\tan^{-1} \omega\tau$$

In Fig. 32.14, the frequency response of $e^{-\tau s}$ and the linear approximation of Eq. (32.14) are compared. Notice that the frequency is plotted as

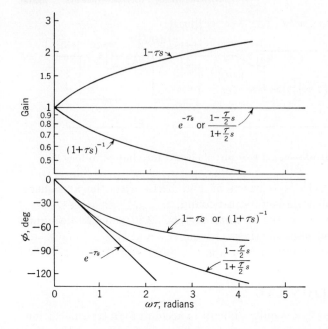

Fig. 32.14 **Frequency response of approximations to $e^{-\tau s}$.**

$\omega\tau$, the dimensionless frequency in radians. We see immediately that the gain of the approximation is not constant at 1 as required by $e^{-\tau s}$ and that the phase angle is in close agreement only for small values of $\omega\tau$.

If $e^{-\tau s}$ is written as $1/e^{\tau s}$ and the denominator expanded in a Taylor series, the result is

$$e^{-\tau s} \cong \frac{1}{e^{\tau s}} = \frac{1}{1 + \tau s + \tau^2 s^2/2 + \tau^3 s^3/3! + \cdots}$$

Keeping only the first two terms in the denominator gives

$$e^{-\tau s} \cong \frac{1}{1 + \tau s} \tag{32.15}$$

The frequency response for this transfer function is

$$\text{Gain} = \frac{1}{\sqrt{1 + \omega^2 \tau^2}}$$
$$\phi = -\tan^{-1} \omega\tau$$

and is also shown on Fig. 32.14. It can be seen that this approximation is no better than $1 - \tau s$.

Another approach is to write

$$e^{-\tau s} = \frac{e^{-\tau s/2}}{e^{\tau s/2}}$$

Expanding numerator and denominator of this last expression in a Taylor series and keeping only terms of first order give

$$e^{-\tau s} \cong \frac{1 - (\tau/2)s}{1 + (\tau/2)s} \tag{32.16}$$

The frequency response for this transfer function is

Gain = 1

$$\phi = -2\tan^{-1}\frac{\omega\tau}{2}$$

and is plotted in Fig. 32.14. This approximation provides the correct gain, but the phase-angle curve agrees with the true curve only for small values of $\omega\tau$; however, the agreement is better than in the previous cases.

An approximation which is more general than the ones just described is the Padé approximation

$$e^x = \lim_{(u+v)\to\infty} \frac{F_{uv}(x)}{G_{uv}(x)}$$

where

$$F_{uv} = 1 - \frac{vx}{u+v} + \frac{v(v-1)x^2}{(u+v)(u+v-1)2!} - \cdots$$
$$+ \frac{(-1)^v v(v-1)\cdots 2\cdot 1\cdot x^v}{(u+v)(u+v-1)\cdots(u+1)v!}$$

$$G_{uv} = 1 + \frac{ux}{v+u} + \frac{u(u-1)x^2}{(v+u)(v+u-1)2!} + \cdots$$
$$+ \frac{u(u-1)\cdots 2\cdot 1\cdot x^u}{(v+u)(v+u-1)\cdots(v+1)u!}$$

For $u = v = 1$, we obtain the first-order Padé approximation

$$e^{-\tau s} = \frac{1 - (\tau/2)s}{1 + (\tau/2)s}$$

This is identical with the Taylor expansion of Eq. (32.16). For $u = v = 2$, we obtain the second-order Padé approximation.

$$\frac{Y}{X} = e^{-\tau s} \cong \frac{1 - \tau s/2 + \tau^2 s^2/12}{1 + \tau s/2 + \tau^2 s^2/12} \tag{32.17}$$

for which the frequency response is

Gain = 1

$$\phi = -2\tan^{-1}\frac{6\omega\tau}{12 - \omega^2\tau^2}$$

The Bode diagram for Eq. (32.17) is shown in Fig. 32.15. Notice that the phase angle is correct up to a dimensionless frequency $\omega\tau$ of about

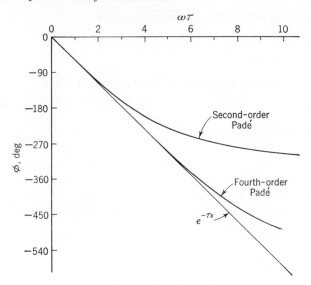

Fig. 32.15   Padé approximations to $e^{-\tau s}$.

2 rad.   This is a considerable improvement over the Taylor-series approximations of Fig. 32.14.   A computer circuit for simulating Eq. (32.17) is shown in Fig. 32.16.   The reader can verify this circuit by cross-multiplying the terms of Eq. (32.17) and rearranging to yield

$$Y = X - \frac{1}{s}\left(\frac{6}{\tau}X + \frac{6}{\tau}Y\right) + \frac{1}{s^2}\left(\frac{12}{\tau^2}X - \frac{12}{\tau^2}Y\right)$$

Inspection of Fig. 32.16 shows that the circuit diagram does solve this equation.   The frequency response of a fourth-order Padé approximation ($u = v = 4$) represents $e^{-\tau s}$ quite closely up to $\omega\tau$ of about 6 as shown in Fig. 32.15.[1]

A number of other circuits are discussed in the literature.   Single

[1] It is not necessary that $u = v$ in the Padé approximation, but we have chosen such cases as examples because they are the approximations most commonly used.

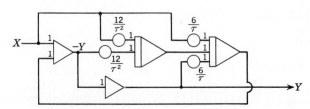

Fig. 32.16   Circuit for second-order Padé approximation to $e^{-\tau s}$.

and Stubbs[1] have suggested an approximation requiring a four-amplifier circuit that gives correct gain and a maximum phase error of 1.09° up to a phase shift of 7.44 rad, or 427°. This means that, for $\tau = 1$ sec, the error in phase is less than 1.09° for the frequency range of 0 to 7.44 rad/sec. These results are also reported in Jackson.[2] Another approach used to simulate the transportation lag is based on a magnetic tape recorder. In principle, the input signal is recorded by a "write" head on moving tape. The output is obtained from a "read" head at some distance from the write head location. For a fixed distance of separation between the write head and the read head, the transportation lag parameter τ is inversely proportional to the speed of the tape. Since this mechanical simulator is not based on standard components of an analog computer, it is not yet in wide use.

Another transportation lag simulator, which is similar in principle to the tape unit just described, consists of a series of capacitors on which the input signal is stored sequentially as discrete voltages. The output is obtained after a desired delay time by feeding these stored voltages back to the circuit at the same rate used in storing them. The response from this type of simulator, which is stepped, can be smoothed by appropriate filtering. As many as 100 high-quality capacitors may be needed to produce a satisfactory simulation.

To illustrate the use of the Padé approximation, consider the control system shown in Fig. 32.17, which is a two-tank chemical reactor with transportation lag. This same system was discussed in Chaps. 11 and 17. The block diagram for the system is shown in Fig. 32.17b with the numerical values of the parameters included in the transfer functions. The details of obtaining the transfer functions and constructing the block diagram have been considered in Chap. 11 and will not be repeated here. The dynamic lag of the measuring element is assumed to be negligible.

The computer circuit for simulating this control system is shown in Fig. 32.18. Scaling has not been considered. Notice that the gains of amplifiers 1, 2, and 3 have been made 2.5, 2, and 2, respectively, and the setting of potentiometer 1 is therefore $0.03K_c/10$. This arbitrary selection of gains makes it possible to vary $0.03K_c$ from 0 to 10.

The gains in the Padé circuit have been arranged so that the transportation lag τ can be reduced to 0.35 if desired. For this example in which $\tau = 0.5$, we obtain the following potentiometer settings from the information on Fig. 32.18:

$$P_2 = P_3 = \frac{0.12}{(0.5)^2} = 0.48$$

$$P_4 = P_5 = \frac{0.3}{0.5} = 0.60$$

[1] Single and Stubbs, Transport Delay Simulation Circuits, WAPD-T-38, *Tech. Rept. Dept.* 38, Westinghouse Atomic Power Division, May, 1952.

[2] *Ibid.*, p. 256.

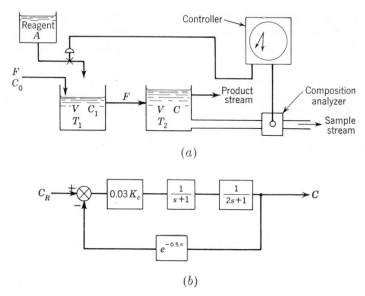

(a)

(b)

Fig. 32.17   Control of composition in a chemical reactor.

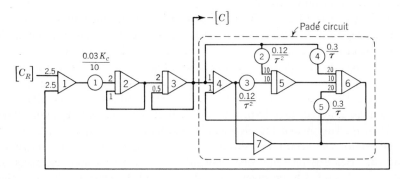

Fig. 32.18   Computer circuit for control of composition.

To introduce a step change in C_R, a constant voltage is applied to amplifier 1. The output of amplifier 3 is then recorded to obtain the transient response $C(t)$. The time-scale factor has been taken as 1 so that 1 sec of computer time represents 1 min of time in the physical problem.

Some recordings from the circuit of Fig. 32.18 for several values of K_c and τ are shown in Fig. 32.19.   If curve 1 is compared with the curve of Fig. 17.16, which is a plot of Eq. (17.7) that is based on the root-locus diagram of Fig. 17.15, it is seen that the response predicted by the root-locus diagram is very close to the corresponding curve of Fig. 32.19.   In fact, if values of C are calculated from Eq. (17.7) and compared with values of C

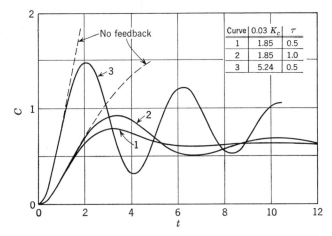

Fig. 32.19 **Response curves from computer circuit of Fig. 32.18.**

from Fig. 32.19 at corresponding times, it is found that the difference between the values is less than 0.01, which is less than 2 percent of the final value. This small deviation is probably well within the error of the computer and x-y recorder. The root-locus diagram of Fig. 17.15 used Eq. (32.16) as an approximation to $e^{-\tau s}$; therefore, for this particular example, the simpler approximation of Eq. (32.16) would also give satisfactory results.

In order to compare the closed-loop response of the control system with the open-loop response during the first τ min (which must be simply the second-order response for the two tanks), the step input was introduced with the feedback loop disconnected. These open-loop response curves are shown dotted in Fig. 32.19. From the computer results, no difference was seen between open- and closed-loop responses up to $t = \tau$. This is further indication that the Padé circuit is correctly delaying the feedback signal by τ units of time.

SIMULATION DIRECTLY FROM DIFFERENTIAL EQUATIONS

In this section, we consider the simulation of nonlinear systems, for which case the transfer function formulation of the problem is not valid. The procedure is to solve the differential equations directly. To illustrate the approach, we shall first develop an analog computer circuit to simulate the dynamic behavior of the gas absorber described in Chap. 25.

Simulation of a Gas Absorber In the simulation, we wish to provide for changes in both inlet gas concentration and liquid flow rate for the two-plate column shown in Fig. 25.5. The pertinent equations which describe the dynamics are Eqs. (25.26) to (25.29), which are repeated here for convenience:

$$\frac{dx_1}{dt} = \frac{1}{H}(L_2x_2 - L_1x_1) + \frac{Vm}{H}(x_0 - x_1) \tag{25.26}$$

$$\frac{dx_2}{dt} = \frac{Vm}{H}(x_1 - x_2) - \frac{1}{H}L_2x_2 \tag{25.27}$$

$$\frac{dL_2}{dt} = \frac{L_3}{\tau_2} - \frac{L_2}{\tau_2} \tag{25.28}$$

$$\frac{dL_1}{dt} = \frac{L_2}{\tau_1} - \frac{L_1}{\tau_1} \tag{25.29}$$

There are four dependent variables (L_1, L_2, x_1, and x_2) and two forcing functions (x_0 and L_3) in this set of simultaneous equations. Recall that x_0 is related to the inlet gas concentration y_0 by

$$x_0 = \frac{y_0 - b}{m}$$

In other words, x_0 is the concentration of a liquid that would be in equilibrium with the entering gas.

The computer circuit will be developed for the following problem:[1] A two-plate absorber as shown in Fig. 25.5 is operating at steady state at 25°C and 1 atm total pressure. An air-SO_2 mixture containing 2 mole percent SO_2 enters the column at a flow rate of $V = 0.051$ lb moles/min of gas mixture. Pure water enters the top of the column at a rate of $L = 0.90$ lb moles/min. The equilibrium relation is approximated by

$$y = 27x - 0.00324 \tag{32.18}$$

The holdup, which is the same for each plate, is $H = 0.11$ lb moles. The liquid dynamics time constant, which is also the same for each plate, is $\tau = 4$ sec. From this information, it is desired to devise a computer circuit which will simulate the following transient runs:

Run *a*. A step change from 0.02 to 0.043 mole fr. SO_2 is made in y_0, all other conditions remaining the same.

Run *b*. A step change from 0.90 to 0.45 lb moles/min is made in L_3, all other conditions remaining the same.

Before programming Eqs. (25.26) to (25.29) for the computer, scale factors must be selected for x_1, x_2, L_1, and L_2. From the statement of the

---

[1] The values of H and τ used in this problem were obtained in an experimental study on the dynamics of a bubble-cap absorber by L. B. Nobbe, "Transient Response of a Bubble-cap Plate Absorber," M.S. Thesis, Purdue University, January, 1961. The plates, which were 8 in. in diameter, each contained three bubble caps having a diameter of 2½ in.

problem, $L_{1_{max}} = L_{2_{max}} = 0.90$. An estimate of x_1 can be obtained by assuming that the liquid leaving is in equilibrium with the entering gas, for which case substitution into Eq. (32.18) gives $x_{1_{max}} = 0.0016$. Since $x_2 < x_1$, it is permissible and convenient to also take $x_{2_{max}} = 0.0016$. Then, since the scale factors for these variables will be the same, the corresponding voltages may be compared directly. From these estimates, the scale factors in the following table are obtained:

Variable	Max value	$\dfrac{100}{\text{max value}}$	Scale factor	Computer variable
x_1	0.0016	6.25×10^4	5×10^4	$[5 \times 10^4 x_1]$
x_2	0.0016	6.25×10^4	5×10^4	$[5 \times 10^4 x_2]$
L_1	0.90	111	100	$[100 L_1]$
L_2	0.90	111	100	$[100 L_2]$

For convenience, the scale factors have been rounded off as shown in the table.

Introducing these parameters and operating conditions into Eqs. (25.26) to (25.29) and writing the equations in terms of computer variables give the following set of equations:

$$\frac{d[5 \times 10^4 x_1]}{dt} = 9.1 \frac{[100 L_2][5 \times 10^4 x_2]}{100} - 9.1 \frac{[100 L_1][5 \times 10^4 x_1]}{100}$$
$$- 12.4[5 \times 10^4 x_1] + 12.4[5 \times 10^4 x_0] \quad (32.19)$$

$$\frac{d[5 \times 10^4 x_2]}{dt} = -9.1 \frac{[100 L_2][5 \times 10^4 x_2]}{100} - 12.4[5 \times 10^4 x_2]$$
$$+ 12.4[5 \times 10^4 x_1] \quad (32.20)$$

$$\frac{d[100 L_2]}{dt} = 15[100 L_3] - 15[100 L_2] \quad (32.21)$$

$$\frac{d[100 L_1]}{dt} = 15[100 L_2] - 15[100 L_1] \quad (32.22)$$

A computer diagram for solving these equations is shown in Fig. 32.20. The settings of the coefficient potentiometers are shown in Table 32.2. Note that the time-scale factor β, discussed in Chap. 30, has been included in the setting of the potentiometers on the inputs to integrators.[1] For this problem the steady-state conditions ($L_{3_s} = 0.9$ and $x_{0_s} = 0.00086$) are obtained when switches A and B are open. Closing switch A introduces a positive step change in gas concentration; the magnitude of the change

---

[1] It was found convenient to let $\beta = 20$ in the actual operation of the computer.

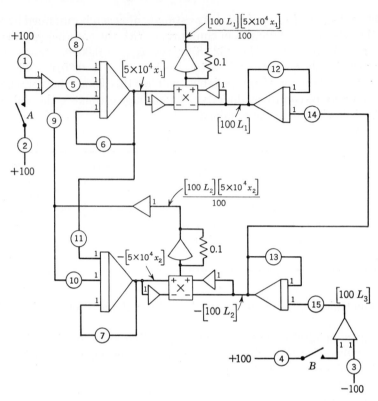

Fig. 32.20   Computer diagram for gas absorber.

Δx_0 is determined by potentiometer 2. Closing switch B introduces a negative step change in flow of water; the magnitude of the change is determined by potentiometer 4.

The operation of the circuit occurs when the computer is in the operate mode. The procedure for getting the circuit into operation is as follows:

Table 32.2

Potentiometer number	Parameter represented	Setting	Remarks
1	$[5 \times 10^4 x_{0_s}]/100$	0.432	
2	$[5 \times 10^4 \Delta x_0]/100$	0.368	Used in Run *a* only
8, 9, 10	$(1/H)\, 1/\beta$	$9.1/\beta$	
5, 6, 7, 11	$(Vm/H)\, 1/\beta$	$12.4/\beta$	
12, 13, 14, 15	$(1/\tau)\, 1/\beta$	$15/\beta$	
3	L_{3_s}	0.9	
4	ΔL_3	0.45	Used in Run *b* only

Set the potentiometers according to Table 32.2, open the switches A and B, and place the computer in the operate mode. After a transient period, the voltages from the amplifiers approach steady-state values, which correspond to the steady-state operation of the tower. A step change in concentration or flow can now be introduced by closing the appropriate switch.

Figure 32.21 shows the response of x_1 and x_2 for a change in inlet gas composition as specified in Run a. [The time scale in Fig. 32.21 is in terms of the time (in minutes) for the physical system, not the computer time.] Notice that the response of x_2 (for which deviations are proportional to those in y_2) resembles an overdamped second-order response. This is in agreement with the theoretical transfer function given by Eq. (25.34). Also note that the response of x_1 appears as a first-order response.

Figure 32.22 shows the change in x_1 and x_2 for a change in liquid flow rate as described by Run b. Since the changes in x_1 and x_2 were small, Fig. 32.22 is plotted in terms of the deviations of x_1 and x_2 with time. Notice that the x_1 drops slightly before rising to the new steady-state concentration. This is a different type of response from any we have previously observed and is a consequence of the nonlinearity in Eqs. (25.26) and (25.27) when L_1 and L_2 are variable.

Having obtained a computer simulation of the gas absorber, we could now study the control of the absorber by adding a few more circuits to account for the controller, valve, and concentration-measuring element. Figure 32.23 illustrates the modification needed in Fig. 32.20 to provide for proportional control of x_1 when the valve has first-order dynamics with a time constant τ_v and the measuring element is assumed to have no dynamic lag. Amplifiers 1 and 2, along with potentiometers 16 and 17, simulate the proportional controller. Potentiometer 16 represents the set point and is set to $5 \times 10^4 x_{1_s}/100$, so that the output of $A1$, which is proportional

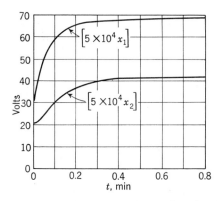

Fig. 32.21 Response of gas absorber to step change in inlet gas concentration.

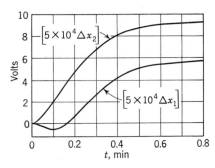

Fig. 32.22 Response of gas absorber to step change in liquid flow rate.

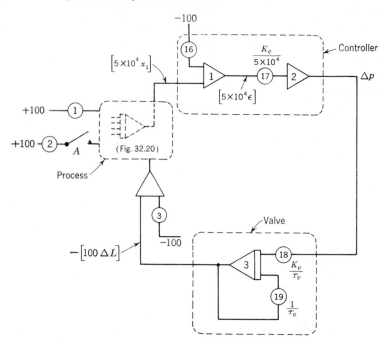

Fig. 32.23 Computer simulation of control of a gas absorber.

to the error, is zero when x_1 is at the steady-state value. Potentiometer 17 is used to vary the controller gain. The output of $A2$ is the deviation in pneumatic pressure which is applied to the valve top. (In a computer operation, this signal would probably be scaled; however, the scale factor here is taken to be 1 for simplicity.) Note that the units on K_c are pounds per square inch per mole fraction SO_2 in liquid. The valve is simulated by the circuit containing $A3$. This circuit, which represents a first-order process, was given in Fig. 32.2. The output from $A3$ is the deviation in flow needed for corrective action. This signal replaces the step disturbance that was introduced by means of switch B in Fig. 32.20 and provides a continuous adjustment in liquid flow rate. With the circuit of Fig. 32.23, one can study the effect of parameters such as K_c, τ_v, and τ on the transient response of the controlled absorber to step changes in inlet gas concentration.

Simulation of Nonlinear Chemical Reactor In Chap. 28, the phase-plane analysis of the control of an exothermic chemical reactor was presented. The analysis led to Eqs. (28.13), which are repeated here for convenience:

$$\frac{dy}{d\tau} = 1 - y - yr \tag{32.23}$$

$$\frac{d\theta}{d\tau} = 1.75 - \theta + yr - (\theta - 1.75)[1 + K_c(\theta - 2)] \tag{32.24}$$

where

$$r = e^{50(1/2 - 1/\theta)},$$
(32.25)

These equations are highly nonlinear because of the term r. However, the analog computer can be programmed in a straightforward manner to solve this set of equations.

One of the most tedious aspects of this type of problem is scaling; however, the numerical results of Aris and Amundson,[1] obtained on a digital computer, were useful in obtaining the scale factors shown in the following table:

Problem variable	Range	Max value	Scale factor	Computer variable
y	0–1	1	100	$[100y]$
θ	1.7–2.3			
r		10	10	$[10r]$
θ_1	(−0.05)–0.55	0.60	160	$[160\theta_1]$
$1/\theta$	0.44–0.59			
θ_1/θ	0–0.24	0.25	400	$[400\theta_1/\theta]$

The terms θ_1, $1/\theta$, θ_1/θ are derived quantities used in the computation and are explained below.

It is convenient for computer programming to rewrite Eq. (32.24) in the form

$$\frac{d\theta}{d\tau} = -2(\theta - 1.75) - K_c(\theta - 1.75)(\theta - 2) + yr$$
(32.26)

The scale factor table shows that the range of θ is not very large. It was found convenient to define a new variable

$$\theta_1 = \theta - 1.75$$

Introducing θ_1 into Eqs. (32.23), (32.25), and (32.26) gives

$$\dot{y} = 1 - y - yr$$
(32.27)
$$\dot{\theta}_1 = -2\theta_1 - K_c\theta_1(\theta_1 - 0.25) + yr$$
(32.28)
$$r = e^{50(\frac{1}{2} - 1/1.75 + 1/1.75 - 1/\theta)}$$
$$= 0.0281e^{28.6\,\theta_1/\theta}$$
(32.29)

where $\dot{y} = dy/d\tau$ and $\dot{\theta}_1 = d\theta_1/d\tau$. The reason for writing the exponent in Eq. (32.29) in this form is that θ_1/θ has a greater range than $1/\theta$, as shown by the above table. This will make it possible to generate this exponential

---

[1] Rutherford Aris and N. R. Amundson, An Analysis of Chemical Reactor Stability and Control, I, II, and III, *Chem. Eng. Sci.*, 7:121–149 (1958).

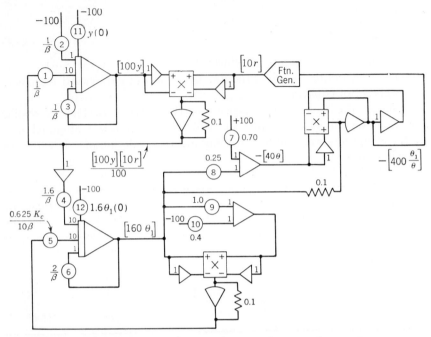

Fig. 32.24 **Computer simulation of an exothermic chemical reactor.**

term more accurately on the analog computer, which uses a division circuit based on a quarter-square multiplier card. These rearrangements of equations, which may seem unusual, are quite important in increasing the accuracy of the analog computer solution.

Equations (32.27) to (32.29) can now be written in terms of computer variables to give

$$[100\dot{y}] = 100 - [100y] - 10\frac{[10r][100y]}{100} \tag{32.30}$$

$$[160\dot{\theta}_1] = -2[160\theta_1] - 0.625K_c\frac{[160\theta_1][160\theta_1 - 40]}{100} + 16\frac{[100y][10r]}{100} \tag{32.31}$$

$$[10r] = 0.281e^{0.0715[400\theta_1/\theta]} \tag{32.32}$$

These equations can be programmed directly to give the circuit shown in Fig. 32.24. The input-output relation of the function generator[1] which produced Eq. (32.32) is shown in Fig. 32.25 to illustrate the highly nonlinear nature of the function r. Notice that potentiometer 5 is used to adjust the controller gain K_c and potentiometers 11 and 12 are used to adjust the initial condition on y and θ_1. It was found convenient to use $\beta = 2$.

[1] The actual circuit used to produce the exponential function made use of log cards, which are standard components available for computers and are used in a manner similar to the quarter-square card discussed in Chap. 31. An alternate device for producing the desired function would be a diode function generator.

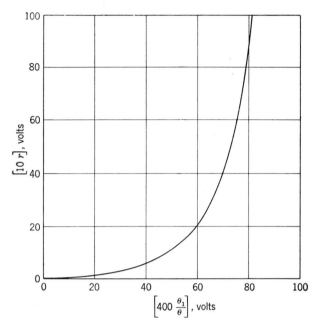

Fig. 32.25 **Nonlinear reaction rate.**

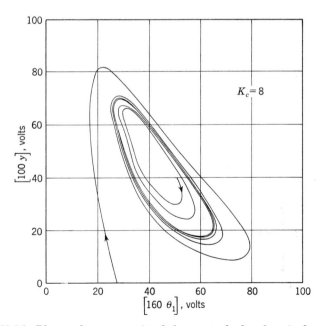

Fig. 32.26 **Phase-plane portrait of the control of a chemical reactor.**

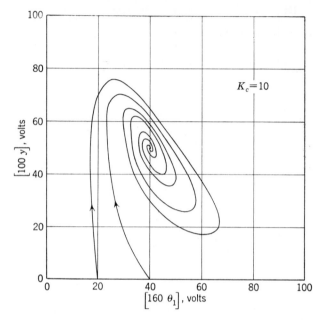

Fig. 32.27   Phase-plane portrait of the control of a chemical reactor.

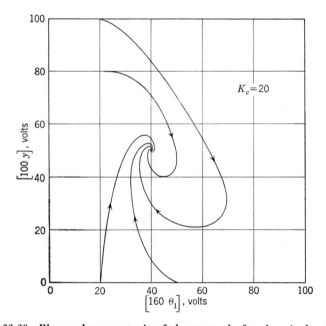

Fig. 32.28   Phase-plane portrait of the control of a chemical reactor.

The phase plane of y versus θ_1 can be obtained by connecting the outputs of amplifiers generating these variables to an x-y recorder. Phase-plane portraits which were actually obtained from the computer are shown in Figs. 32.26 to 32.28. The reader should study these figures along with the discussion of Chap. 28. Notice that the computer results agree with the predictions of the analysis. For $K_c > 9$, the response is stable and approaches the critical point $y = 0.5$, $\theta = 2$. (This corresponds to $[100y] = 50$, $[160\theta_1] = 40$ in Figs. 32.27 and 32.28.) For $K_c < 9$, the system cannot remain at this critical point. If $K_c = 8$, Fig. 32.26 shows that the system will spiral away from the critical point to a stable limit cycle. This results in continuous oscillation in the composition and temperature, clearly an undesirable operating condition.

PROBLEMS

32.1 Show that the computer circuit of Fig. P32.1 simulates the transfer function

$$\frac{Y}{X} = \frac{K(\tau_1 s + 1)}{\tau^2 s^2 + 2\zeta\tau s + 1}$$

Determine the potentiometer settings in terms of the parameters K, τ_1, τ, and ζ.

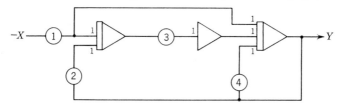

Fig. P32.1

32.2 The computer circuit of Fig. P32.2 simulates a transfer function Y/X which involves the parameters τ_1, τ_2, τ_3, and K. Determine the transfer function if the potentiometer settings are

$$A = \frac{1}{\tau_3} \qquad B = \frac{K\tau_1}{\tau_3} \qquad C = \frac{K\tau_1}{\tau_2\tau_3}$$

$$D = \frac{1}{\tau_2\tau_3} - \frac{1}{\tau_1\tau_2} \qquad E = \frac{1}{\tau_2}$$

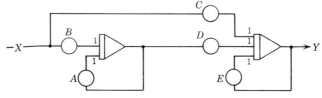

Fig. P32.2

32.3 Construct a computer diagram that will simulate the control system of Prob. 13.6. The circuit should provide for variation in K_c from 0 to 10 and τ_D from 0 to 1. The gains of the inputs to summers and integrators are either 1 or 10.

Appendix   *Proof of Root-locus Rules of Chapter 15*

In the discussion to follow, we let

$$A(s) = \frac{KN}{D} = \frac{K(s - z_1)(s - z_2) \cdots (s - z_m)}{(s - p_1)(s - p_2) \cdots (s - p_n)} \tag{A.1}$$

and

$$B(s) = \frac{A(s)}{K} \tag{A.2}$$

A root locus is therefore a locus of values of s for which $A(s) = -1$ and $B(s)$ is real.   It is also assumed that $n \geq m$.

Since the equation

$$A(s) = -1 \tag{A.3}$$

is an nth-order polynomial, it follows that to each value of K there correspond n values of s satisfying Eq. (A.3).   This proves Rule 1.

Consider the behavior of $A(s)$ in the vicinity of a typical pole p_k.   We test a small circle centered at p_k,

$$s = p_k + \delta e^{j\theta} \qquad \delta \text{ very small} \tag{A.4}$$

for satisfaction of the angle criterion, Eq. (15.11). Substituting Eq. (A.4) into Eq. (15.11) yields, after discarding negligible terms in δ,

$$\sum_{l=1}^{m} \angle (p_k - z_l) - \sum_{\substack{l=1 \\ l \neq k}}^{n} \angle (p_k - p_l) = (2i + 1)\pi + q\theta \tag{A.5}$$

where q is the order of the pole at p_k. Equation (A.5) always has the solution

$$\theta = \frac{1}{q} \left[\sum_{l=1}^{m} \angle (p_k - z_l) - \sum_{\substack{l=1 \\ l \neq k}}^{n} \angle (p_k - p_l) - (2i + 1)\pi \right] \tag{A.6}$$

which is equivalent to Eq. (15.14). The number of distinct values of θ, which can be generated from Eq. (A.6) by taking different integers for i, is equal to q. Thus we have demonstrated that there are q points on the circle of Eq. (A.4) which satisfy the angle criterion and hence lie on the root locus. The angles of these points are given by Eq. (A.6), which is true only as $\delta \to 0$, that is, right at the pole. A similar study can be made for points on the circle

$$s = z_k + \delta e^{j\theta}$$

to derive Eq. (15.15) and verify the behavior near a zero. This then proves both Rules 2 and 6, with the exception of the behavior near an asymptote.

Substituting Eq. (A.4) into Eq. (15.10) and solving for K yield, after discarding negligible terms in δ,

$$K = \frac{\delta^q \displaystyle\prod_{\substack{l=1 \\ l \neq k}}^{n} |(p_k - p_l)|}{\displaystyle\prod_{l=1}^{m} |(p_k - z_l)|} \tag{A.7}$$

Equation (A.7) shows that $K \to 0$ as $\delta \to 0$, or that the loci begin at the poles. Similar arguments show that the loci terminate at the zeros, where $K \to \infty$.

Rule 3 follows directly from the angle criterion for any point on the real axis satisfying the conditions of the rule. Its proof is suggested as an exercise for the reader.

If $n = m$, there are no asymptotes and all loci terminate at finite zeros. Hence, we consider only the case for $n > m$ in deriving the asymptotic behavior. Expanding the numerator and denominator of $A(s)$ yields

$$A(s) = K \frac{s^m + a_1 s^{m-1} + \cdots + a_{m-1}s + a_m}{s^n + b_1 s^{n-1} + \cdots + b_{n-1}s + b_n} \tag{A.8}$$

where

$$a_1 = - \sum_{k=1}^{m} z_k$$

$$b_1 = - \sum_{k=1}^{n} p_k$$

and the other a's and b's are other functions of the zeros and poles. Dividing numerator and denominator of Eq. (A.8) by the numerator results in

$$A(s) = \frac{K}{s^{n-m} + (b_1 - a_1)s^{n-m-1} + \text{(terms in lesser powers of } s)} \tag{A.9}$$

As s becomes very large in magnitude, Eq. (A.9) behaves essentially as

$$A_\infty(s) = \frac{K}{[s + (b_1 - a_1)/(n - m)]^{n-m}} \tag{A.10}$$

The solution to the equation

$$A_x(s) = -1$$

is

$$s + \frac{b_1 - a_1}{n - m} = K^{1/(n-m)} e^{j\pi \frac{2i+1}{n-m}} \qquad i = 0, 1, 2, \ldots, n - m - 1 \tag{A.11}$$

Equation (A.11) describes a group of $(n - m)$ straight lines radiating from the center of gravity defined by Eq. (15.12) at the specified angles and thus verifies Rule 4.

To prove Rule 5, we proceed as follows: Consider the neighborhood of a typical point (not a pole) on the root locus, s_0. Since $B(s)$ is the ratio of polynomials, it is differentiable and may be expanded in a Taylor series around s_0. Retaining only the linear terms,

$$B(s) - B(s_0) = \frac{dB}{ds}\bigg|_{s=s_0} (s - s_0) \tag{A.12}$$

Now for s on the root locus *both* $B(s)$ and $B(s_0)$ are real. Therefore, taking the argument of both sides of Eq. (A.12) yields

$$\measuredangle[B(s) - B(s_0)] = i\pi = \measuredangle\left(\frac{dB}{ds}\bigg|_{s=s_0}\right) + \measuredangle(s - s_0) \tag{A.13}$$

The slope of the locus,

$$\theta = \measuredangle(s - s_0)$$

is thus determined by Eq. (A.13) as

$$\theta = \begin{cases} - \measuredangle \dfrac{dB}{ds}\Big|_{s=s_0} \\ \pi - \measuredangle \dfrac{dB}{ds}\Big|_{s=s_0} \end{cases} \tag{A.14}$$

unless

$$\frac{dB}{ds}\Big|_{s=s_0} = 0 \tag{A.15}$$

in which case Eq. (A.14) has no unique meaning.   Therefore, Eq. (A.14) shows that only one locus may pass through a point s_0 which does not satisfy Eq. (A.15) or conversely that an intersection of root loci necessitates that Eq. (A.15) be satisfied.   If $dB/ds\big|_{s=s_0} \neq 0$, the two angles of Eq. (A.14), which differ by π, correspond to moving in different directions along the locus.

Now consider intersections of two root loci on the real axis.   Other intersections do not occur with sufficient frequency to be of practical significance.   As we have shown, Eq. (A.15) must be true at the intersection. At any point s_0 on the root locus, where Eq. (A.15) is satisfied, we can write a Taylor series

$$B(s) = B(s_0) + (s - s_0)^2 D(s)$$

where $D(s)$ is a polynomial in s.   In the *immediate vicinity* of s_0,

$$s - s_0 = \delta e^{j\theta} \qquad \delta \text{ very small}$$

$B(s)$ may be differentiated to yield approximately (neglecting terms of order δ^2)

$$\frac{dB}{ds}\Big|_{s\approx s_0} \approx 2\delta e^{j\theta} D(s_0) \approx 0 \tag{A.16}$$

Hence,

$$\measuredangle \frac{dB}{ds}\Big|_{s\approx s_0} \approx \theta + \measuredangle D(s_0) \tag{A.17}$$

This assigns values to the angle θ in the vicinity of the intersection.   Thus, combination of Eqs. (A.14) and (A.17) yields

$$\theta = \begin{cases} \dfrac{-\measuredangle D(s_0)}{2} \\ \dfrac{\pi}{2} - \dfrac{\measuredangle D(s_0)}{2} \end{cases} \tag{A.18}$$

Since θ is 0° for the loci entering the intersection from the real axis,

Eq. (A.18) shows that $\not\perp D(s_0)$ is either 0 or 90° and that the loci leaving the axis must leave at 90°.

The breakaway point can be found by direct application of Eq. (A.15). To do this, it is easiest to write $B(s)$ in logarithmic form. Thus,

$$\ln B(s) = \sum_{l=1}^{m} \ln (s - z_l) - \sum_{l=1}^{n} \ln (s - p_l) \tag{A.19}$$

Differentiating,

$$\frac{d \ln B(s)}{ds} = \frac{1}{B(s)} \left[\frac{dB(s)}{ds} \right] = \left(\sum_{l=1}^{m} \frac{1}{s - z_l} - \sum_{l=1}^{n} \frac{1}{s - p_l} \right) \tag{A.20}$$

Since on the root locus $B(s) \neq 0$, Eqs. (A.20) and (A.15) yield Eq. (15.13) directly. This concludes the proof of Rule 5.